Plate 1. Tree architecture and habit. (a, b) Trees from R. Envira with erect plagiotropic branches, sparse foliage and narrow canopies; (c) young adult tree with horizontal branches; (d) *Criollo* tree with single plagiotropic branch.

2a

2b

2c

2d

Plate 2. Tree architecture and habit. (a) Multiple-stemmed tree with arching trunk; (b) tall tree in forest with typical habit resulting from competition; (c) tall tree in Ecuador apparently devoid of inflorescences on the stem; (d) fruits produced on ends of branches.

Plate 3. Variation in flush leaf colours. (a) Pale green flush leaf of genotype from the R. Villano (R. Napo system); (b) red flush leaf of red-fruited genotype; (c) pale green flush leaf of progenies from cross-fertilization of pigmented leaved individuals of two F$_1$ families with a common parent from the R. Napo; (d) pale green flush leaves of *Criollo* genotype; (e) pale flush leaf of genotype from the R. Javarí.

Plate 4. Variations in leaf characters. (a) Flush leaf colour segregation in R. Japurá family; (b) pigmented petioles in R. Japurá genotype; (c) leaves with serrated edges from R. Japurá; (d) round leaves of '*Jaca*' mutant.

5a

5b

5c

Plate 5. Examples of leaf type segregants in families of Amazonian varieties.

6a

6b

6c

Plate 6. Variability of leaf characters. (a) 'Crinkle-leaf' mutation; (b) variation in leaf shape in families of accessions from the R. Japurá; (c) a chimerical plant from the root of a normal plant.

7a

7b

7c

SIAL
93

7d

Plate 7. Examples of fruits in the spherical extremity of the range of fruit shapes. (a) Spherical fruit from Amazonian population with thick husk (x0.37); (b) green and red fruits of an '*amelonado*' variety in São Tomé; (c) São Tomé, the '*Laranja*' mutant-type of *Theobroma sphaerocarpum* Chev.; (d) Brazil, Bahia, representative of fruit type attributable to the '*Comum*' variety (x0.33).

8a

8c

8d

8b

Plate 8. Variations in flower characteristics. (a) Examples of flowers from different varieties; (b) flowers with reflexed sepals; (c) mutation in which the staminodes are converted to functional anthers; (d) branch bearing deeply pigmented inflorescences of a red-fruited genotype.

9a

9c

9b

9d

Plate 9. Variations in flower characteristics. (a) Examples of flowers from different varieties; (b) flower with long staminodes; (c, d) flowers and fruits and seeds, respectively, of genotype 'BE 2' showing relationship of cotyledon colour to the short rose-coloured staminodes and guide-lines.

10a

10c

10b

10d

10e

Plate 10. (a–c, e) Examples of fruits from the Ecuadorian segments of tributaries of the R. Marañon. (d) Large fruit from the R. Santiago, x0.30.

11a

11b

11c

11d

Plate 11. Examples of fruits from the drainage system of the R. Napo. (a) Tall tree with narrow canopy and sparse fruiting on trunk from the R. Payamino; (c) x0.34.

Plate 12. Examples of fruits with surface pigment from the Amazonian Region. (a) Brazil, R. Javarí; (b) Ecuador, R. Conambo (R. Tigre drainage system); (c) Ecuador, R. Sucumbios (San Miguel) R. Putumayo system; (d) Brazil, upper Solimões; (e) Brazil, R. Içá (Putumayo).

12a

12b

12c

12d

12e

13a

C.SuL 1   C.SuL 3   C.SuL 4   7   C.SuL 8   C.SuL 9

13b

13c

Plate 13. Examples of fruits from the R. Ucayali and R. Juruá systems. (a) Peru, Loreto, tree from the R. Tapiche near junction with the R. Ucayali; (b) Brazil, R. Juruá, fruits of some existing collections from areas located immediately below Cruzeiro do Sul; (c) Brazil, R. Juruá, other fruit types from areas adjacent to that of the specimens in (b).

14a

14b

14c

Plate 14. Fruits from the R. Envira and R. Solimões to show special characteristics. (a, b) fruits of the trees depicted in Plate 1a and 1b demonstrating the range of variability; (c) striking fruit type from the mid-Solimões region.

15a

15b

15c

15d

Plate 15. Brazil, examples of variability in fruit types of the populations of the upper R. Solimões.

16a

16b

16c

16d

Plate 16. Brazil, fruits from tributaries of the upper R. Solimões. (a) Typical fruit from the R. Itacoai, tributary of the R. Javarí; (b–d) some of the variability of fruit characters of genotypes from the R. Içá (R. Putumayo).

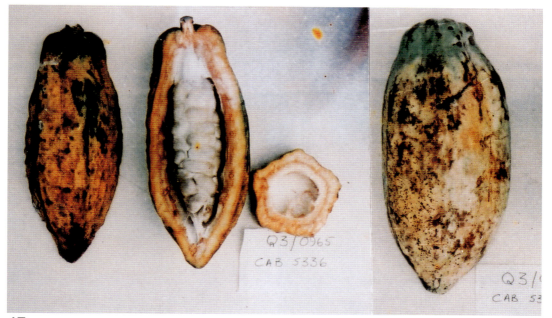

17a

17b

17c

Plate 17. Brazil, fruits of trees from the R. Solimões exhibiting suggestions of pentagonal shapes. (a) Fruit with definite pentagonal characteristics (left) with a fruit from another individual of the same family (right) (x0.285); (b) fruit of the parent of the above family; (c) fruit of a progeny of a tree said to produce fruits of pentagonal shape.

18a

18b

18c

18d

Plate 18. Brazil, examples of tree and fruit types from the R. Solimões. (a) Tree with low branching habit and dark green fruits; (b) an outstanding tree from an island population; (c) another tree with a tendency for a low branching habit; (d) fruits of members of a family, denominated 'SPA', presumed to have originated from the R. Napo, Peru (x0.364).

19a

19b

19c

19d

Plate 19. Brazil, R. Purús valley. (a, b) R. Acre, the range of fruit type variation from genotypes sampled near R. Branco; (c, d) examples of the extremes of the fruit variability of the population of the R. Iaco.

20a

20b

20d

20c

Plate 20. Colombia, R. Caquetá. Variability of fruit characteristics in the progenies of trees sampled in the population located below the Falls of Araracuara, in order as river is descended – fruits of individual trees. (a) 'EBC 5' (x0.243) and (b) 'EBC 10' (x0.306) from same island; (c) 'EBC 76', 'EBC 77' and 'EBC 79'; (d) 'EBC 85' and 'EBC 91' (x0.36).

21a

21b

21c

21d

Plate 21. Colombia, R. Caquetá. Samples of fruits from the stretch of the river from La Pedrera to the frontier with Brazil. (a) 'EBC 125' fruit type with pentagonal appearance, perhaps related to the trees in Plate 20 (x0.294); (d) fruit of a tree ('EBC 148') from frontier area with Brazil (x0.30).

22a

22b

22c

22d

22e

Plate 22. Examples of fruits from the R. Branco system, Brazil. (a) Fruit of a large tree, Taiano; (b) fruit of inbred progeny of 22a; (c, d) fruits from the family of a tree at Caracaraí; (e) near pentagonal shaped fruit from a tree at Taiano.

23a

23b

23c

23d

Plate 23. Some of the genotypes grown on the Island of Careiro, in R. Amazon near Manaus. (a) 'MA 11';
(b) 'MA 12' and 'MA 15'; (c) the genotypes collected in 1967 with identifier 'CA'; (d) inbred progenies of 'CA 4'.

24a

24b

24c

24d

Plate 24. Examples of fruit types from the R. Jí-Paraná system, Rondonia, Brazil.

Plate 25. Various fruit types from the Amazonian Region. (a) Near pentagonal shaped fruit from tree cultivated in eastern Pará State (x0.36); (b) fruit of a progeny of tree of (a) (x0.43); (c) fruit of 'SPEC. 54' homozygous genotype from R. Papurí (R. Negro system); (d) fruit of a progeny of a hybrid of two Amazonian genotypes.

26a

26b

26c

26d

Plate 26. Fruit types of a range of genotypes possibly descended from Amazonian Region populations. (a) Fruits of three collections from the lower R. Jarí valley, Amapá, Brazil; (b) Amazonian type (right) cultivated in Dominica, West Indies, with variant fruit shape (left); (c) uniform variety cultivated in Miranda State, Venezuela, perhaps derived from the 'Calabacillo' of Trinidad; (d) fruit without ridges of M137, Trinidad (x0.34).

Plate 27. Ecuador, fruits of types attributable to the '*Nacional*' variety. (a) From Manabí Province; (b) from Hacienda Clementina; (c) type known in Manabí Province as '*Balao*', from southern Ecuador.

Plate 28. Fruit types of the *Criollo* group. (a, b) Examples of '*Pentagonum*' shaped fruits, Costa Rica and Soconusco, Mexico, respectively; (c, d) green and red fruits, respectively, of the '*Porcelana*' variety, R. Encontrados, Venezuela.

Plate 29. Examples of fruits of the *Criollo* group. (a) Progenies of the Chuao valley population, Venezuela; (b) fruits from the Dominican Republic possessing some *Criollo* characteristics; (c) a uniform *Criollo* type cultivated in the Dominican Republic from seed introduced from Venezuela, early 20th century; (d, e) fruits and distinctive leaves, respectively, of tree of *Criollo* ancestry, St Lucia, West Indies.

30a

30b

30c

Plate 30. Fruits of *Criollo* group varieties. (a) A sample of fruits from the Lacandon forest, Mexico; (b, c) from two locations in Esmeraldas Province, Ecuador.

31a

31b

31c

Plate 31. Leaf type mutations. (a) An example of the segregation for a recessive leaf abnormality; (b) inbred progeny (left) of 'CA 3' compared with hybrids (right) of the clone at the same age; (c) plants with the 'Luteus-PA' leaf colour mutation (right) compared with plants of the same family with normal leaves (left).

32a

32c

32b

32d

Plate 32. Examples of genotypes possessing the anthocyanin inhibitor gene and polyembryony. (a) 'Witte Kakao', a *Criollo* type with the gene which occurred in Java, illustrated in 1863; (b) fruits of a tree homozygous for the anthocyanin inhibitor gene in Alenquer district, Brazil; (c) 'twin' embryo from the genotype '*Catongo*', showing selfed- (right) and cross- (left) fertilized zygotes; (d) seedlings grown from polyembryonic seeds of the '*Nacional*' variety, Ecuador.

THE GENETIC DIVERSITY OF CACAO AND ITS UTILIZATION

Botany, Production and Uses

THE GENETIC DIVERSITY OF CACAO AND ITS UTILIZATION

B.G.D. Bartley

CABI Publishing

CABI Publishing is a division of CAB International

CABI Publishing
CAB International
Wallingford
Oxfordshire OX10 8DE
UK

CABI Publishing
875 Massachusetts Avenue
7th Floor
Cambridge, MA 02139
USA

Tel: +44 (0)1491 832111
Fax: +44 (0)1491 833508
E-mail: cabi@cabi.org
Website: www.cabi-publishing.org

Tel: +1 617 395 4056
Fax: +1 617 354 6875
E-mail: cabi-nao@cabi.org

A catalogue record for this book is available from the British Library, London, UK.

Library of Congress Cataloging-in-Publication Data
Bartley, B. G. D. (Basil G. D.)
    The genetic diversity of cacao and its utilization / B. G. D. Bartley.
        p. cm.
Includes bibliographical references (p. ??).
    ISBN 0-85199-619-1 (alk. paper)
    1. Cacao--Germplasm resources. 2. Cacao--Utilization. I. Title.
    SB267.B27 2004
    633.7'423--dc22

                                                        2003021400

ISBN 0 85199 619 1

Typeset in Souvenir Light by Columns Design Ltd, Reading
Printed and bound in the UK by Biddles Ltd, King's Lynn

Contents

About the Author

Dr Basil Bartley was born in New Delhi in 1927 and obtained the degree of Bachelor of Science (Agriculture) from the University of Allahabad in 1947. In the same year he was admitted to Iowa State University, Ames, Iowa, USA, to undertake graduate studies in Plant Breeding. He was granted the Master of Science and Doctor of Philosophy degrees in 1949 and 1950, respectively. For the latter thesis his work involved the analysis of path coefficients in progenies of soybean which was the first application of the method in plants

Dr Bartley was appointed to the Cacao Research Scheme at the Imperial College of Tropical Agriculture (which later became the Faculty of Agriculture of the University of the West Indies) in Trinidad. Among the activities undertaken was the establishment of breeding systems applicable to cacao, studies on incompatibility and disease resistance and the collection of species of *Theobroma* and *Herrania* and inter-specific and inter-generic hybridization. He was responsible for introducing the concept of hybrid cultivars and worked on the development of parents for the purpose including the use of inbreeding.

Dr Bartley's first contact with the Amazonian Region was as a member of the Anglo-Colombian Cacao Collecting Expedition which led to an interest in the study and conservation of the species' genetic resources. This resulted in the formation of the first gene bank in Trinidad specifically designed for the purpose. In 1970 he went to São Tomé and Príncipe as an adviser to the cacao programme, which involved various activities including investigations into practical methods of evaluation of cultivars.

In 1975 Dr Bartley was contracted by the Interamerican Institute for Cooperation in Agriculture to work with CEPLAC in Brazil as an adviser to the cacao improvement programme. Working at the Cacao Research Centre and the Department for the Amazon until 1990 he was involved in the development of various breeding activities including disease resistance, compatibility studies and research on quality factors. The CEPEC gene bank was set up under his guidance. In the Amazon he drew up the plan of action for the genetic resources conservation programme and coordinated its activities, playing a leading role in the characterization of the germplasm and its utilization. From extended visits to Venezuela, Ecuador (where he worked for 6 months in 1992) and the eastern Caribbean he obtained a fundamental knowledge of the cacao populations in those countries and advised in their research programmes.

Dr Bartley has maintained a constant interest in the development of genetic resources programmes carried out worldwide and participated actively in all aspects including the exchange of germplasm. Through his studies of the characteristics of the plants he has played a leading role in the development of the Descriptor Lists. Since 1994 he has served as a consultant to the International Cocoa Germplasm Database.

Preface

This summary of our knowledge concerning the genetic diversity of cacao is the result of 53 years of activities related to the genetic resources of the species. During this period the author has been active in all aspects of the subject from collecting to utilization. The period has also seen considerable advances in the knowledge of the diversity. It will be evident from the information contained in the book that we are still a long way from knowing everything about the species' diversity and that the subject is still evolving. In view of the absence of a generalized treatment of the subject it has been felt that the time has come to present a survey of the available knowledge in order to provide an orientation for future research programmes that will use the diversity.

It should be made clear that, although there may be expectations that such information would be included, no details are given regarding the origins and characteristics of the selections available. Some of these selections are mentioned only to explain various aspects of the subject.

Although much of the information contained in this volume is derived from the author's personal observations and interpretations, we must not fail to acknowledge the part played by numerous persons, past and present, who have also participated in genetic resources programmes. These persons have contributed their knowledge to the general accumulation of information about the species. In particular we must mention the farmers, farm workers and residents in areas where the species occurs who, by daily contact with the trees, have provided many invaluable observations and ideas.

It is not pretended that the account presented here contains the complete knowledge of the species' diversity. For one thing, the limitations of the size of the volume prevent us from including all of the information and illustrations at our disposal. Even then, where large and heterogeneous populations are concerned, the description of the range of variability contained in them would be beyond our capabilities. A large quantity of data is locked up in reports and files to which we have not been able to obtain access. The guardians of these data do not always appreciate the contents of the files and other records and their importance.

In addition, it has not been possible to consult all of the literature that is known to exist from which relevant information may be obtained, or even to cite all of the available works that are concerned with the subject of genetic resources.

The most appropriate manner in which to analyse the diversity would be to study in detail the variability that occurs in the individual populations. The results of these studies would be published in which the diversity existing in each component population would be described. Wherever possible information on the inheritance of the characters obtained from breeding experiments would be included. It is realized, as explained in the succeeding chapters, that this would imply a considerable amount of time and effort. It is hoped that this volume would stimulate interest in conducting further research on the subject.

In this book an attempt has been made to correct a deficiency that occurs in most books written in the English language. This refers to the disregard of literature written in other languages, much of which is indispensable to an understanding of the nature of the diversity.

Various persons have contributed to the preparation of the book by providing photographic material. The following persons, to whom the author expresses his gratitude, have collaborated in providing the images listed: J.M. Ferrão, Plates 2d, 7b, 7c, 8d; W. Phillips Mora, Plates 9b, 25b; H.C. Evans, Plates 30a, 30c and Figs 10, 13 and 14; P.C. Holliday, Fig. 22. In addition, the Royal Botanic Gardens, Kew, England, kindly made available the historical photographs used in Figs 31 and 41 as well as Plate 32a. Plate 32a was reproduced from the illustration in van Nooten (1863). The permission to publish these illustrations is gratefully acknowledged. Special acknowledgements are due to the staff of the Almirante Centro de Estudos de Cacau, Bahia, Brazil, for their collaboration in supplying the images on which Figs 8, 9 and 15 are based. Figure 24 was created from a photograph taken by R. Pereira at the author's request. Figure 29 was also composed from photographs presented to the author by A.O. Silva Santos.

The remaining photographs are the property of the author and should not be reproduced without permission.

The maps are the result of painstaking work by Maria Isabel Carvalho whose contribution must also be acknowledged. Owing to various circumstances, the maps do not include all of the points of reference mentioned in the text. With the aid of the maps the reader could fill in the details from complete maps that show the areas concerned. However, to help readers locate geographical features that are named in the text, references to the respective maps in which they are named are provided in the index.

To Dr Claire Lanaud, the author wishes to express his thanks for providing copies of much of the literature relating to the molecular analyses undertaken with cacao. The restrictions of space and time have prevented a complete analysis of the results from being presented and the interpretations involve only the studies that are relevant to the situations described in the book. The survey of diversity is intended to be a guide to persons undertaking analyses of the genetic structure of cacao regarding the origins and relationships among the genotypes subjected to analysis.

Special acknowledgements are due to the staff of various archives who have permitted access to their holdings and given the author incalculable assistance. The archives of the Royal Botanic Gardens are of particular importance for the wealth of material relating to the early history of research of cacao varieties for which the Gardens played a fundamental role.

The author wishes to express his thanks to Tony Lass for his constructive suggestions and support for the publication. Particular thanks are due to the Biscuit, Cake, Chocolate and Confectionery Association (BCCCA) of the UK, which financially supported the printing of this book.

It is the custom in cacao to use abbreviations and acronyms to designate genotypes, clones and families. Although the reader is advised to consult other sources for the meanings of variety names, in order to facilitate the search for this information the equivalents of the abbreviations for the material mentioned in the text are given in the Appendix.

It is hoped that the information in this book will show the unique characteristics of the diversity of cacao and that it is an interesting subject for research at all levels. As such, it merits the attention of scientists in fields other than those specifically concerned with cacao. The treatment of the subject would also be a valuable aid for researchers in the areas of botany and environmental studies by providing an understanding of the characteristics of the species.

Foreword

The future success and sustainability of cocoa (cacao) cultivation will depend largely on the ready availability of improved cocoa planting materials for the next generations of cocoa growers. Improvements are required in pest and disease resistance, in bean quality parameters, in early bearing, productivity and ease of establishment; making progress on any one of these requirements presents the modern cocoa plant breeder with a series of significant challenges.

To obtain good, secure results, cocoa plant breeders have a need for a number of essential building blocks. These include reliable long-term funding, readily available data on potential parents, access to these parents and related data on them as well as an efficient system of intermediate cocoa quarantine to permit the introduction of parental material without the introduction of any pests or diseases. Over recent decades the Biscuit, Cake, Chocolate and Confectionery Association (BCCCA), which represents the UK consumers of cocoa and manufacturers of chocolate, has had the vision to fund these essential building blocks on behalf of the world cocoa industry. Amongst other projects, BCCCA has supported the conservation and characterization of wild cocoa trees in the collection at the Cocoa Research Unit in Trinidad and to fund the International Cocoa Germplasm Database as well as the Intermediate Cocoa Quarantine Facility – both these latter being based at the University of Reading in the UK. It is very clear that without this visionary investment, cocoa breeders would have made very much less progress than they have done in recent years. BCCCA are proud to have been associated with developing and supporting these essential elements for progress towards the future prosperity of cocoa growing.

In addition to these activities, BCCCA has financially supported and encouraged expeditions to collect wild cocoa germplasm from various parts of northern South America – in and around the centre of diversity of the crop in the Amazonian Region. It is therefore entirely appropriate that BCCCA should have given financial support to the publication of this book that describes in considerable detail known wild cocoa populations and the surrounding environments.

Dr Basil Bartley is very well qualified to produce such a work having been active as a cocoa breeder in the Caribbean, West Africa and South America during a distinguished career lasting over 40 years. Furthermore, his work has included participation in a number of wild cocoa-collecting expeditions in the Amazonian Region and the first of these was participation in the Anglo-Colombian Cacao Collecting Expedition in 1952. He was also responsible for the creation of the cocoa genetic resource conservation programme for the Government of Brazil.

A large part of the information in this book actually comes from Basil's personal recollections and notes from his participation in these wild cocoa-collecting expeditions. We feel that capturing this wealth of information would be of great value to future workers in this field. It is our sincere hope that this book will become an essential reference resource for all current and future generations of cocoa breeders, giving them vital background information on the characteristics of the wild cocoas, the sites from which they were collected and the surrounding environment of these sites.

We wish to record our real appreciation to Dr Basil Bartley for having taken the time to collate and present such a wealth of information on the wild cocoa populations and for his considerable contribution to our knowledge of cocoa breeding over the last four decades or so.

Tony Lass
Chairman
Cocoa Research Committee
BCCCA
September 2004

1 The Background to the Subject: Concepts and a Brief History

The answer to the question 'Why should we be interested in the genetic diversity of the species?' is, that from the practical point of view, its existence is essential to achieving progress in its improvement in terms of the development of cultivars that satisfy certain established objectives. In addition to this principal reason the diversity of cacao is of such a nature that it is interesting in terms of its prospects for use in various lines of research. The genetic variability is contained in the aggregate of the species' genetic resources. For this purpose it is necessary to understand that the term 'genetic resources' includes all of the individual plants that belong to the species.

The existence of genetic diversity is indicated by the differences exhibited by its individual components with regard to the phenotypic expressions that are observed. In this sense all existing plants and the future generations produced by natural and artificial recombination are considered to be components of the species' diversity. On this fundamental principal no individual plant can be excluded from the universe of the diversity merely on the basis of its phenotypic expression resembling that of another individual, until it can be shown that their genotypes are identical. The genetic compositions of individuals or genotypes can only be determined through the application of the procedures for the purpose. In Chapter 9 various examples are given of the potential use of genotypes that would perhaps not be considered capable of producing useful cultivars.

The Organization of the Study and Conservation of the Genetic Resources

The effective achievement of the aims of any programme of genetic improvement depends not only on the quantity of genetic resources, as measured by the number of individual genotypes available, but also on the qualities of the diversity that they represent. The use of the term 'qualities' in this context refers to the genes carried by the individual genotypes, especially those that will contribute to the fulfilment of the aims of the programme. It goes without saying that an improvement programme whose objectives are concerned with the development of cultivars that possess certain qualities cannot be met if the genetic resources that are available do not contain the genes that will satisfy the desired objectives.

Consequently, the usefulness of any given component of the species' genetic resources will depend on the ability to identify in it the factors that are required. The need to obtain an adequate knowledge of the composition of the genetic resources extends beyond existing requirements and objectives. The possible future uses of the genotypes must also be taken into consideration. The long-term objectives of a breeding programme, especially for a perennial species such as cacao, should be subject to constant revision in order to cope with the appearance of new and additional circumstances. These include the genetic control of new diseases and pests, alterations to product use and quality requirements and the development of cultivars adapted to specific environmental situations and cultural systems.

In order to obtain the diversity and determine the genetic composition of the individuals of which it is composed the best procedure is to establish and conduct genetic resources programmes. Since cacao is a perennial species with specific intrinsic characteristics it is necessary to understand that a genetic resources programme, which will produce results, has to be conducted on a long-term and dynamic basis. A programme of this nature is made up of the following sequential and interdependent stages:

1. The acquisition of the genetic diversity, normally through collection, selection and exchange.
2. Propagation.
3. Conservation, by establishment in the field.
4. Characterization and evaluation of the individual genotypes involved.
5. Manipulation of the genotypes by breeding processes designed to release and make available the factors of the diversity that are contained in the genotypes.
6. Documentation of the information obtained.

The Geographical Basis of the Distribution of the Natural and Cultivated Populations in the Americas and their History

Theobroma cacao is a neotropical species occurring in primitive and secondary regions of distribution that are to be found in the Americas within a range between latitudes 20° N and 20° S. The cacao populations that would have existed in the pre-Columbian era could be considered to have occupied two geographical regions of distribution. These regions coincide with what, for the purposes of this account, would have been the primary distribution region and with a secondary region that can be considered as an entity in terms of genetic

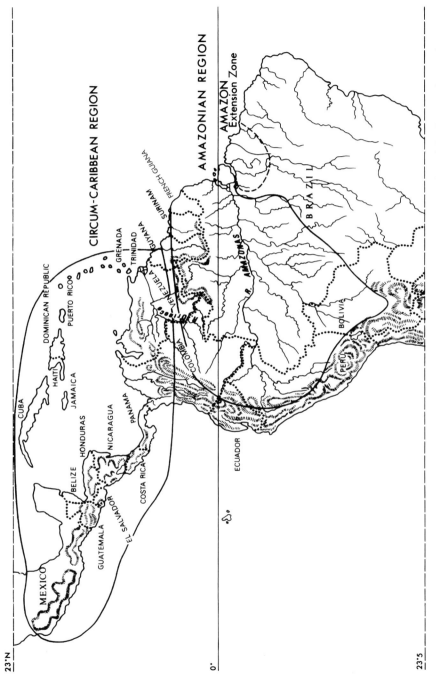

Fig. 1. Map of the range of cacao distribution in the Americas showing the approximate boundaries of the Amazonian and Circum-Caribbean Regions, and the Amazon Extension Zone.

resources on account of the occurrence of a single, unique variety. The approximate boundaries of these regions are shown in the map of the Americas in Fig. 1 and they are defined and described as follows.

The Circum-Caribbean Region

For the purposes of the analysis of diversity this title will be applied to what is considered to be the region of secondary distribution. It embraces particularly the Mesoamerican region and the hinterland of the northern coast of South America. It is in this region that the species had reached the highest level of domestication in pre-Columbian times. These were the populations in which the Spanish conquerors had their first contacts with the plant and its products. The varieties that were cultivated or otherwise occurred in the region are considered to belong to a single group on the basis of the possession of a conjunction of certain distinctive characteristics.

Although it was reported that Christopher Columbus was the first European to encounter cacao, in the form of its dried seeds, along the coast of what is now Belize, the first genuine report of the occurrence of cacao was given by Oviedo y Valdez (1855). In this, he related that Alonso Pinzón found cacao being cultivated in the vicinity of Chetumal, in southern Yucatán (which is contiguous to northern Belize) when he explored the Caribbean coast in 1510. It is obvious from this account that the Spaniards already knew all about cacao and its commercial and social importance to the inhabitants.

In fact, it was Oviedo y Valdez (1855), who commenced writing his monumental work in 1536, less than 50 years after the first voyage of Columbus to the New World, which the Spaniards called 'Las Indias', in which he provided us with the first complete description of the trees, their cultivation and the uses made of their products as well as the ceremonies associated with cacao. This was based on the observations he made during his stay in western Nicaragua. It is probably one of the most accurate descriptions of the crop. If the illustration of a cacao field that is included in the edition of his work, cited above, was published in the original edition it would be the first accurate picture of the plant and its cultivation.

The knowledge of the scale of cultivation of cacao in Mesoamerica was enlarged as other territories were incorporated by the Spanish conquests. It was the cacao exported from these territories that provided the first samples known in Europe and during a considerable period they were the main suppliers of cacao to satisfy the increasing demand for this exotic product. The history of cacao consumption outside the Americas has been related by various authors, such as Coe and Coe (1996). These authors also give a fairly extensive account of the social and practical aspects of the situation regarding the crop in pre-Columbian and colonial times.

Although other Spaniards arrived at and settled the lands of South America bordering the Caribbean Sea shortly after the voyages of Columbus, it was not until several decades later that the occurrence of cacao seems to have been recognized. From all accounts the species would have been distributed in the region to the southwest of Lake Maracaibo. From this point the exploitation and cultivation of the local varieties commenced and the area cultivated expanded towards the east of what became Venezuela. The product from the plantations was exported to Europe and in time supplanted that from Central America as production declined in the latter region.

Up to the time that the countries gained their independence from Spain, the whole of the coastal and inland areas along the south of the Caribbean Sea were under the same government, that of the Viceroyalty of Nueva Granada. For this reason it is necessary to consider within the area that supplied the cocoa exported under the name of 'Caracas' the regions of the present day Republic of Colombia adjacent to the Venezuelan States of Zulia, Mérida and Trujillo.

Although cacao probably did not occur on the islands of the Caribbean in pre-Columbian times these will be included in the Circum-Caribbean Region because of the relationships of the varieties that were introduced on the islands and similarities in the subsequent evolution of their populations.

The Amazonian Region

The principal area covered by this title includes all of the immense area drained by the Amazon River Basin. However, the definition of the Region will include the areas contiguous to it in which cacao is found. Most of the adjacent areas included in this definition are situated to the north of the Amazon Basin. The approximate northern limit of the region will be considered to be latitude 6° N. The contiguous area is drained by rivers that are independent of the R. Amazon and flow in north and northeasterly directions into the Atlantic Ocean. Precisely, the areas to the north of the Amazon Basin include the southern part of the Orinoco River Basin and the Guianas.

Also, to simplify the demarcation of the distribution area the lands in the humid tropical belt embracing the lower R. Tocantins and towards the east are considered to be part of the Amazonian Region. Although these areas evidently do not fall within the area of natural distribution of the species in virtue of all the cacao being cultivated the genetic resources contained in them are, by their derivation, related to the varieties that occur in the Amazon Basin. The area of cultivated cacao will be identified as the Amazon Extension Zone.

Soon after Francisco Pizarro conquered the Incas he sent Peranzures de Campo Redondo to explore the land of the 'Chunchos' in the forests to the east of the Andes. This expedition, which took place in 1538–1539, reported that on one river a large quantity of cacao was encountered (see Patiño, 1963, p. 289). This is the first account of cacao being found in the Amazonian Region.

Although Patiño (1963) refers to documents alluding to the existence of cacao in the Amazon Basin during the century after the Spanish occupation the existence of sizeable stands of cacao and the prospects for its commercial utilization did not become generally known until 1639. In that year Father Cristobal d'Acuña accompanied the expedition of Pedro Teixeira on its return journey from Quito to Belém. In the report of his journey Acuña (1641) wrote about the magnitude of the occurrence of cacao on the river in the following terms:

The second product is the cacao of which the banks (of the river) are so full that sometimes when tree-trunks were felled to make the camps for the troops very few other trees were used besides those which produce this fruit so esteemed in New Spain and the countries where the chocolate is known. This cacao is so profitable that each tree gives an annual yield, discounting all expenses, of eight silver reals. On this river the same kind of trees can be grown with great facility and the minimum of skill because nature by itself ladens them with abundant quantities of fruits.

It would not only have been this report but the information given by Acuña and Father Domingos Brieva, the expedition's guide on its ascent of the river, to their companions on the expedition that sparked off the rush to exploit the wealth offered by harvesting the produce. Cacao then became the major contributor to the economy of the Amazon Basin during a period that lasted for over 200 years.

The reason why the occurrence of the species in the Amazon Basin had not been noticed previously can be attributed to the fact that the colonizers of the river up to the year of the journey did not know the name 'cacao' given to the plant in Central America. It would have been the presence of a person who had been in Central America and knew the plant who would have been able to connect the trees seen on the Amazon with those known as cacao in Mesoamerica. Acuña probably would have known the cacao already being cultivated on the coastal plains of Ecuador bordering the Pacific Ocean.

In the early accounts of the Amazon and adjacent regions, as well as the Caribbean coast of South America, there are frequent references to the occurrences of various fruits, and cacao would probably have been included in this category where it occurred. The plant would have been known by the different names applied by the indigenous tribes, as is still the case today. In Central America the tribes descended from the original inhabitants also have names for cacao that are specific to them.

Once the species had been identified as a producer of a commercial product of potential value resulting from the growing demand for it in Europe its exploitation in the Amazon got under way on a fairly large scale. The first exploitation was based on harvesting the fruits from natural stands along the river banks. This practice was followed for over 200 years and the volume of the harvests that was exported was responsible for cacao being the major foundation of the economy of the Amazon Region for many decades, until supplanted by rubber in the late 19th century.

The details contained in reports of travellers in the Amazon Basin and data regarding the cacao harvests are important sources of information concerning the locations of the occurrences of natural stands of cacao. From these documents it is also possible to obtain some ideas about the characteristics of the cacao trees in these locations and the uses made by the indigenous inhabitants of certain parts of the trees as well as the seeds.

One of the reasons why the European colonists of the Amazon did not connect the species there with the Mesoamerican plants was the absence of systematic cultivation and the use of the seeds for the same purposes. Several writers have interpreted the absence of signs of domestication as indicating that the Amazon Indians were not interested in cacao and had no use for it. The truth is that the contrary is the case. The Amazon aborigines, that is, the tribes that existed at the time of the European invasion, would not have had a need to cultivate a tree that occurred in some abundance. There is evidence that the Indians did plant cacao near their villages and this practice is followed up to the present time. The purpose of the trees grown near villages would have been to provide material for the various social and other uses to which the plant parts were put. There is also enough evidence to show that cacao seeds in the fresh form were consumed as food by certain tribes who occupied areas in which the species was plentiful.

Not many years passed before cultivation was taken up, particularly in the areas in which the species was not found in the natural state, these extended eastwards as far as near the mouth of the R. Amazon. The Jesuit missionaries played an important role in the expansion of the species' range through the plantations they established. The cultivation systems adopted apparently followed those in Mesoamerica that had been brought to Brazil by persons who had visited Mexico and Central America.

About 1720, French explorers discovered substantial areas of cacao beyond the headwaters of the R. Oiapoque in French Guiana and neighbouring areas of Brazil. The first reports of the occurrence of the species in the south of the Dutch Colonies of Essequibo and Demerara (in present day Guyana) were made about 1760. Cacao has since been reported as occurring in a spontaneous or sub-spontaneous state in other locations. All of these belong to valleys that eventually drain into the R. Amazon or in areas on the periphery of the Amazon Basin. In 1922 areas of what could be considered to be natural stands of cacao were discovered in the south of Surinam, thus completing the connection along the lands adjacent to the Amazon Basin beyond its northern watershed. The details of these discoveries will be given in the appropriate places in Chapter 5 when the cacao populations of the respective areas will be described.

Few of the accounts relating to cacao in these regions give any information regarding the nature of the plants that occur in the countries and areas concerned. This lack of information does not allow an analysis to be made of the diversity existing at any given time or place. The first person to observe differences among the cacao varieties he encountered in the Circum-Caribbean and Amazonian Regions was Alexander von Humboldt (1821). He described the contrast between the types on the coast of Venezuela and those of the upper Orinoco valley as follows:

> The tree that produces the cacao is not at present found wild in the forests of Terra Firma to the north of the Orinoco; we began to find it only beyond the cataracts of Ature and Maypure. It abounds particularly near the banks of the Ventuari, and on the Upper Orinoco, between the Padamo and the Gehette. This scarcity of wild cacao trees in South America, north of the latitude of 6 degrees, is a very curious phenomenon of botanical geography and yet little known. This phenomenon appears the more surprising, as according to the annual produce of the harvest, the number of trees in full bearing in the cacao plantations of Caracas, Nueva Barcelona, Venezuela, Varinas and Maracaibo is estimated at more than 16 million. The wild cacao tree has many branches, and is covered with tufted and dark

foliage. It bears a very small fruit, like that variety which the ancient Mexicans call 'thalcacahualtl'. Transported into the conucos of the Indians of Cassiquiare and the Rio Negro, the wild tree preserves for several generations that force of vegetable life which makes it bear fruit in the fourth year.

Secondary Areas of Cacao Diversity in the Americas

In addition to the two regions of the distribution of cacao that existed in the Americas at the end of the 15th century and that can be considered to be the sources of the genetic diversity of the species, several other areas have developed. The germplasm established in these supplementary areas would be derived from the intentional transfer of propagating material out of the principal regions. The transfer of this material would have involved small quantities, such as a single fruit or a few fruits and, possibly in rare cases, of plants. The result would have been the creation of populations that, through having small beginnings, developed with a small genetic base.

The oldest of these secondary populations is that occurring in the Pacific littoral of Ecuador that probably had its origin in the cacao varieties native to the eastern flanks of the Andes. After the discovery of cacao in the Amazon Basin and its exploitation, germplasm was taken to other areas of Brazil. The first was the area of tropical forest in the then 'Capitania' and later the State of Maranhão. Various attempts were made to establish cacao in regions along the coast and it is possible to find cacao planted in several places as far south as the Tropic of Capricorn. However, the main area of cultivation developed in the State of Bahia, resulting from various introductions made during a period extending over 100 years from 1746, the year when, it is claimed, the first planting took place in Bahia.

The varieties cultivated in the continental areas of the Circum-Caribbean Region were also important sources of planting material in other territories. Presumably, Jamaica was the first important beneficiary of such material, its incipient cacao industry being established when the island was a Spanish colony. The most likely source of this material would have been the indigenous plantations along the coast of the Caribbean. However, histories of the island written during the British occupation refer to cacao being introduced from Caracas.

Other islands would also have received planting material of a similar kind over a period of several centuries corresponding to the various attempts to establish cacao plantations in these islands. The most important of the islands was Trinidad, where cultivation appears to have begun soon after it became a Spanish colony. During the Spanish occupation the island was geographically and administratively connected with Venezuela and it would be logical to suppose that the planting material used came from that country.

The establishment of cacao cultivation in the areas of the Reino de Nueva Granada, which became the nation of Colombia, would have been based on seed brought from the, presumably, original home of the *Criollo* variety associated with the Lake Maracaibo region. A possible migration of cacao could be envisaged starting with the cultivation area of the upper Catatumbo area, around the town of San Faustino. Cultivation would have spread to the west, such as the area centred on San Vicente, then to the Magdalena valley and to Antioquia, the Cauca valley and Tumaco.

However, the widespread habit in Antioquia of drinking a beverage made from cacao, in a fashion similar to that in Mesoamerica, suggests a connection between this part of Colombia and the Circum-Caribbean Region. This may have occurred through the diffusion of customs and cacao through Panama and the Atrato valley.

One area that merits special mention on account of its 'indigenous' cacao population is that comprising the Province of Esmeraldas and the Department of Nariño in Ecuador and Colombia, respectively. Although politically divided these contiguous areas could be considered as being one cultural region and they would share some common elements as regards the history of cacao germplasm. The traditional cacao variety of Esmeraldas belongs to the *Criollo* group and has been described by Preuss (1901). The earliest cacao cultivated in Nariño would also have been similar but it is not known if any connection exists between the varieties originally planted in both areas. The Nariño cacao has been contaminated by the introduction of other varieties, making the comparison difficult. The problem arises in determining whether this variety is related to the ancient cacaos of the Pacific slopes of the Mesoamerican region or whether it derives from migration from the Maracaibo region. Other connections are possible, consideration of which will be made in the appropriate account of the variety. This region is of interest because of the use made of the fruits of *Theobroma gileri*, a species that may be endemic to this region, and which possesses some traits that are similar to those of cacao.

Patiño (1963), who made a compilation of the documentation relating to cacao in the various countries of the New World, provides a great deal of important and valuable detail fundamental to an understanding of the history of the species in its natural distribution range. Although it is the most valuable source of documentary evidence its scope is limited to the documents and publications available to the author when he undertook the review of the literature. Since the subject matter is divided according to modern political divisions, the threads that bind together the components of the regions in terms of their cacao diversity are obscured. Also, the treatment of Brazil is scanty, in terms of its size and importance, but this reflects the author's inability to consult the source literature.

One of the failings of a review of the literature is the absence of descriptions based on personal observations. One work that goes some distance to filling this lack, as it concerns comparative observations made in several cultivated populations, is that of Preuss (1901).

Cultivated Populations in the Old World and Their Origins

Nearly all of the tropical regions in the Old World, which possess environments that are similar to those of the original habitats of cacao, have, at one time or another, experimented with the introduction and cultivation of the crop. The history of cacao in these regions, involving as it does a large number of political and geographical entities and varied circumstances, would be difficult to describe in the space available. Each of these entities has its own history of cacao introduction and development of cultivation. The populations that have evolved all have their origins in germplasm introduced from its primary sources. Some of these are quite distinct. However, the vast majority of Old World producing countries owe their existing germplasm to introductions, either from the same primary source or from other Old World countries.

Consequently, as will be concluded, the genetic bases are similar but the composition of the populations that evolved varies according to the success or failure of establishing specific genotypes belonging to the introductions. Artificial selection, such as that emphasizing certain varieties on account of their qualities, would also influence the composition of the surviving germplasm.

It is necessary to emphasize that the transport of planting material to the Old World would have involved small quantities, primarily in the form of fruits and seeds in the same way as was described for the secondary areas of cultivation in the New World. Therefore, the genetic base of the populations of each country would be an insignificant sample of the germplasm that existed in the source country at the time the introduction was made. When selected varieties were involved in germplasm movement the genetic base would have been even more restricted in composition.

The first recorded migration of cacao from its New World habitat appears to be the arrival in the Philippines in 1666 of a single plant. The probable origin of this plant would have been located in the valleys where the crop was cultivated southeast of Acapulco, the port from which the Manila galleons sailed for Asia. Other sources have claimed that cacao was first established in the Celebes or the Sunda Islands. In the absence of any concrete evidence to support the claim, the most likely means of transport would have been the Manila galleons, as they regularly carried cacao to prepare the chocolate consumed by the crews. The possibility of an earlier, undocumented, shipment cannot be ruled out.

From the Philippines the crop was probably carried westwards in various stages to the Indonesian archipelago, the Malay Peninsula, South India, Ceylon (Sri Lanka) and other Indian Ocean islands. The presence of cacao in the Pamplemousses Garden on Mauritius was noted in the late 17th century, that is, not long after cacao was established in the Philippines. In the absence of the facts of the introduction to Mauritius other origins of the plant could be considered, but seem unlikely.

The first recorded movement of cacao belonging to the Amazonian Region populations took place in 1822. This resulted in the establishment of 30 plants on the island of Príncipe. These plants were also the first known introduction of the crop to the African continent. While cultivation on Príncipe was being expanded, cacao from this island was established on the neighbouring island of São Tomé. This development accompanied the increasing demand for the raw product so that the neighbouring European colonies on the West African coast began to be interested in the crop and in due course the variety that had been introduced from Brazil was being planted in most of the countries. The variety, on account of its adopted home in São Tomé, was there known as 'São Tomé Creolou', a name given, not because of the variety being native to the island, but to distinguish it from other varieties that had been introduced subsequently.

This variety constituted the large majority of the cacao trees grown in the other West African countries, because of which it began to be known as 'West African Amelonado'. The variety was later transported to other colonies of the same powers. In view of the fact that the West African countries became the largest producers of cocoa and its cultivation was adopted in other parts of the globe it constituted the variety of leading importance in world cacao production for many decades. It is still important in the context of the improvement of the species.

Evolution of the Activities Related to the Development of Genetic Resources Conservation

The first introduction of a variety from the Amazonian Region to a potential producer in the Circum-Caribbean Region was made about 1660. However, the introduction from the Amazonian Region to Trinidad in 1757 should be considered to have had a larger effect on development of the cultivated varieties. Hybridization between the two varieties and recombination during the succeeding generations created an extremely heterogeneous population with a vast range of phenotypes.

The origins of a serious interest in the variability exhibited in the heterogeneous population may be attributed to the establishment, on the island of Trinidad in about 1870, of the first collection of genotypes that were considered to represent the 'varieties' that were identified within the population. This collection had an important influence on cacao development far beyond the confines of the island, since it was the planting material supplied from this collection that contributed to cultivation in the Old World. The last decades of the 19th century and first decades of the 20th century were marked by an enormous activity in the exchange of planting material between the older producing countries. The result was the creation of new genetic variability through hybridization and subsequent recombination during succeeding generations.

During the subsequent decades, accompanied by the establishment and development of research activities related to genetic improvement, mainly stimulated by work in Trinidad, additional varieties were added and increasingly greater information became available from cacao populations in other countries. Although such activities were conducted on a relatively small scale they did make important contributions to the development of knowledge about the phenotypic and genotypic differences within and between varieties from different origins. The early years of the 20th century were important in establishing that the genetic variability that the species possessed was essential for variety improvement and that this variability could be harnessed for the development of superior cultivars. The interest in cultivar improvement was accompanied by the development of methods of vegetative propagation in a species that had, hitherto, been propagated by seed.

At the same time that methods of vegetative propagation were being developed, programmes of individual tree selection were initiated, first in Trinidad and Java and later adopted in other countries, particularly in Latin America. During the 30 years following the example of Trinidad several countries had established clones from the selections made in the local populations and the aggregations of these formed the first germplasm collections. Later the idea was put forward to exchange the selections among the countries and this resulted in the formation of collections with a wider range of diversity.

However, it was not until the 1930s that information became available about the diversity that existed in the Amazon valley. The results of collections made at that time gave an impetus to interest in conserving and utilizing (particularly with regard to disease resistance) the genetic diversity of this large region. Subsequently, various activities that were concerned with collecting and conserving the germplasm that occurs in the individual countries were undertaken. The first of such activities was carried out in the late 1940s, during which time a few collections were made in the Amazonian area of Ecuador.

Activities of this kind started later in Brazil. The first collecting activity made in the Brazilian Amazon took place in 1965 and collecting continued sporadically on a small scale during the next 5 years. In 1976 a programme designed to survey the entire territory was initiated. The objective of the programme was to undertake systematic collections in the various drainage systems and was carried out between 1979 and 1987. In total these activities resulted in the establishment of the large number of genotypes that were acquired in a network of collections that embrace the largest collection of diversity in the world.

The last half of the 20th century has seen much of the large pool of diversity that exists in the region becoming available for research. The stimulus and motivation for conducting germplasm conservation programmes has been due in part to the pressure on the natural habitats of the species by human factors responsible for development and population growth. During the course of these programmes, experience has been accumulated regarding the management of germplasm collections. The need for undertaking the characterization and description of the individual accessions at all levels of the genetic resources programmes was accepted. In order to provide a guide for persons involved in the descriptions of the characteristics of the genotypes a comprehensive descriptor list was created in 1979, in which the characteristics are presented in a logical order.

As a result of these activities it has become an established fact that the Amazonian Region is the storehouse of virtually all of the diversity of the species. The scale of diversity and the state of availability of information about it are such that no single person has a conception of the total range of the genetic variability that the species contains. In view of this situation it becomes necessary to attempt to gather together as much as is known about how the diversity is constituted and its nature in order to provide the users of the genetic variability with a wider conception of the subject than has hitherto been at their disposal. Because of the limitations and space, this large subject cannot be treated in its entirety. The most that can be done in the circumstances is to present as much of an overview as possible, which would be of benefit to persons with an interest in the subject and with the need to utilize the germplasm for various purposes.

In this way, it cannot be claimed that we have reached the stage at which we possess a comprehensive knowledge about the species' genetic diversity. Documents such as travellers' reports and official reports and herbarium specimens provide information that shows that many areas remain to be explored. In the regions of natural dispersion there is a strong imbalance with regard to the efforts expended by the countries concerned in germplasm conservation. Even in the countries in which most collecting and conservation has taken place it is not possible, for physical and other reasons, to sample all of the variability that exists in a given location, especially where the population is large and heterogeneous. Although conservation of the diversity has proceeded to the stage where the genotypes have been established in the field, these have later been lost because of a variety of reasons.

The state of availability of information is also far from complete and probably unsatisfactory. The nature of the species is a factor that makes characterization and evaluation a slow and usually costly business. In addition to the species being a perennial tree development of plants can often be slow so that time is required before characterization can be undertaken. The various organs of the plants are produced in stages and during separate seasons requiring that the plants be kept under observation for several years in order to obtain the information necessary for the complete characterization of their attributes. In the majority of circumstances, although the germplasm has been collected and established, observations have not been made because of the inadequate conditions for conducting the research that usually are the rule. This applies particularly to the observations that would be made in the field. There are accessions that derive from collecting activities carried out 20, and perhaps, 50 years ago, about which we still know little or nothing of the characteristics and utility of the genotypes represented in these collections.

2 The Terminology Specific to Cacao

One of the features that is special about *Theobroma cacao* that a person who is connected with the species needs to understand is the nomenclature that has evolved during the history of its cultivation. This specifically applies to all aspects concerning names of plant parts and varieties. Most of the terms that are frequently encountered in the literature of the species are Spanish in origin, some of which derive from the first contacts the conquerors had with the countries in which the crop was cultivated at the time of the conquest. Most of the countries in which cultivation was practised were under Spanish rule during the greater part of the past 500 years and Spanish is still the language of most of the producing countries in Latin America. In many cases the Spaniards applied to the various American plants and their parts terms that derived from their similarities to those of plants with which they were familiar in Spain. Accordingly, the original meanings of several of the terms used in connection with cacao must be sought for in the common Spanish vernacular. Some of these have undergone alterations when applied in countries where other languages are spoken in order to conform to the pronunciation or spelling used in those countries. There are, of course, cases in which the words are directly translated into the languages used in other countries.

The most important terms used presumably can be attributed to their origin in Trinidad where the crop was cultivated when the island was a colony of Spain. However, there are expressions used relating to cacao that may be specific to some Mesoamerican countries and regions as well as to other major producers, such as Venezuela. These expressions may be encountered in publications from the countries concerned.

The Common Name of the Species

The word 'cacao' is accepted as being a Spanish corruption of the Nahuatl name for the plant and its product. This entered the vocabulary almost as soon as the Mesoamerican region had been subjugated. It continues to be the name used in all of the languages in the countries of the American continent. (In Brazil the word Cacáo, which should be the correct spelling of the name, was used until at least the end of the 19th century, although the modern spelling is cacau.) Cacao is, of course, used in the specific name of the botanical combination attributed to Linnaeus.

At some time in the past the word 'cocoa' entered the English language, generally accepted to have been a mistake. However, the term has been adopted, unfortunately, in English-speaking countries as a name for the crop and also occurs in titles of various organizations. When the word 'cocoa' was introduced it was considered to be inappropriate, especially since it was confused with similar names of other plants. Since the word 'cocoa' had already become common in trade and industrial contexts, the policy was adopted by the early research workers of using cacao with reference to the plant and cocoa where the commercial product is concerned.

Since my personal preference is to use a terminology that is universally practised, the word cacao will be entirely used in this book, with the exception of quotations and titles in which 'cocoa' occurs.

Plant Parts

It is possible to find various descriptive words used locally in several of the producing countries for organs of the plants that are most distinctive. Most of these terms are of Spanish, Portuguese or French origin. Few of them perhaps have entered into cacao literature in English except in occasional travellers' reports or botanical descriptions. Among the most important terms that can be found universally is the Spanish word '*horqueta*' or in its corrupted form '*jorquette*' (again derived from Trinidad and/or Venezuela). This name is given to the whorl of plagiotropic branches that are produced at the end of an orthotropic stem. '*Horqueta*' translates as the fork or crotch of a tree but there does not appear to be an English or botanical equivalent for the particular morphological character in cacao.

In English usage the word 'pod' is applied to the fruit. In old documents other words of similar implication can be found such as 'cod' or 'purse'. However, the word 'pod' is the incorrect botanical term for the fruit. For this reason, and to use a term applicable in all circumstances, the word 'fruit' is preferred. The fruit is known by other names in various countries, for example, '*mazorca*' is the term most commonly used in Spanish-speaking countries.

Following from the use of 'pod' for the fruit, the seed has been called a 'bean'. Since this, again, does not define correctly the type of seed it will be avoided in this account. In the past the seed was also called an 'almond' (Spanish – *Almendra*, Portuguese – *Amendôa*) from its resemblance to the seed of that species.

In order to establish definite, unambiguous terms for the respective parts and to conform to the practice in industry, the words 'husk' and 'shell' will be used, respectively, to denote the outer covering of the fruit and the dried testa (covering) of the seed.

Names Used in Relation to Cacao Varieties

It is in this area that words derived from Spanish predominate, but not exclusively. The manner in which these terms are applied can be extremely confusing since there are no exact definitions as to what they mean. The result is that these terms are applied differently by various persons and in a variety of circumstances. On account of this lack of definition the same variety type can receive two different names or the same name is applied to different genotypes that are usually distinct in most of their other characteristics. The several ways in which the various terms are applied have been used to construct hierarchical classifications of 'varieties' or 'populations'. It will be seen that, when the geographical origins, genetic constitutions and relationships of the genotypes are considered, there is no justification for maintaining these terms and classifications.

Many of the difficulties encountered in interpreting such classifications and variety descriptions are the result of the applications of knowledge derived from the small range of diversity that was available to the early writers on the subject or to certain observers. The terminology adapted during the early period does not apply to the much larger range of diversity that is currently at our disposal and in the context of the enhanced knowledge concerning the genetics of the species.

The fruit of cacao is the part of the tree that is most usually visible and it, together with its component parts, is most closely connected with the commercial product. As a consequence, it has been customary to use the fruit characters to indicate differences between trees and thus most of the names adopted to describe cacao varietal types and populations are based on the fruit characters. These are usually Spanish words, frequently relating the fruit types to those of other species of plants. Other terminology applied to fruit characters has been based on botanical terms or with the intention of establishing a botanical classification for the diversity.

There are several instances involving the attribution of 'taxonomic' epithets to distinctive types of plants that have attracted attention, many of these being based on the fruit characters. In this context it should be observed that the accepted ending for taxonomic names in the genus *Theobroma*, at species or sub-species levels, is the neuter form *-um* and not the feminine form *-a*. For reasons of correctness and uniformity, the ending *-um* will be used throughout the book even though the feminine form may have been given in the original applications of the names concerned and repeated in the literature.

The unsatisfactory state of affairs relating to the classification of genetic resources requires that a rational system must be developed in which the genotypic distinctions are clearly defined and universally applicable. In the following sections an attempt will be made to explain the meanings of the terms that have been and are currently used.

There are two words derived from Spanish that are basic to any discussion of variety types and the classifications that have been proposed in the past. These words are 'Criollo' and 'Forastero' and they appear frequently in the literature.

'*Criollo*' is translated as 'native'. It can be used in a variety of circumstances to distinguish the cacao types that existed, or exist, in a given geographical location or region from the varieties that have been introduced from other regions. As a result, the word can have different meanings in connection with variety identification according to the circumstances in which a particular variety occurs. It is used in the original form or translated into the equivalent words in other languages in the indigenous distribution range, Portuguese *Creoulo*, French *Créole*. The word is also applied to the first variety cultivated in a geographical unit outside the indigenous range of the species to distinguish it from later introductions, e.g. Guadeloupe Créole. The variety of ways in which the word '*Criollo*', or equivalents, can be employed is a source of much confusion to persons who lack a proper perception of the variety situation in the species.

For this reason it is necessary to establish a specific definition for the genetic type to which the word is normally applied and referred to in classifications. This use of '*Criollo*' as a variety name may have originated in Trinidad to distinguish the variety that had been exclusively cultivated there from the variety introduced from the Amazon Basin in 1757. As will be explained later, the first variety cultivated in Trinidad probably derived from the populations of cacao that had been brought into cultivation in northern Venezuela. In view of this genetic association it became permissible to include the Venezuelan types under the same generic name of '*Criollo*'. As an extension to this use and the fact that the ancient or classic varieties cultivated in Mesoamerica possessed similar characteristics it became the practice to include them in the term '*Criollo*'. The name could also be extended to include the varieties cultivated in other areas of South and Central America that were descendants of the above varieties.

In recognition of the similarities expressed by the diverse populations to which the word 'Criollo' has been applied historically when referring to cacao varieties it is proposed to continue this practice and, accordingly, apply the term in the text in the form of *Criollo*. However, it must be appreciated that the diversity covered by the term could comprise local and regional variations in respect of one or several characters. Some of these variations are given specific names that will be described below. In applying the term, Cheesman's (1944) proposal will be followed in that we should consider the existence of a group of varieties to which the term *Criollo* is applied and not to establish a hard and fast definition of the term. In this way it will be sufficient to give names to each of the various populations belonging to the *Criollo* group, when it is appropriate to do so.

The word '*Forastero*' signifies foreign, that is, a variety not connected with the indigenous (or first cultivated) variety of a region. It could be assumed that the name was first applied to the variety that was introduced into Trinidad from the Amazon Basin in 1757. However, this particular variety was given the name '*Calabacillo*' (= Little Calabash) with reference to the shape and size of the fruit. The term '*Forastero*' began to be used as a name for the mixture of types that segregated in the generations following hybridization of '*Calabacillo*' with the original '*Criollo*'. From Hart's description of '*Forastero*' in Bailey

(1947) it is abundantly clear that the name was applied to the hybrid forms. (It could be added that some authors have spelled the term as 'Forestero' assuming that it means 'from the forest'.)

Following the attempts to produce a rational classification of the diversity that developed in the Trinidad population as well as those of other Caribbean island growers, 'Forastero' was applied to any type that could not be identified as belonging to the *Criollo* group as defined at the time. In this way 'Forastero' came to be used to cover all of the variability that existed in the Amazon Basin. As will be apparent from the descriptions of the cacao varieties indigenous to the Amazon Basin no single term could cover all of the diverse forms that occur. Although *Forastero* would be a suitable term to apply in countries outside the Amazonian Region to genotypes from the region it is unacceptable that the word would apply to varieties that would be native (and, therefore, Criollo) to the Region. Hence, it is considered that the word *Forastero* having this implication in variety nomenclature is inadmissible and will not be used in this book.

The names applied to local variations in the *Criollo* group are often of a descriptive nature, usually referring to fruit characters. The most noteworthy of these names are 'Porcelana' and 'Pentagona', which will be defined in the appropriate place when the indicators of variability are treated. Regional names also occur, such as 'Caracas' and 'Soconusco', but as these names were applied to populations that formed part of a wider distribution of the same varieties they cannot be considered to be variety names and should be confined to their use for identifying the origins of commercial cacao.

It is obvious that the concept of the name 'Forastero' as used in the early literature relating to cacao varieties in Trinidad was inadequately defined and was applied to several genotypes or groups of them in a haphazard manner. In fact, no established system of names existed and, as Preuss (1901) observed at the time of his visit to Trinidad, no two persons agreed on the names of 'varieties'. Some of the published accounts of cacao in Trinidad dating from about the beginning of the 20th century do not use 'Forastero' in connection with variety names.

In several publications, especially those relating to specific populations, the words 'Angoleta' and 'Cundeamor' occur. Normally, they would be applied to individuals belonging to *Criollo* populations or descendants of their hybrids. Both names concern fruits of an elongate shape, the difference being that a 'Cundeamor' fruit, as the term is presently applied, has a basal constriction (bottle-neck) and a rough surface, while the 'Angoleta' fruits have a broad base. Often the distinction is slight and, in some trees, the form of the base of the fruits can vary to such an extent that even the most ardent supporters of these names cannot determine the class to which a particular tree belongs. In addition, these terms cannot be applied universally to all of the genotypes with elongate fruits that occur in the universe of the species' diversity as the various illustrations in this book attest. Accordingly, these words, along with *Forastero*, will be ignored unless used in quotations.

Cundeamor is the name given to a vine-like plant belonging to the *Cucurbitaceae*, the fruits of which are small with a rough surface. Some observers are of the opinion that the current application of the name is wrong.

The fruit type called '*Cundeamor*' in the Dominican Republic is illustrated in Ciferri (1929, 1933), from which it can be seen that it differs from the current definition.

Other adjectives associated with variety names include those relating to fruit colour, basically involving shades of greens and reds, where unripe fruits are concerned. The practice of distinguishing the fruits with red colour has been particularly prominent, which means that such 'varieties' received epithets that included the word 'red' or its synonyms in other languages; e.g. *colorado*, *rojo*, *roxo*, *vermelho*, *morado*. 'Green' appears to be less frequently used in this context and, if so, it is usually in combinations with the Spanish form *verde* (e.g. 'verdilico'). The Spanish *blanco* (= white) may be found applied to light shades of green. The colours of the ripe fruits are usually shades of yellow and names of types having 'green', unripe fruits may also include the adjective 'yellow' or 'amarillo' to distinguish them from the otherwise similar variants with red fruits. Presumably, the name 'Golden Caracas', given to a variety grown in Guyana, applied to trees with green fruits.

The large range of expressions of fruit shapes and other characters has resulted in the application of a diversity of descriptive names to distinguish between the forms. These names are, again, mostly derived from Spanish equivalents and have entered into the process of variety nomenclature. The origins and the meanings of these terms will be discussed in greater detail in Chapter 3.

The emphasis placed on fruit characters results in the vernacular names of fruit shapes being treated as variety names, a practice that is not always justified on the basis of the genetic composition of the genotypes to which the name is applied. It has, unfortunately, become a practice, in certain circles, to use the word '*Amelonado*' (given due to a similarity to the shape of melons) as the name of a population or variety. In this context, genotypes possessing fruits of the same appearance are attributed indiscriminately, regardless of their origins and genetic composition, as though they are descended from the same variety.

The use of such terminology in the names of varieties is acceptable when they refer to specific genotypes or varieties and when they are components of a name that includes another identifier that distinguishes them on the basis of the specific attribute possessed by each. Even combinations of terms in names can result in inaccurate and ambiguous nomenclature. For example, the type introduced into Trinidad from Brazil and named 'Red Amelonado' is by no means the same as the 'Amelonado Roxo' (= 'Red Amelonado') of São Tomé (depicted in Plate 7b). In each case in which the name, or its variants, has been applied, it identifies a specific element of the diversity and not as a general identifier as erroneously supposed.

Bernoulli (1869), in an attempt at developing a classification of *Theobroma*, proposed several 'sub-specific' combinations within *Theobroma cacao* based on the occurrence of distinctive fruit types within the *Criollo* population that was cultivated in Guatemala at the time. These 'subspecies' were given the names *Pentagonum*, describing the distinctive five-ribbed fruit and *Leiocarpum* applied to fruits with smooth husks that had no prominent ridges. Neither of these names could be acceptable as variety names in their own right, as will be explained in subsequent chapters, unless they form part of a name in which a specific component is included that describes the origin or relationship of the genotype concerned.

Since the term 'Leiocarpum' describes smooth fruits a tendency arose to attribute this name to varieties from other populations, the fruit husks of which are also smooth (or nearly so). Among the varieties to which the name 'Leiocarpum' began to be attached was the Calabacillo of the West Indies (which would perhaps be called Amelonado according to other definitions of these terms). The indiscriminate use of the word 'Leiocarpum' for genotypes from the Amazonian Region populations or hybrids is invalid because the name proposed by Bernoulli was originally based on the Criollo varieties he studied. The name Leiocarpum has generally fallen out of use but occurs in the literature, for which reason it is necessary to explain its application.

Another name that is applied to fruits with smooth surfaces is 'Porcelaine', normally associated with former Dutch colonies. The term used in this context should not be confused with the similar name 'Porcelana' although the meaning is the same, since the 'Porcelaine' of Suriname applied to a variety from the Amazon Basin. However, the name 'Java Porcelaine' was given to a Criollo type on that island.

A 'taxonomic' name that was introduced and has caused a great deal of confusion through the improper use not connected with the initial application is the word 'Sphaerocarpum' (= spherical fruit). The term was proposed by Chevalier (1908) for a type he encountered on São Tomé with spherical fruits (locally named 'laranja' = orange), which is depicted in Plate 7c. Later, in his revision of the genus Theobroma, Cuatrecasas (1964) proposed creating a sub-specific form of cacao called Sphaerocarpum, under which all cacao occurring in the Amazonian Region would be placed. Such a practice is unacceptable but, unfortunately, has been adopted by some. In the first place, there is no botanical basis for the proposed classification since the genotypes of the Amazon Basin embrace a large range of fruit shapes and it is a fallacy to assume that the typical fruit of the region has a spherical, or almost spherical, shape. In fact, the type illustrated in Cuatrecasas (1964) as a representative of his subspecies sphaerocarpum type is anything but spherical in shape. The Herbarium of the Royal Botanic Gardens at Kew in England has specimens of dried fruits from São Tomé (Lima e Gama, 1911). These are perfectly spherical and undoubtedly represent the 'Laranja' mutation. These specimens are identified as 'Caracas'. This indicates that the mutation would have occurred in the Criollo variety introduced into the island, a conclusion confirmed by the fact that red fruits occur, such as those of the tree in Plate 7c. Although Cuatrecasas would have seen these specimens he ignored the connection with Criollo when he created the subspecies sphaerocarpum to designate the Amazonian varieties that have entirely green fruits. Consequently, it is necessary to ignore Cuatrecasas' proposed classification.

Local and regional names abound in cacao. One of the more important sources of such names is Trinidad. At the Botanic Garden in Port-of-Spain a collection of 'named' varieties was established in the late 19th century. The creation of the collection has some significance since seed from these 'varieties' was sent to countries such as Fiji, Ceylon (Sri Lanka) and Singapore and the names may still be encountered in these places. The names applied to the 'varieties' are of interest in connection with nomenclature. They include: 'Cundeamor' (of different types) and 'Criollo' under the name 'Ochroe' with red fruits. ('Ochroe' is the Trinidad equivalent of okra *Hibiscus esculentus*, which suggests that the name was applied to a narrow elongate fruit with prominent ridges.) Other names are 'Forastero' – red; 'Cayenne' – red and yellow (small, elongate, perhaps having a smooth surface); 'Verdilico' – green (perhaps correctly 'Verde Liso' = green/smooth); 'Sangre Toro' (= bull's blood, but this is ambiguous since the list has 'only green'). Perhaps 'Sangre Toro' implies a name for something else, with a less 'genteel' association, for example, being confused with the name 'Cojón de Toro' (= bull's testicle).

The three types under the name *Cundeamor* include one whose fruits had a smooth surface and another with a broad shoulder. Both of these diverge from the present concept of the name and show that the concept that existed in 1880 was very different to that applied during the past decades. From the separation of varieties in the groups called *Cundeamor* and *Forastero* it is evident that these names referred to different types.

Although no accurate descriptions of the types in the Trinidad Botanic Garden collection have been found, the details accompanying the variety names permit some comparison to be made with types still existing in Trinidad. The Trinidad Botanic Gardens' reports refer to a set of paintings of the fruits of these 'varieties', which had been made and kept for exhibition. Discovery of the whereabouts of these paintings would provide a means of defining the characteristics of the types concerned.

Later, another collection of Trinidad 'varieties' was made involving 18 named types. Planting material from them was supplied from this collection to neighbouring countries. The 18 'varieties' included ten *Forastero* of various fruit colours. Four of these were 'verrugosa' (rough husk) and two were 'Amelonado'. This demonstrates the large range of variability that occurred within the concept of *Forastero*. Three of the other types were called 'Creole' but two of these types were described as 'smooth', which may indicate a similarity to some of the Chuao population or that they did not belong to the *Criollo* group. There is only one *Cundeamor*, there being no hint as to its characteristics.

The differences in the names used in the two lists are symptomatic of the confused state of nomenclature that existed. Various attempts at classifying this variability were attempted. At that stage no consideration was given to the origins of the Trinidad cacao populations. Also, since the subject of inheritance was unknown, the genetics of the characters that distinguished the varieties could not be applied in the development of classifications. Morris (1887) made the first attempt to classify the cacao varieties in Trinidad and this influenced the variety nomenclature followed by several authors. Cheesman (1944) proposed an extended classification that included the Amazonian Region types that had been added to the existing germplasm in Trinidad. It was in this classification that the Amazonian varieties were included in the *Forastero* class.

As will be described in Chapter 6, the introduction into Venezuela in the 1830s of planting material from Trinidad brought about a change in the composition of the cacao populations in certain parts of that country, where the cultivated types had exclusively belonged to the *Criollo* group. Either the introduced types or the progenies descended from hybridization began to be distinguished and given names.

Preuss (1901) provided an account of the names applied to different fruit types under the general title of 'Carúpano' (the region closest to Trinidad where the introduction process presumably started). The names are sufficiently descriptive to enable some comparison to be made to the Trinidad varieties but few of them have any significance as they do not appear in the literature and may have fallen out of use. Perhaps one Venezuelan name is of interest since it occurs in later literature. This is 'Cojón de Toro', which is somewhat descriptive of the fruit shape that corresponds to the 'Calabacillo' named above. Another name current in (parts of) Venezuela is 'Legón', which may be the same variety or something related to it.

Another name that probably arose in Venezuela is 'Trinitario'. Presumably this name was applied to material introduced from Trinidad. Pittier (see Schnee, 1960) reported that the word was used to designate the type planted in eastern Venezuela, which would correspond to the original 'Calabacillo' of Trinidad. A tree of this type in the Barlovento region of Venezuela is illustrated in Plate 26c.

In the course of time the word '*Trinitario*' came to designate the various elements of the Trinidad populations, which could not be identified as *Criollo* or any type recognizable as being of Amazon origin. In other words, the term covers all of the products of hybridization and recombination through various generations. In this sense it has the same connotation that '*Forastero*' had previously.

In some senses 'Trinitario' may be a valid designation for the components of the diversity of phenotypic types of the Trinidad population prior to the more recent introductions. Through the common parentage the offspring of planting material that was supplied from Trinidad to several producing countries, in the New World as well as the Old World, would also be included.

However, an unfortunate tendency has developed to label as 'Trinitario' all genotypes of hybrid origin regardless of parentage and other genotypes that are not attributable to other variety categories. In other words, the term is used as a sort of mental waste bin for unclassifiable material. This policy is inappropriate in that the expression is used to include varieties that are not related in any way to the genotypes from which the Trinidad population evolved.

Among the regions in which specific variety or population names evolved is the cacao area on the Pacific coast of Ecuador. In times past the cacao of the region – at least in other countries – was identified as 'Guayaquil'. Following the introduction of other germplasm from about 1890 the name 'Nacional' came to be used to distinguish the indigenous varieties from the introductions to which the name 'Venezolano' is usually applied. However, some farmers use the term 'Criollo' for the indigenous varieties. Another term associated with the Ecuadorian population is *Arriba*. *Arriba* is a Spanish word meaning 'above' and refers to the region 'above' (north of) the part of Guayaquil. It is synonymous with the area marked as 'Guayas Basin' on the map in Fig. 5. It is a name used in the trade to distinguish the product of the region from those of other areas in Ecuador. The variety concerned would have been the traditional 'Guayaquil' or 'Nacional', but *Arriba* is not a variety name.

In Jamaica at the beginning of the 20th century several names appear to have been used for different varieties but it is probable that they are no longer used. These names were given by Morris (1887) but no descriptions accompany the list. The incorporation of the word 'Spanish' in some names indicates that they identify descendants of the *Criollo* types cultivated during the Spanish period. 'San Domingo' was given to another variety. This would have been the 'Amelonado' type introduced from Haiti by the French settlers who fled the country during the revolution of 1804.

In Brazil variety designations are not common and in rare cases names that were applied to specific types can be found in the districts of the middle Amazon where cacao is cultivated. The few references on the subject, such as Le Cointe (1934), recognize two extreme types of the spectrum of diversity based on fruit shapes. One was called 'Jacaré' (= Alligator) on account of the elongate fruit with a rough husk. This name is equivalent to the 'Lagarto' (translated as 'Alligator') of Central America, where it is applied to the fruit of the '*Pentagonum*' with its knobby surface. Preuss' (1901) description of the cacao in Nicaragua suggests that the word 'Lagarto' was applied to any fruit with a very rough surface. In the two regions the allusion to 'Alligator' implies a similarity to the appearance of the fruits but by no means that the trees belong to the same genetic population.

At this point the use by Pound (1938) of the term 'lagarta' for descriptions of fruit shapes could be mentioned. It this context the word has no relation to the *Criollo* 'Lagarta'. His use is rather ambiguous and, as described in Chapter 5, the term was applied to elongate fruits regardless of the roughness of the surface.

Another type in the range of fruit shapes referred to by Le Cointe (1934) was called 'Cabacinho', which is the same as 'Calabacillo' and the name could be applied to some of the varieties in the State of Bahia.

The other types introduced into Bahia and elsewhere from the Amazon were distinguished according to the respective introductions. In Bahia they were named 'Comum' (= common or usual) or from their origins 'Para' and 'Maranhão'. The origins and characteristics of these types will be described more fully in Chapter 6.

In Mexico (and, in the absence of indications as to the occurrence of such varieties, these names may apply to other Central American countries) the names 'Guayaquil' (descended from seed introduced from Ecuador) and 'Ceilán' or 'Ceylán' are found. Although the latter are described as being cultivated no satisfactory definition has been found to enable the type to be identified in relation to the other germplasm.

With regard to the varieties exported from the Amazon prior to the 19th century the name 'Martinique Créole' was used, as previously mentioned, to distinguish the first variety introduced into that island from the later introductions.

In this volume this name will be retained to distinguish the variety concerned from the Trinidad 'Calabacillo'. As suggested later the 'Martinique Créole' was probably the ancestor of the Amazonian type that is found in Costa Rica, to which the name '*Matina*' is given (name of place). A similar variety that was introduced to Colombia in about 1890 is known there as '*Pajarito*' (= little bird), which may have been the same as the Trinidad '*Calabacillo*'; apparently, the name was given on account of the small seeds.

The extent of the possible variety names, based on geographical, personal and descriptive designations, is limitless but the above are those that are more common in the literature and could be referred to in the text. However, a number of distinct types have arisen in various populations and have received designations based on the specific characters expressed in each genotype. Such types include 'Jaca' (leaves resembling those of jackfruit), 'Crinkled-leaf Dwarf' (leaf abnormality) and 'Maracujá' (flowers without staminodes, resulting from their supposed similarity to the flowers of passion fruit; Plate 8c). There are numerous other designations derived from location and personal names.

In addition to some cases of place names, usually found in abbreviated forms, proper nouns are rare as variety names. A few examples of these are 'Catongo', 'Almeida' and 'Mocorongo', all three genotypes possessing the anthocyanin inhibitor gene.

When the selection of individual trees began, an alpha-numeric system was introduced to identify the individual selections. Some of the alphabetic components of the identifiers are complete words such as the names of locations. More commonly, the custom has developed to use abbreviations and acronyms. A plethora of these is now in existence, the alphabetic components referring to countries, locations, geographical features, institutions and programmes, to cite but a few. The same system has been applied in the identification of families resulting from inbreeding or hybridization. The description of the meanings of the alphabetic components of variety or clone names is beyond the scope of this book. Explanations of the meanings and origins of these designations are to be found in the International Cocoa Germplasm Database (www.icgd.reading.ac.uk). The reader who requires details about any of the identifications that may be referred to in the text should find the details in the appropriate publication of the database. However, to enable readers who are unfamiliar with the alpha-numeric system of nomenclature of cacao genotypes and families to obtain a notion of the meanings of the abbreviations and acronyms, those of the genotypes named in the book are listed in an Appendix with their derivations and origins.

3   The Indicators of Variability

An individual plant's phenotype is a composite of the expressions of the various organs of which it is composed. In the universe of plants that make up the species each organ is observed to exhibit a variety of forms that differ according to the individual plants. The variation in the phenotypic expressions is the result of the action of the different alleles that occur in the genes that control the specific characteristics and the total of the alleles make up the plant's genotype. The degree of variability that is observed in some organs is exemplified by the vast array of forms and characters of the fruits that were generated in the descendants of the hybrid between the *Criollo* variety originally cultivated in Trinidad and the '*Calabacillo*' introduction. The scale of this variability attains significant importance when it is considered that the complex population created derived from two genotypes. This would have been especially important when account is taken of the fact that they would have been homozygous, or nearly so, for all characters.

The range of variability exhibited in terms of the forms and colours observed becomes larger and more complex with the exploration of new populations. Studies carried out on the available resources or in the progenies developed by inbreeding and hybridization show the extent to which the genetic variability is developed.

As indicated in Chapter 2 a significant emphasis has been placed on the appearance of the fruits for purposes of measuring variability and differentiating the genotypes into varieties or populations. This emphasis on the fruit characters has resulted in the variability expressed in the other organs of the trees being ignored, except for rare examples in which differences or specific attributes of one or more parts of the plants have been described.

Information about the expressions of the various organs of the plants has been accumulated over many years of observation. The data obtained have been important in widening the knowledge concerning all of the variable characters exhibited by the genotypes observed. The knowledge acquired demonstrates that the individual genotypes and the populations to which they belong are distinguished in many ways other than on the basis of fruit characters alone.

Attention has been paid to determining the procedures for describing the character expressions of the individual genotypes ever since the first attempts at variety nomenclature and selection were made. In Chapter 2 it was stated that the first known attempt to demonstrate the differences expressed in various plants concerned the illustrations made of selected 'varieties' in Trinidad. The subject of describing the distinguishing characteristics of cacao varieties has been raised from time to time. The availability of an organized system of descriptions of cacao genotypes became a necessity in order to identify the large number of selections that had been made and were being distributed to various countries. In the 1960s steps were taken to develop such a system on the basis of investigations on the variations shown in certain characteristics.

The subsequent years witnessed a great expansion, not only in the quantity of genetic variability that had become available, but also information about the characters involved and their modes of expression. The first phase in the development of the descriptors for cacao was concerned with the identification of the selections that had been made and were exchanged between research stations. The accumulation of a larger amount of information on the character expressions from the extensive range of germplasm that was becoming available prompted the development of a comprehensive list of the descriptors that applied to cacao. This list involved the systematic arrangement of the various organs of the plants and the expressions of the character states about which observations had been made.

Since it deals mainly with the characteristics that are easily observed and measured, the descriptor list that was published (Engels *et al.*, 1980) does not include some traits in which genetic variation occurs but which are not readily determined. In general, these traits include those related to the composition of the plant organs and others of a quantitative nature whose expression is influenced by non-genetic factors. Among the traits in the latter category are those involved with growth and productivity.

From the list of descriptors it can be appreciated that many elements comprise the genetic variability shown by the genotypes that make up the diversity. The list provides a guide to the individual traits that will be observed in determining the constitution of a genotype and the variability among genotypes. The knowledge contained in the descriptor list can be applied in all stages of the genetic resources programme. A person who is involved in collecting germplasm in a population of trees would use the list to make observations on the individual trees from which the degree of heterogeneity in the population would be estimated. This estimate would be used to determine the sampling procedure to be adopted in order to obtain a representative sample of the variability in the population.

During the management of a germplasm collection, observations of the characters, or morphological markers, expressed in the accessions are essential for purposes of identification. The description of the characteristics of the individuals belonging to a population provides the means of recording the extent of variability in the population sampled and the differences among populations.

In the following pages some of the more important attributes that have been observed to produce the differences among plants will be described and the various expressions that have been noted in each case will be discussed. It should be appreciated that, given the fact that the process of observing the variability is still developing, there will always be the possibility that other expressions may be encountered as the range of diversity is enlarged. Another point that should be understood is that several characters may occur in conjunction to constitute the genotype of an individual. This is particularly the case when the genes involved have a pleiotropic action in which the expressions of several characters are the product of a single gene.

The Structure and Properties of a Cacao Tree

The knowledge concerning the structure and development of cacao trees is essential to obtain a comprehension of the details presented in this chapter. Although most readers may possess a basic knowledge of the subject it is appreciated that others may not be familiar with the appearance of cacao trees and the manner in which they grow. The following brief account of their structure will set the scene for obtaining an understanding of the descriptions of the expressions of the traits exhibited by cacao trees.

Germination is of the epigeal type with the cotyledons being raised above the soil and the two cotyledonary leaves being opposite. Vertical, or orthotropic growth, continues with successive new growth. The leaves on the main stem are arranged in a spiral in a five-eighths phyllotaxy. These leaves differ from those of the lateral branches by their long petioles. Additional vegetative buds are contained in the axils of the cotyledonary nodes which may be stimulated to develop if the terminal bud is suppressed.

At a certain height the terminal bud of the main stem ceases to be active and is replaced by buds which form the lateral branches. These arise in whorls usually consisting of five, unequal, branches which result from three successive divisions of the terminal bud. Growth of the lateral branches continues through elongation and successive new growths. The leaves of the lateral, or plagiotropic, branches are arranged in a one-half phyllotaxy being alternate and opposite. The large leaves have short petioles that may be thick with prominent pulvini. Additional branches are formed from initials in the leaf nodes of the lateral branches. The leaves normally are short-lived and their shedding depends on climatic conditions. Adult trees may become almost leafless under severe conditions.

In addition to the initial buds from which plagiotropic branches are produced the leaf axils also possess buds which can give rise to vertical stems which are similar to the main stem. These adventitious stems differ from the main stem by having plagiotropic initials. They are also capable of producing their own aerial adventitious roots when conditions are suitable for this to take place.

Further vertical growth continues on the main stem by the activation of the orthotropic buds in the leaf axils below the whorl of lateral branches. In the course of time a further whorl of lateral branches is produced. Further growth occurs with successive development of vertical stems and lateral branches. In theory, there is no limit to the height a cacao tree can attain under natural conditions when there is no competition from other vegetation and it is free from mechanical and other damage. Such trees may present a symmetrical development with the lateral branches of the succeeding whorls becoming progressively shorter as the trees grow. On the other hand, when the canopy is dense the lower branches may be shed, leaving an elongate main stem that may attain a height of over 10 m.

Inflorescences are produced from all leaf axils. These are most commonly found on old wood from which the leaves have been shed. At times, inflorescences may be formed on new flushes where the leaves are still present. An inflorescence normally produces many flowers in succession. Although these appear to arise from the same spot the inflorescence is really a cyme, the base of which is concealed below the bark. Sometimes the inflorescence is extruded beyond the bark; in which case it is possible to observe its structure as a modified branch with rudimentary leaves and stipules. Extrusion of the inflorescence occurs under certain circumstances in trees of the *Criollo* group but also may be seen in cases where the buds have been infested by pathogens and pests. Flowering may take place throughout the year but the timing and intensity may be controlled by climatic factors and production of fruits.

After fertilization of the flower the ovary develops into a fruit. In cacao, embryogenesis follows an unusual pattern. The zygote develops for a few divisions. After this it remains dormant for about 50 days after which the normal process of fruit development is resumed.

The fruit is an indehiscent drupe which remains attached to the tree by its peduncle after it matures. The husk of the fruit is woody of varying thickness according to variety. The seeds are contained in the fruit in five rows according to the five loculi in the ovary. Besides the embryo, the seed contains two cotyledons which are convoluted. When the fruits become ripe the seeds are enclosed in a white or yellow pulp which is produced from the ovary wall. The seeds are recalcitrant, having no dormancy period, and germinate when they are exposed to a moist environment.

Growth Habit

This heading refers in the most part to the characteristics seen in the developing plants and young trees. When the plants are in the juvenile stage of development their structure and growth characteristics are easier to observe and measurements are made with greater accuracy than is possible in the case of adult trees. The intrinsic characteristics of older trees are often indiscernible owing to the alterations to their architecture and canopy form that result from external factors. In cultivated fields, particularly, pruning and harvesting practices and competition create significant changes in tree structure rendering it difficult to detect genetic differences. Most of the attributes of varieties described in this section concern those of plants derived from seed propagation or developed from orthotropic shoots. Plants propagated vegetatively from plagiotropic branches do not form the natural architecture and the differences between varieties are less distinct. However, in some cases some specific characteristics may be observed in such plants.

The identification of stable genetic differences between plants in families with regard to vegetative development in the early stages of juvenile growth is usually subject to uncertainty because of the various factors that determine growth. However, some varieties have exhibited marked differences among their progenies in growth variables during early and later development. Examples of these cases are shown in Plate 1a, b and c.

Differences in overall tree size at a given stage of development of the plants can be measured by such attributes as trunk size and rate of growth, comparative tree height and canopy size. Some of these attributes can exhibit a considerable degree of intra- and inter-plant variability.

The height at which the first jorquette (whorl of plagiotropic branches) is produced would normally depend on environmental conditions, such as light, soil fertility and the age of a plant at establishment in the field. Consequently, the determination of genetic effects regarding this characteristic could be inconclusive. However, some genotypes have been prominent in their ability to produce the first whorl of branches at heights well below 1.0 m. The observations made on their progeny have shown that this trait is inherited. Plate 18a shows a seedling plant with branching at about 50 cm from the ground and exhibiting a high reproductive potential for a plant of this size. Interest has been shown for a long time in the identification of genotypes with a 'dwarf' habit that could be utilized as rootstocks for the purpose of reducing the height of otherwise vigorous plants.

Significant and interesting disparities between individuals from different populations regarding the angles at which the plagiotropic branches develop from the whorl have become evident. The extreme branch angles known range from 90° to almost vertical. Plate 1a and b shows plants with almost erect branches and narrow canopies while Plate 1c shows a 90° branching habit. In both cases the traits are inherited. The most outstanding case of specialized branch development is a tree in which the branches are elongate and geotropic, producing an effect in which the trunk is enclosed by the branches. A growth expression that has been found only in *Criollo* populations is the formation of a single plagiotropic branch instead of the normal five branches, an example being shown in Plate 1d.

There are observable differences between genotypes that are probably inherited with regard to such traits as the distance between internodes and the number of plagiotropic branches per whorl. However, these expressions may be subject to modification by environmental effects such as the cultural treatments the plants receive. It is possible that the differences observed may apply to a definite period of growth or be of a transitory nature.

All of the variations of growth factors combined with those of the leaves produce remarkable differences between plants in the characteristics of the canopies. These are most evident in plants that have attained their adult structure at the time reproduction starts; that is, prior to the changes that would be brought about by cultivation practices. Canopy volume and density of vegetation of different population groups are very noticeably variable as well as the appearance of the foliage. The attributes of the genotypes in this respect could have an important relationship to the reproductive capacity of the individual genotypes.

Leaf Characters

One of the most obvious differences between plants that has been encountered, becoming especially prominent as a distinguishing feature of varieties, is the range of colours exhibited by the flush (young, developing) leaves. The colours range from the lightest green (sometimes referred to as white or unpigmented) to various shades of red. The light green shades are most evident in certain Amazon populations but also are found in the *Criollo* group. The red shades of flush leaves are produced by the presence of anthocyanins in the tissues and are often associated with red fruits, as expressions of the same allele, being more intense when the individual is homozygous for the fruit colour allele.

A few examples of flush leaves without anthocyanin pigment are shown in Plate 3, in which they are contrasted with leaves having red pigment from a red-fruited genotype (Plate 3b). However, several genotypes with green fruits also exhibit significant variation in the degree of intensity of the presence of anthocyanin. The intensity of pigmentation produced by an individual is subject to effects such as the age of the flush as it develops and environmental factors. As a result casual observations may not detect the true expression of pigmentation in the leaves of a given plant and it is likely that the position of the genotype in the range of variation may not be correctly determined.

Few observations have been reported concerning the shape and texture of the flush leaves. However, there is one notable character that appears in some varieties, and that may be associated with a given range of populations. This relates to the blades and pedicels of the flush leaves that are thick and soft, producing a 'velvety' appearance.

On account of the dimorphic habit of cacao, differences in the characteristics of the leaves of orthotropic and plagiotropic branches may exist that are related to genetic effects but although these may be suspected their identification may prove to be difficult. Differences in the appearance of the leaf pedicel, such as the thickness of the pulvini, attributed to genetic effects have been reported but this trait has not been sufficiently studied for the existing observations to be applied for purposes of characterization.

Extremely large differences are found in the leaf blades of genotypes, embracing all populations in the range of diversity. In terms of shape, leaf blades vary from narrow elongate to broad. While the shape of most leaves is elliptic, genotypes are found with obovate or ovate shaped leaves, which are distinguished according to the position of the widest part of the blade. While the edges of most leaves are entire, certain genotypes are found whose orthotropic leaves have serrated edges, such as that illustrated in Plate 4c, which shows juvenile plants from the R. Japurá. Other plants from the same area had leaves with noticeably indented edges. Additional variations based on genetic differences occur in respect of the shape of the base of the leaves in terms of the angles formed by the blade edges when they meet the petiole. Apices may also exhibit significant variability. Figure 2 shows two distinct leaf types from the range of variability exhibited in this organ.

As stated above, in genotypes with red fruits the flush leaves generally have red pigment. This association is also expressed as pigment in the petiole and leaf axil (known as axil spot) in developing leaves. In this way the presence of red pigment alleles can be identified in young plants, since the petiole pigment is normally absent in genotypes with green fruits. However, an exception to this relationship was exhibited in another genotype with green fruits from the R. Japurá (Plate 4b), the petioles of which were distinctly pigmented.

Considerable variation also occurs in the texture of the blade. Some leaves are smooth, thin (with the consistency of paper) and shiny, while at the other extreme the blade is broad, thick and rough to the touch. Various combinations of these characteristics can be used to differentiate varieties or are specific to certain genotypes.

Several mutations have been identified and conserved. Many of these are distinguished by possessing anomalous leaf characters. The divergent types include crispate leaves such as those shown in Plate 6a, of which the 'Crinkle Leaf Dwarf' is another example and the almost circular leaves of 'Jaca' (Plate 4d and Fig. 50). Both leaf abnormalities have appeared in several genotypes. The 'Jaca' leaf type mutation has occurred in at least three distinct populations located in very distant regions. Each genotype has been preserved as herbarium specimens or as living plants. A few examples of leaf abnormalities or forms that have segregated in the progenies of various genotypes are

Fig. 2. Two contrasting leaf types represent the scale of variability encountered in this organ.

depicted in Plate 5. The range of genotypes that are characterized by leaves that are very distinct is large and, as additional germplasm becomes known, many more forms could be added.

Flowers

The flower of cacao is a complex organ and possesses some features that make it unique and distinctive. The entire complex of the reproductive organs and processes of the species is especially important with regard to genotypic differentiation.

Figure 3 illustrates the various organs of a typical flower, the relationships among them and the names of the individual parts. The genera *Theobroma* and *Herrania* are somewhat unique in possessing two unusual organs; the infertile staminodes and the ligule attached to the petal pouch. Another feature that is almost universally present refers to the coloured lines that occur in the inside of the petal hoods. In cacao terms they are known as 'guide-lines' on the assumption that they facilitate the orientation of the pollinating insects to reach the anthers, in which case it may be more accurate to use the botanical term 'nectar guides'.

Although cacao is notable for possessing the characteristic of cauliflory, not all genotypes exhibit this trait. Members of some varieties seem to lack this behaviour, producing their flowers entirely on the branches and, therefore, are distinguished by the lack of fruits on the orthotropic shoots. Plate 2c depicts a tall tree, which, on account of the pale flush leaves, appears to be related to the populations of the Oriente of Ecuador, and does not show any signs of flowering having taken place on the trunk. The most extreme form of fruit

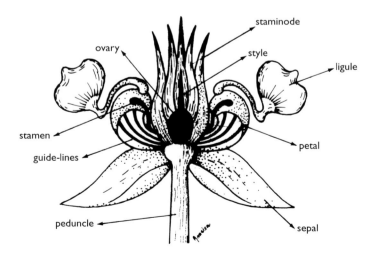

Fig. 3. Sketch of a cacao flower with the names of the individual parts.

distribution in the trees is that of certain genotypes that produce flowers and fruits on the newest flushes, an example of which is shown in Plate 2d.

The intensity of flowering is observed to vary considerably according to the variety. This trait may be influenced by environmental factors and certainly can be influenced by the reproductive capacity of a tree as determined by its compatibility status. Consequently, it is often not possible to attribute a genetic cause to the differences that occur. However, there are some genotypes that consistently produce smaller quantities of flowers during a given period. Genotypic variation of this nature can be ascertained only through comparative observations and when they are made over a period of time.

Differences between genotypes in regard to seasonal distribution of flowering and fruiting have been demonstrated on the basis of observations and data recording carried out during several years on a comparative basis.

Variation in the colour of the peduncle is noticeable, the range being green or white to red. However, the intensity of pigmentation is subject to non-genetic influences, such as the quality of light received, which make its determination uncertain.

The flower buds exhibit considerable variation with regard to their shape and pigmentation. The shapes are determined by the characteristics of the sepals, which are discussed below, the narrow elongate sepals resulting in elongate buds while the short sepals produce buds of a rounded shape. Pigmentation varies according to the population concerned. Some buds are entirely white while the buds of other genotypes exhibit differences in the degree of pigment. In general, there appears to be no relationship between the pigment expressed in the buds and the colours of the fruits of the same genotypes. The buds of some varieties are notable in having pigment along the joins of the sepals.

The flowers of different genotypes exhibit considerable diversity in respect of their general appearance. Differences are notable among certain populations, which often can be used to attribute a given plant or variety to the population group to which it belongs. The most notable differences that are observed concern size and colour. Plates 8 and 9 depict a few examples (out of thousands of samples) to provide an idea of the range of variability that occurs in the species. The over-all appearance of a flower is produced by the combination of its individual parts. Variation in the sizes of the individual parts resulting from conditions such as adverse climatic factors may produce abnormal variations in flower size.

One of the most observable organs of the flower is the calyx. Frequently, in an open flower some of the sepals are joined. This tendency may be associated with certain genotypes but the variation is not yet fully understood. Sepals vary with regard to size and shape, the former normally being correlated with the genetic differences in total flower size. The shape of the calyx can be measured as the ratio of length to the width at the widest part. The form of the apex possibly has a major contribution to the length. The basic colour of a calyx is white but various degrees of anthocyanin pigmentation may be superimposed on it, on either the interior or exterior surfaces, or on both of them and also distributed in relation to the adjacent edges of the sepals. One of the characters of the calyx that is most outstanding as regards genotypic differences is the position of the sepals in relation to the other organs of the flower. The positions vary from 'reflexed', in which the sepals are parallel to the peduncle (as in Plate 8b), to horizontal or curved slightly upwards forming a saucer shape.

The staminodes are specialized structures, characteristic of the genus *Theobroma*, which have developed as sterile stamens, perhaps with a specific function as the whorl of staminodes protects the gynoeceum. It is possible that the ancestral plants of the genus possessed a whorl of ten stamens. Certain genotypes, especially a few mutations, are notable for the absence of the normal staminodes, each of which is replaced by a functional stamen as depicted in Plate 8c. In such cases the stamens are exposed, a feature that is also found in a few genotypes with a normal flower structure, this being an additional factor in recognizing genotypic variability.

The staminodes of *T. cacao* are awl-shaped and present few variations in terms of shape. However, variations in colour patterns occur according to the variety to which the genotype belongs. There is one feature that is of interest in connection with the variations of staminode characters and their relation to the structure of populations. Most genotypes have flowers with erect staminodes. A variation is the occurrence of staminodes that bend outwards along the upper half of their length. This trait (centre of lower row of Plate 8d) is a constant feature of certain varieties. In some plants there is a tendency for the staminodes to be closed together at the extremity in a tent-like fashion. The expression of the trait appears to be inconsistent and, at the present time, may be considered to be less useful as an indicator of variability.

The petal presents some complexities related to its form and function. The petal hood by itself does not appear to exhibit any notable variation in appearance that can be measured. Some genotypes are distinctive regarding the intensity of pigmentation that can be observed from the outside of the flower. The guide-lines exhibit a significant amount of variability, which is rather complicated to describe because of the diverse combinations of effects that are found. Variability occurs in relation to the colour of the guide-lines, which, in plants with a normal genetic mechanism, can range from pink (Plate 9c) to very dark red or purple. In addition to the intensity of colour there are variations concerning the distribution of pigment along the guide-lines and the differences between the inner and outer lines. The guide-lines also exhibit variation in respect of their thickness.

Another characteristic feature of the genus is that most species have an appendage to the top of the petal hood that is termed a 'ligule'. The ligule is usually a flat organ that is attached to the petal hood by a ribbon-like structure, designated a 'strap'. The strap is an organ that carries vascular tissues that may be coloured and varying in their prominence. The length of the strap, when considered in terms of the sizes of the other parts of the petal, may be seen to differ among varieties, some varieties being notable on account of the long straps in their flowers. Strap length may play a part in determining the difference in the positions of the ligules in relation to the rest of the flower. The positions of the ligules appear to be determined genetically; in some genotypes they are pendulous, hanging between the sepals, in others they are held horizontally. The latter condition gives the flower the appearance of being larger than flowers of types with pendulous ligules.

The scale of variability expressed in the ligules is fairly large, notable differences being observed in terms of size, shape and colour. Most ligules are flat organs but the variety-related differences in the shape attributes might be so subtle as to require detailed observations to obtain adequate descriptions of them. The general variation in ligule shape ranges from narrow elongate to one in which the breadth is greater than the length. Not all varieties have flat ligules. Notable exceptions are found in certain populations where the outer edges of the ligules are curved upward, producing an appearance similar to that of a spoon.

The basic colour of the ligule is yellow and is expressed in different varieties on a scale of tints. At one end are the ligules in which the shade of yellow is so light that the colour could be described as being white. At the other extreme the ligule colour is a dark yellow. Intermediate shades are encountered that indicate the presence of anthocyanin pigment. Red pigmentation is superimposed on the basic colour or in combination with it, depending on the genotype being observed. In the case of genotypes in which pigmentation occurs, the range of expressions varies from the flowers in which pigment is confined to a few of the veins on the ligule surface, through increasing intensity to the rare cases where the entire ligule blade is red.

The combinations of the different scales of pigmentation with the variation in the base colour of the ligule result in variations in the appearances produced. For example, a fairly intense red combined with dark yellow produces an effect that, for lack of a precise name, can be described as a shade of bronze. A base colour with a lighter shade of yellow combined with an intermediate intensity of red produces an effect in the range of orange. In such cases, it is necessary not only to attempt to describe the scale of the effect of the colour combinations, but to describe the scale of each of the colour components concerned.

The presence of anthocyanin pigmentation in the stamen filament is one of the most important indicators of genotypic and varietal differences. In some populations only unpigmented filaments are found. In the varieties in which the stamens possess pigmentation, a great deal of variability is exhibited in its intensity and the patterns of its distribution. At one end of the range only slight amounts of pigment occur and variations are expressed as to whether the pigment is distributed evenly or irregularly. At the other end of the scale some varieties are distinguished by having intensely pigmented stamens, some of which are outstanding on account of the pigment being glossy in appearance. Several varieties are notable for possessing specific expressions with regard to the stamen colour states and on this basis they are easily identified.

The stamen pigmentation may be extended to the anthers and claims have been made for determining genotypic differences on the basis of pigmentation at the base of the anthers. However, this characteristic does not appear to be sufficiently consistent to be useful in distinguishing genotypes.

Some of the anomalous character states that occur in the androeceum have been described above and others probably occur. At least one genotype exists in which the stamens have been converted to a leaf-like structure; the genotype, of course, is male sterile.

Because of the small size of the organ, differences in the shape of the ovary are usually difficult to detect. It is possible that the shape of the ovary may be related to the shape of the fruit of an individual genotype but definite observations to determine if such a relationship exists have not been made. The most notable variations in the ovaries are those relating to pigmentation in the same manner as found in the other organs of the flower. Ovaries are entirely white in colour or exhibit variations in the presence of pigment. Variations in pigment expression include the presence of red coloration at the ends of the ovary hairs, a collar of red at the base of the style or uniform pigment throughout the ovary. To a certain degree pigment in the ovary is determined by the allele that produces the red colour of some fruits. The intensity of pigment varies according to the red-fruited genotype and its homozygosity status.

The ovary of cacao contains a number of ovules, the fertilization of which results in the production of the seeds in the fruits. The numbers of ovules per ovary varies considerably and is conditioned by the genetic composition of the plants belonging to a variety. Accordingly, it can be used as an indicator of the variety or population group to which an individual belongs and the relationships among populations. When the numbers of ovules are accurately determined they are found to vary within small limits within a given genotype. The ovule numbers determine the maximum numbers of seeds that can be produced in a fruit. Maximum seed numbers counted per fruit have a range of 40 to 77 according to the genotype. Since the ovary is made up of five loculi this works out to a range of eight to 16 ovules per loculus. It appears that some genotypes resulting from mutation have lower numbers of ovules. The determination of the correct ovule numbers is a somewhat tedious procedure that requires a certain amount of skill and patience. Attempts have been made to devise other methods of counting. All of the substitute methods lead to inaccuracies in counting, which result in illogical situations in which genotypes have been reported as having fewer ovules than the numbers of seeds counted in the fruits.

Fruits

The fruit characters have played an important role in differentiating individual genotypes with regard to their position in the diversity of the species. Primarily, the reason for the emphasis that has been placed on the fruits is a consequence of the fruits being perhaps the most conspicuous organs of the plants, allied with the fact that the fruits provide the usable product of the species. However, the importance of the fruit as an indicator of diversity lies in the enormous range of expressions of its parts and contents.

The most notable aspects of the fruits are those of shape and colour. Shapes range from spherical (sometimes oblate – length less than diameter), as in Plate 7a, to very narrow elongate, where the length:diameter ratio is >3

(Plate 15a). In most cases the form of the apex of a fruit is important in determining its measurable external length. Other dimensions being the same, the relatively short fruits have rounded or obtuse apices while the elongate fruits have attenuate apices. The characteristics of the base of a fruit also influence the dimensions and the ratios resulting in specific values for the fruit shapes of the varieties that possess these characters.

The surfaces of the fruits exhibit an ample diversity of expressions for various aspects. Some of the variations in the characters involved in creating the appearance of fruits can be specific to certain varieties. Most fruits have superficial ridges that vary in their prominence, width and arrangement on the fruit. Basically there are five pairs of ridges, each pair being associated with and marking the locular divisions.

Considerable variation occurs among genotypes and varieties in terms of the range of expressions relating to the pairing of the ridges. When there is no space between ridges of a pair the fruit assumes a five-sided appearance. The pentagonal shape of the cross section of a fruit of this type was the basis for the name *Pentagonum* that was applied to some fruits (Plate 28a, b) as mentioned in Chapter 2.

At the other end of the range of expressions of ridge types are the fruits in which the distance within the pairs of ridges is equal to the distance between the ridges of adjoining pairs. This results in the ten ridges of a fruit being equidistant. Principally, in rounded fruits with imperceptible ridges the appearance of the fruit surface is smooth. The extreme of the expressions of the ridge traits is the occurrence of fruits that appear to have no ridges, such as those in Plate 26d. Such visual appearances could be misleading since the presence of ridges can be ascertained by touch. It should be pointed out that the smoothness of the surface is not exclusive to fruits that are nearly spherical in shape.

One specialized character of fruits of certain varieties is the covering of the surfaces with small protuberances, or lenticels, that produces an appearance that, in the absence of an adequate term describing this condition, has been called 'mealy' or 'pimply'. A more appropriate description would be 'granular'.

The illustrations that accompany this account of the diversity of cacao provide examples of the many forms of variation of fruit shape and surface characters that are encountered in the species. A striking demonstration of the scale of variability that occurs in one producing country, Venezuela, representing the segregation in fruit characters that was produced by hybridization is to be found in an illustration in Pittier (1926).

Less visible attributes of the fruits concern those of the husk. Large differences in thickness are observed. The husk also exhibits a significant range of expressions regarding its hardness or rigidity, which are genetically determined. The scale of hardness is determined by the degree of lignification of the mesocarp. Some varieties possess husks of such hardness that rodents have difficulty in penetrating to the interior of the fruit. At the other extreme, some genotypes have husks that can be compressed by hand pressure and a wire could be passed through the fruit without encountering resistance.

As mentioned above, there is an enormous range of variability in respect of fruit colours. The basic fruit colours are shades of green. These vary from light green or whitish, for which reason it has been customary to describe the lack of colour by the Spanish word *blanco*. The extreme shade is a very dark green exemplified by the fruits in Plate 18a.

Red pigmentation due to anthocyanins is superimposed on the green base. The expression and distribution of pigmentation on the fruit surface are influenced by the amount of light that a fruit has received during development. In the absence of firm evidence concerning the mechanisms by which pigmentation is developed on the fruit surfaces, we may recognize the existence of two types of pigment expression depending on the genetic constitution of the genotype concerned. One type is produced by the action of a specific gene or combination of genes responsible for the development of anthocyanin, which are dominant in their effect. The other situation occurs in genotypes that are homozygous for the recessive gene that is responsible for the basic green colour. In these genotypes the pigmentation on the surface results from the action of supplementary genes. The inheritance of these pigment patterns will be described in Chapter 7.

The dominant gene for the expression of the red fruit colour occurs in some populations of the *Criollo* group and is, therefore, inherited in the offspring of hybrids with these genotypes. This gene has a single effect but the expression of pigment in the fruits of a given genotype would be modified by the intensity of the basic green colour of the surface, which is intrinsic to the genotype concerned. The modifying effect would explain the amplitude of shades and tints in plants that possess the pigment gene. The variation in shades is also determined by the homozygosity status of the gene responsible for pigmentation.

Considerable variability also exists with regard to the sizes of the fruits. The actual sizes of the fruits of a genotype are influenced by several genetic and non-genetic factors. In these circumstances the quantification of fruit size, whether by measurement of the dimensions, weight or volume, is subject to considerable error. However, real genetic differences can be detected when fruits of different varieties are compared. The range of average fruit lengths varies from 120 mm to 300 mm. In view of the lesser degree of variation in fruit diameter and the fact that various details of the fruit shape are, as described above, implicated in determining fruit length, this measurement is not an exact indicator of size. Fruit weights are less reliable indicators of variability on account of the uncontrollable non-genetic influences, such as the variation in seed numbers and environmental effects during development, which are responsible for the expression of this variable.

As the peduncles of the flowers vary in length according to the genotype there is a similar degree of variability in respect of the peduncles of the fruits. Perhaps, in fruits, the diameters of the peduncles provide more reliable indicators of genotypic variability. The peduncles of some genotypes are long and thin so that the fruits are pendulous. At the other extreme are those genotypes whose peduncles are short and thick, as those of the tree shown in Fig. 4, so that the fruits can be held horizontally.

Fig. 4. Fruits with short peduncles. Note the scars left after the fruits have been removed.

Seeds

In relation to the sizes of the seeds considerable variation is also encountered. However, seed size is determined by several non-genetic factors, thus it is difficult to obtain accurate estimates of genetic values for this character. Relative seed sizes can be estimated by weight or volume and also by the measurement of their dimensions.

The weight of seeds in their 'fresh' condition in recently harvested fruits provides a rough estimate of the sizes of the seeds since it is the value that is most easily determined. However, the values obtained are subject to many sources of variation, of which the number of seeds in a fruit, the time elapsed between fertilization and maturity of the fruit and the length of time between maturity and harvesting are a few of the most important factors. One of the characteristics of the 'fresh' seeds that results in inaccurate estimates of the seed weight of genotypes is that inherited differences exist among them regarding the moisture content. This situation leads to seeds with high moisture contents having low dry seed weights.

Accordingly, the most accurate indicator of seed variability is that based on dry seeds, especially the cotyledons, where the variation among fruits of a genotype is smaller than that from whole seeds. However, non-genetic factors such as climate can determine the actual and relative values that can be measured. The seeds of some genotypes are very small, perhaps having average weights of 0.5 g when dry and the largest weigh about 2.0 g.

The shapes of seed vary from almost spherical to flat (depth less than breadth), being broad at the hilum and tapering towards the distal end. Some varieties are prominent in possessing seeds that are described as cylindrical; that is, having an almost circular cross section. It is possible that associations exist between the fruit shape and the seed shape, a situation that may be important in classifying the diversity.

The pulp surrounding the seed in its fresh state should not be ignored as an indicator of variability. The colour of the pulp, perhaps occurring in most varieties, is white and of such a nature as to make it impossible to discern differences among genotypes. However, some varieties are outstanding in possessing pulp of a yellow colour. The taste of the pulp of genotypes belonging to different populations can reveal interesting variation in this component of the fruit. A few genotypes are notable in possessing pulps that have a sweet and pleasant taste. Other varieties, notably members of the *Criollo* group, are notable for their acid or even disagreeable pulp. The identification of such variations would be made on a subjective basis and could depend on the observer assessing the trait in the material concerned.

Other Discriminating Characteristics

One trait by which the different varieties are distinguished is the period of time that elapses between fertilization of a flower and the initiation of maturity of the fruit that develops. The known range of this period is in the order of 135–180 days. Genotypic distinctions are fairly stable but the exact length of the period measured depends on the environmental conditions prevailing during the periods of observations. The most accurate method of determining the length of time to maturity is to produce fruits by artificial pollination; in these cases the dates of fertilization can be known exactly.

Self-incompatibility is common in the species and occurs across a wide range of genotypes. Individual genotypes and populations possess alleles that determine the function of this character that are specific to them. As a more detailed discussion of the subject of compatibility will be given in Chapter 7 it will suffice to state here that the individual alleles and the states in which they occur are some of the most important and useful indicators of inter-population differences. In subsequent chapters it will be explained how the determination of the self-incompatibility status in various populations may be used to analyse the relationships among their members' population and obtain information about their origins. However, the identification of the alleles and the determination of their states requires a considerable amount of effort and, sometimes specialized, genetic and laboratory techniques. On account of the specialized nature of such determinations the subject has not received the attention it requires, resulting in the state of knowledge being incomplete.

Combinations of Descriptor States as Variability Indicators

It is obvious that a difference exists between the descriptors that are observed on the trees and those that are measured on the organs after they have been removed. The descriptors in the former group comprise the characteristics that are observed on the trees in their natural conditions of occurrence, which give an idea of the variability contained in a population. Regarding the other descriptors an unfortunate notion persists among certain observers that only some measurements need to be made on various organs in order to determine the differences between genotypes and populations. This notion leads to important characteristics being ignored. In order to obtain a comprehensive picture of the diversity and the characterization of the individual elements of which it is composed it is necessary to include all fonts of variation.

The fact is that the diversity will have to be described in terms of the combinations of the different traits that are variously distributed throughout the species. In this sense, each genotype is not a collection of statistics but an interpretation of how the traits combine and interact with each other.

There is a general misunderstanding about the purposes for which certain measurements are obtained. For example, the measurement of individual dimensions of an organ is not the end of the matter. In the case of organs such as leaves, sepals, ligules, fruits and seeds, length and breadth by themselves have little meaning. The dimensions are determined with the objective of obtaining the ratios from which the shape of the part or organ is based. It is necessary to go further than this by including other attributes that make up the shapes. In the case of the fruit, for example, the length:breadth ratio alone has little meaning unless the apex characteristics are also considered.

The sizes of the individual parts of the flower are naturally related to its over-all size so that individual measurements are of minor importance for the purposes of comparison of genotypes. Hence, it will be logical to find that the parts of a small flower would be smaller than the corresponding parts of a large flower. The circumstances in which individual flower parts differ from the normal range are those that could be considered to be important for distinguishing between genotypes.

The variations in the values of the lengths of the staminodes and the gynoeceum (ovary and pistil) may individually be indicators of diversity among genotypes. However, it can be seen that the genotypes also differ with regard to the comparative lengths of the two parts of the flower. Thus, the ratios of the lengths of the gynoeceum and staminodes also serve to discriminate between varieties.

One attribute that is frequently estimated but that has no value in determining the genetic differences among genotypes is the number of seeds contained in fruits resulting from natural fertilization. Since the effects of natural fertilization are random they lead to large variations among the fruits produced by individual trees with regard to the numbers of seeds that develop in each fruit. As explained above the ovule numbers determine the maximum seed number that can be produced in a genotype's fruits; hence the only valid basis for determining genotypic differences is given by the maximum numbers of seeds, provided that the determinations of these characteristics are based on sufficiently large samples.

The sizes of the seeds of genotypes representing the range of diversity of the species are determined by various factors in addition to any direct genetic effect. Non-genetic circumstances play an important role in determining the seed size in a given sample of fruits of a genotype. For example, the availability of moisture during the development of the fruits is directly related to the weights of the seeds in fruits harvested in different periods. Other factors that may determine seed size include the seed number, as determined by the number of ovules of the genotype, and the shape and size of the individual fruits. Genotypes with large fruits and low ovule–seed numbers tend to have larger seeds than the genotypes having intrinsically small fruits and high ovule numbers. Some association may be found between the shape of a fruit and the shape and size of the seeds it contains but it does not hold in all varieties.

When the colours of the various parts of the plant are considered the situation is quite complex. In some populations the colours of the cotyledons determine the colours of the flush leaves in the plants that grow from them. For example, white cotyledons produce plants with light green flush leaves and pigmented flush leaves are found in plants that have grown from purple cotyledons. This association is not generalized throughout the range of diversity and varies according to the population concerned. Although one of the assumed characteristics of the *Criollo* group is the possession of white cotyledons, few of the plants belonging to this group have unpigmented flush leaves. The expressions in the *Criollo* group probably are dissimilar to those of the genotypes described above.

Another example of the relationship between cotyledon colour and flush leaf pigment is the case of the population to which the clone 'BE 2' belongs. As shown in Plate 9 the cotyledons of this genotype have a pink colour but the leaves of their progeny are not pigmented. A few instances have been observed where plants with unpigmented flush leaves developed from seeds with purple cotyledons.

Among the combinations connected with the mode of distribution of anthocyanin pigment in the flower parts, are the notable relationships observed of the expressions of pigmentation in the staminodes and the guide-lines. Various combinations occur in this complex relationship, which include both parts with similar shades of purple pigment (any differences being difficult to detect), purple staminodes and pink guide-lines and both staminodes and guide-lines being light purple or pink. The combinations expressed are also related to the cotyledon–flush leaf pigment complexes. The genotype 'BE 2', mentioned above, in addition to the pink cotyledon colour has flowers with pink staminodes and guide-lines. Other genotypes with unpigmented flush leaves and cotyledons have purple staminodes and guide-lines.

The allele producing the anthocyanin pigmentation, conspicuous in fruits, that is, those derived from the *Criollo* group, has important effects in the individuals that carry the allele. For one thing, the allele has pleiotropic effects. The presence of the allele can be observed as pigmentation in the hypocotyls of the germinating seeds and, as described above, in the pedicels of the leaves and as the 'axil spot' character, especially observable at the stage when the flush leaves are hardening. The presence of pigment in the hypocotyls and pedicels of a plant indicates that it will produce red fruits.

The pigmentation allele is frequently associated with the red coloration on the inner surfaces of the sepals. Preuss (1901) was the first to refer to the effect of the allele in the cotyledons of red-fruited trees of the *Criollo* group that are slightly coloured, in contrast to the white cotyledons of green-fruited trees.

The anthocyanin pigment that occurs in the stamen filaments is almost exclusive to genotypes with green fruits. This character is rarely observed in red-fruited genotypes. In fact, only one definite case is known where this combination is stable, a situation that could indicate an unknown origin of the parents of the genotype in question.

The phenotypic expressions of genotypes possessing the pigmentation allele are determined by the status of homozygosis in the particular genotype and by its genetic background. In the plant parts in which the red pigment is expressed it is likely to be more intense in the homozygous individuals than in the heterozygous genotypes. However, other factors may modify the expression to such an extent that the pigmentation is less readily observed and sometimes may only be evident under certain conditions.

The allele producing red pigmentation is virtually confined to cultivated populations and to varieties derived from these populations. In this respect the presence of the allele is an indicator of the origin of populations derived from hybridization with *Criollo* varieties. The amplitude of its effects is as yet incompletely known and further observations, especially those based on controlled breeding experiments, may lead to future discoveries about the expression of anthocyanin pigmentation when the allele responsible is combined with genes from other genetic backgrounds.

4 The Manifestation of the Diversity and Its Conservation

Because of the extensive area over which the species occurs and the nature of the trees growing under natural conditions characterization of the diversity is virtually impossible to achieve. The best manner in which to study the diversity is in situations where the genotypes can be adequately characterized and compared so as to provide an ample vision of the differences among them. In this way, study of the diversity would be effectively carried out as an integral part of programmes directed at its conservation. Characterization of the diversity is one of the stages that comprise a genetic resources conservation programme as set out in Chapter 1.

In the present chapter the details of the procedures involved in conducting genetic resources conservation programmes will be discussed. An understanding of how such programmes can be effectively carried out depends on the appreciation of the various circumstances in which the species occurs and the representation of the phenotypic variability in the forms that are specific to the species.

The experience acquired during the 20th century through conservation programmes conducted in the areas where the diversity occurs in natural and cultivated situations has given us an ample knowledge about the species' base of diversity. In addition, programmes of selection, studies of inheritance and the determination of the molecular structure of the basic genetic materials have provided us with an insight into the nature of the diversity and the manner in which it occurs and is distributed.

For the purposes of the description of the variability within and between the various areas of occurrence it is necessary to adopt a definition of the aggregation of the plants that exist in a determined area or location. A particular aggregation of plants will be referred to as a 'population'. The expression 'sub-population' could be employed in certain circumstances to distinguish between aggregations of plants within a given region, which, although possessing certain common traits, also express characteristics that are specific to them. The term 'variety' may be applied in the situations in which a population, or a component of it, possesses distinctive individual characteristics that distinguish it from other populations or aggregations of trees within the population concerned. In such circumstances these varieties would be identified by a name that is related to its location of occurrence or specific traits expressed.

The starting point in a genetic resources programme should be to identify the areas and locations where cacao occurs. A study of the documentary evidence of the existence of cacao within the area of interest for a conservation programme is one method of determining the exact locations where the species occurs. The sources of this evidence include travellers' reports, official documents, accounts of explorations (principally those related to natural history), and histories of the regions concerned or locations within them. A useful source of such information is Patiño (1963). Some of these documents may provide information regarding the characteristics of the plants that exist in specific areas. However, details of this nature that can be found in old accounts are rare or, at most, provide little information of practical use. This is due to the fact that most observers would not have been familiar with the species and would not be able to describe differences among the plants in different locations. Even naturalists who have visited several locations where different varieties of cacao occur have not recognized and described the differences between them.

Collections of botanical specimens in herbaria constitute a potentially useful source of information. Not only do the specimens provide details of locations of cacao trees but often include some description of the tree from which the material was collected and its environment. Where descriptions are lacking, the characteristics of the collected material could be determined by examination of the specimens. Collections made in a region over periods spanning many years can also contribute to a compilation of the history of the cacao populations of the region.

Expressions Displayed by Cacao Trees in Various Habitats

An understanding of the nature of the diverse situations in which cacao populations are found, ranging from their primitive state to that in which they are exploited by man, is fundamental to the development of the story of the genetic diversity of the species. For this purpose the cacao populations can be attributed to the following three basic categories, depending on the degree of human involvement in their establishment:

1. Spontaneous – a situation in which the population had been established and developed naturally.
2. Sub-spontaneous – this situation is defined as one in which the establishment of the population has resulted from human action but where, at the present time, there is no evidence of actual human involvement in its development or exploitation.
3. Cultivated or domesticated – such populations are those formed by human action and have been exploited continuously.

The above classification implies that the spontaneous, or wild, populations are those in which the primitive diversity has been created and perpetuated. The treatment of the subject in this book is based on the premise that the primitive diversity would be confined to the area of natural distribution of the species in the Amazonian Region. Consequently, it is considered that the diversity in the populations attributable to the other categories has been derived from the primitive diversity. Alteration of the genetic composition of the base populations may have occurred during the passage of time through mutation, mixing of genetic elements from various primitive populations and natural and artificial selection.

In practice it is often difficult to judge whether a given population is natural or artificially derived, especially in situations such as in forests and recently cleared lands.

Although claims have been made that certain populations encountered are wild it is highly unlikely today to be able to categorically state that a given population arose through natural development in prehistoric times. For one thing we do not know exactly who were the original inhabitants of an area in the natural distribution range of the species. The human distribution patterns in the Amazon Basin and other regions have changed considerably since the arrival of the Europeans. As a result, a cacao population encountered in a given area in an, apparently, natural state may have been established by a tribe that inhabited the area several centuries ago.

Even today, some tribes have the custom of planting a few cacao trees in their villages. These trees remain in the locality after the family moves to another location. Such abandoned plantings can be identified by the presence of other useful species of plants and, especially, by the colonization of the area by trees of the genus *Cecropia*.

However, it is a general rule that natural populations of cacao are found only on soils of high fertility. As will be described in the discussion of the diversity of the Amazonian Region much of the natural occurrence of the species is found on alluvial soils along river banks. Where natural populations occur at distances from watercourses and at higher elevations these are located on naturally fertile soils, including, in the appropriate regions, soils that have been derived from volcanic activity.

Populations have been observed that, from the absence of any historical proof of human occupation and from the nature of the surrounding forest, would have been expected to be wild. However, in those cases where the size of the population is fairly large the appearance of the trees suggests that they had, at one time, been exposed to a greater incidence of light than they would have received had they developed under natural forest conditions. The arrangement of the trees in these populations also seems to be non-random in contrast with what would be expected if they had been established by natural dispersal mechanisms.

One feature of sub-spontaneous populations is the fact that they are composed of groups of small numbers of trees. These groups can be formed from the survivors of the village plantings, as described above. Other means by which such groups may have been formed include the accidental sowing, from fruits carried by the indigenous tribes during their migrations. One or at most a few fruits may have been harvested in the home village or along the route for some purpose; for example, to suck the pulp, the seeds being deposited at the locations where the fruits were opened. Similar methods of accidental sowing are attributed, perhaps in more recent times, to the collectors of natural products, such as rubber (the latex of *Hevea* spp.), balata (the latex of balata, *Mimusops balata*) and Brazil nuts (the seeds of *Bertholletia excelsa*). These collectors, starting from a place where cacao was present (or passing through one on their journeys) would take a fruit for similar reasons and, after the fruit was opened, the seeds were allowed to fall on the ground where they germinated if the conditions were suitable.

The small amounts of seed involved in these migrations would explain the narrow variability encountered in situations where the human element was responsible for the establishment of the isolated populations. Any heterogeneity encountered would be derived from the fruits harvested from heterozygous individuals.

Various situations have developed since the European colonization of the Amazon Basin and the commencement of the exploitation of cacao, that may also have a bearing on the issue of spontaneous and sub-spontaneous populations. Among the possibilities that should be considered is the deliberate or accidental sowing in the areas where the natural populations were exploited during the extractive period. (In current usage the term 'extractive period' refers to the era in which the product of the natural occurrences of cacao was harvested.) In addition, colonization of lands for the purpose of agriculture on whatever scale where cacao was present may have involved leaving the original trees (of the, perhaps, natural stands) and these populations expanded, deliberately or accidentally. This process continues in modern times. In most of these situations the seed used would have been derived from the existing populations so that the genetic composition would have been maintained even though some segregation may have occurred among the progenies if the parent trees were heterozygous. Through the subsequent generation cycle populations would have been created with variability of a greater scale than that which would have been apparent in the parental natural populations.

The uncertainty of defining which populations are spontaneous or wild presents certain difficulties regarding the methods employed in sampling the diversity in conservation programmes in the regions where the primitive diversity is expected to occur. The collector of the plant material needs to use intuition as to whether the status of the target population can be defined as wild. The mere fact that cacao trees are found in forests is no basis for claiming that they belong to a wild population. Such claims have been made in cases where, although the forest appears to be primitive, the area has been subjected to human occupation in the past, during which time the forest has been able to return to a natural state. In the R. Vaupés valley, an area of soils of lower fertility, it was stated that it takes about 30 years for the forest to return to its original state after a clearing has been abandoned. This means that, even if the forest vegetation is in equilibrium, the cacao trees found in it may derive from fairly recent plantings of introduced seed.

Under natural conditions of development, the cacao trees in spontaneous and sub-spontaneous situations present a variety of habits. The forms of the individual trees that are found in such states result from the influence of various factors. Some of these relate to the environment at the site of occurrence, such as forest density and topography. The ages of the trees and the genetic composition and density of the population as well as its past history contribute to the various tree habits that are encountered.

When the forest is dense a young tree grows towards the source of light. The result is that the trunks tend to be long and thin with long internodes. If branching takes place the canopy formed is usually restricted in area. The lateral location of the light source may cause the trunk to curve towards it according to the positions of the neighbouring trees. This is a common feature when certain varieties develop in some situations and may give the impression that the cacao trees develop a 'vine-like' habit. The occurrence of trees possessing this feature has been the subject of claims of finding a 'new' type of cacao. An example of a tree in dense forest with an attenuated trunk and weakly branched habit is depicted in Plate 2b. Some varieties also produce very tall stems even under more open light conditions (Plate 2c).

While some trees have thin trunks even when they are fairly old, other trees are encountered with tall trunks that have a large diameter. It has been found that the growth of progenies descended from some very large trees has been relatively slow. When such situations occur, it can be concluded that these large trees may have taken several hundred years to reach their present size. These examples prove the longevity of some of the cacao trees that are encountered under spontaneous conditions, but they also show that the normal types of tree habits cannot be implied from the appearance of the trees found under natural conditions.

At the other end of the range of tree forms are those trees, or perhaps better referred to as 'stands' (or thickets), that are composed of several stems. This is the consequence of the natural propensity of most cacao trees to produce suckers from their bases. The number of suckers found in a thicket can vary enormously. Plate 2a shows a stand with relatively few stems but others may have several dozen. Various mechanical causes such as the weight and action of the growth of creepers (lianas) around the stems and fallen branches and trunks of other trees cause the stems to bend forming an arch. This condition initiates the development of more orthotropic shoots from the axillary buds of the main stem, thus increasing the number of stems produced.

Similar situations also apply to older cultivated populations where the trees have been raised under conditions in which they have received the minimum of cultural care, multiple-stemmed stands developing for various reasons. One cause of the existence of trees formed by many tillers (stems arising from the base of a plant) is the practice of allowing them to occupy the open spaces created by progressive reduction in the densities of the original population by deaths and other factors.

From the point of view of collecting germplasm and the sampling procedures to be adopted in the target populations the existence of multiple-stemmed stands and natural vegetative reproduction brings two important problems. One problem concerns the fact that a 'thicket', which may appear to be a 'tree', could be composed of two or more genotypes. In some cases the stand may have been derived from the germination of several seeds. These seeds may derive from the same fruit or from more than one fruit. In the first case the resulting trees would be related but not necessarily phenotypically identical. In the second case, the trees may have been established at various times. Thickets have been encountered where an additional genotype has been added through seeds brought by birds and monkeys germinating within the original stands.

Examples of genetically mixed stands of this type are found in cultivated populations where the practice was to sow two or more seeds in a planting hole. When the trunks representing the progenies of these seeds were allowed to develop together genetically mixed stands resulted. Some of the most obvious examples of this situation are to be found in countries such as Ecuador, where it is common to find fields in which the seeds used for sowing came from parent trees that had red and green fruits. Sowing of pairs of seeds of contrasting colour genotypes produced stands in which the mixtures of genotypes were easy to spot.

In those regions where there is no variability in fruit colour mixed stands are not easily detected. This is especially the case when, on the occasion of a visit to a population at a given location, not all of the trunks have fruits or flowers on which differences between the various trunks can be observed; sometimes these indicators are absent altogether. Consequently, when propagating material is collected from several trunks of a stand and identified as coming from a single tree the progenies obtained may be found to represent two or more genotypes. Another consequence is that progenies of fruits collected from one trunk may be unrelated to the progenies of budwood taken from another trunk. The same situations apply, of course, to progenies of fruits harvested from several trunks but not identified as being different at the time of collection. The disregard for the use of the indicators of variability at the time of collecting to determine whether the trunks belong to the same genotype results in considerable problems in the identification and characterization of the resulting accessions.

The second problem related to vegetative reproduction is the existence of a situation that, in terms of the genetic constitution of the populations concerned, is the opposite of the situations described above. There are two ways in which plots of cacao, seemingly composed of various trees, may have developed from a single individual, other than by means of seed. Where an individual tree has produced numerous basal suckers these would develop into adult trees that may, in time, become independent of the original stem. Subsequently, the original trunk may disappear, leaving a gap between the trees that have developed from the surviving suckers giving the appearance that they are separate trees. Successive events of this nature affecting the older suckers will result in the younger stems becoming spaced further apart. These will, therefore, appear as individual plants but which belong to the same genotype as the ancestral tree.

The other way in which separated trunks will be produced from the original trunk applies to situations in which the arching trunks develop their own orthotropic shoots from their vegetative buds. In certain circumstances, for example, when they are close to or touch the ground, the new shoots produce their own roots, which will grow into the soil. Effectively these are new trees and will become independent of the main trunk when the older part of the 'mother' trunk is severed or decomposes. Progressive reproduction of this kind will, over many years, result in the establishment of many isolated trunks that have a common ancestor.

The consequence of these effects is that in due course a plot will be encountered that appears to be formed of various trunks but that will, however, have identical genotypes. In terms of sampling populations produced in this manner only one trunk will be necessary to provide a sample of the genotype.

The Influence of Ecological Conditions on the Nature of Cacao Populations and their Genetic Constitutions

From the foregoing accounts of the appearances of the tree in the spontaneous and sub-spontaneous states it would be evident that the species is distributed in a wide range of sizes of the aggregations of trees in locations or areas. The concept of 'population', as applied in this book, covers occurrences of solitary trees to large areas containing many thousands of trees at high densities. In the circumstances in which the species is encountered in an area distributed sparsely as individual trees and small groups of trees it could be considered that the aggregation of such trees comprises a population even though the individuals may be genetically distinct. The large variability in population size and situations in which the species occurs have an important influence in the determination of the procedures that are adopted by genetic resources programmes for the collecting and sampling of the diversity present in individual cases.

Solitary trees that produce fruits are expected to be self-compatible and they are almost always homozygous, or nearly so. Hence, collecting a small sample of seeds would suffice to represent the genotype.

Large populations at one site or distributed over a larger area probably contain some genetic variability and it is necessary to estimate the extent of the variability by means of observations based on the appropriate indicators. The sample size should be related to the apparent range of variability rather than the population size and would theoretically be calculated with the objective of obtaining a representative sample of the genetic variability. In practice, however, there are usually limits to the quantity of material that can be collected for reproduction and establishment. These will be considered in the appropriate section.

Another factor that needs to be taken into account in characterizing a population concerns the ages of its individual components and their consanguinity. In cacao, determination of the age of a tree under any circumstances is virtually impossible. As described above, the natural reproduction that may take place during the life of a tree would determine the ages of the individual stems encountered. The consequence of this habit is that the particular stand may have occupied a space for as long as several centuries but the trunks that are encountered may be relatively young.

It would be expected that a tree with an age of several decades would have produced fruits from which the seeds would result in a generation of offspring when they have germinated and the plants are established at the same site. Since there is no alternation of generations in the species this will give rise to populations that, at a given site, may be composed of individuals representing several successive generations, descended from the same ancestor and consisting of trees with a wide range of ages. This situation would be expected to be the norm where large populations are encountered.

However, it must be said that, in the context of spontaneous and sub-spontaneous populations, although quantities of seed may fall on the ground from fruits opened by monkeys, rodents and birds, the occurrence of young plants is rather rare. The only cases known to the author are those where the forest has been cleared. The probable reasons for the absence of developing seedlings in populations under natural vegetation cover are the difficulty of germination and growth under heavy shade and the effects of seasonal flooding. The development of populations composed of several generations would, therefore, be the consequence of special circumstances, such as the opening of the forest canopy by natural or human factors.

Although there may be little evidence of the simultaneous existence of successive generations at a site, the picture may be otherwise when entire regions are considered. Huber (1906a) described in detail the changes to the environment that occur in valleys such as that of the R. Purús, which have important consequences for the constitution of the cacao population in this drainage system. As will be described in Chapter 5, this is one of the most important regions of distribution of the diversity. The river is characterized by its serpentine course, in which constant erosion of the river banks takes place along certain stretches. At the same time building of new banks and beaches occurs through the deposition of alluvium in other stretches. This action of the river has taken place over several centuries as can be verified from the alternate strata of timber and alluvium that are exposed when the high banks of the river and its tributaries are eroded. The erosion results in the loss of older cacao trees from one bank of the river but there has been successive colonization of the new banks through germination of seeds from the old trees transported to the new locations.

The constant loss and renovation means that populations at different locations along the river have been established during various periods of history with trees at some locations being progenies of trees at other locations. The composition of the new populations perhaps represents part of the diversity that may have existed in the older populations, with additional variability being created simultaneously as a consequence of mutation and recombination. Consequently, individual trees of younger populations may express characteristics that are not observed in the parent populations. The process of constant changes to the course of the river has resulted in the formation of lakes, some at a distance from the present course, on the banks of which cacao trees occur. These may represent remnants of older populations but have not been considered in the collecting activities undertaken so far.

The large range of ecological situations in which spontaneous and sub-spontaneous populations are encountered means that local ecological circumstances have a profound influence on the ability of cacao plants to colonize particular habitats. Consequently, the genetic compositions of the populations found in certain habitats would differ according to the ecological conditions specific to each habitat. In regions where the relief is hilly the areas suitable for the establishment and maintenance of cacao are probably limited to certain ecological niches where soil moisture can be maintained. The occurrences are likely to be isolated and, therefore, composed of related individuals with little genetic variability. It is most likely that human agents have been responsible for the establishment of such populations with the consequence that the genetic composition of a given population may be restricted.

In view of the fact that by far the most usual distribution of natural cacao populations is along the river valleys, the effects of the large variety of situations in determining the nature of the diversity have to be taken into consideration. In addition to the variation in the ecological factors affecting certain regions, the characteristics of the rivers can have an important bearing on the composition of the diversity contained in the local populations. These characteristics need to be considered in developing strategies for sampling the diversity of individual river systems for the purposes of conservation.

Factors Involved in Sampling the Diversity in Various Populations

The riparian distribution of the species determines that access to the populations is more likely to be by water. Even when roads exist in the target area they are unlikely to provide access to more than limited areas of distribution. Roads and trails are, of course, indispensable to the sampling of the diversity in those areas where trees occur at distances from navigable watercourses.

Cacao populations along the narrower rivers appear to be more homogeneous in their formation. Travel on these rivers is easier and, since the cacao populations are closer at hand, collecting in such circumstances may be very productive in terms of the population size that can be surveyed and the quantity of material for propagation collected during a defined period of time. The disadvantages presented by the smaller rivers are the increased likelihood of obstructions and the adverse consequences of flash floods. Very small rivers, such as the tributaries, present inconveniences in the form of obstructions by fallen trees and their shallowness, which make passage difficult and onerous. Accordingly, it is likely that exploration in these valleys will be less productive in terms of the determination of diversity and its acquisition. The difficulties of accessibility have resulted in lesser attention being given to the exploration of small tributaries and areas distant from the river banks. As a result the total area of distribution in a valley is imperfectly known and the impression is given that cacao populations occur only on the banks of the major rivers.

At the other extreme are the very broad rivers, often several kilometres wide. The cacao populations tend to be isolated and although probably having similar genetic origins have developed individual variability. Among the main features of the large rivers is the presence of numerous islands. The populations that exist on these islands are restricted in size and genetic composition and, being obligate intra-breeders, contain specific elements of the diversity. From the point of view of the operational side of collecting, large rivers imply that a larger amount of time is spent on travelling, especially across rivers, resulting in less time being available for surveying the diversity in the individual locations and collecting samples of it. As a result, the productivity rate as measured by the quantity of material collected during a specified time period would be lower than that obtained on the narrower rivers. However, as will be explained during the descriptions of the diversity already available for analysis, the lower productivity obtained in sampling the variability in some of these areas is compensated for by the existence of a more ample phenotypic diversity.

The above explanations concerning the situations in which natural cacao populations are encountered demonstrate that diverse factors are responsible for the nature and distribution of the genetic diversity that exists. All of these factors must be taken into account in the planning and execution of the phases of a conservation programme that are concerned with the acquisition of the germplasm. On account of the varied specific situations in which the diversity can occur it is inappropriate to establish rules regarding the operation of collecting activities and sampling procedures. The collecting strategies must be adapted to the particular conditions that exist in the individual target areas.

In addition to the factors that determine the manner in which the cacao diversity is distributed the collecting programmes have, as a rule, to conform to the resources and restrictions that are imposed by conditions that are independent of the nature of the diversity. Owing to the fact that the conservation of genetic resources is a developing subject it is a general rule that collecting activities do not achieve the potential results because of the inexperience of the persons to whom the collecting is entrusted. The influence of resource availability on the planning of the collecting activities and their efficiency in terms of the ability to capture the diversity of the target area can be considered in two levels. One level is related to the actual operation of collecting. The second level concerns the limits to the quantity of material that can be collected imposed by the capacity of the station to which the material is sent to effectively handle the potential quantities involved.

The above factors as determinants of the sampling strategies are also related to the method of propagation that will be employed. In the case of cacao both seed and vegetative propagating material can be used for reproduction of the genotypes sampled. In either case it is important that the destination station possesses adequate facilities for propagation and storage of plants. This would be more uncertain for the use of vegetative propagation since adequate planning is required to ensure that the facilities and details such as ensuring the availability of rootstocks in sufficient quantities would be established well in advance of the collecting activity.

Both methods of propagation have their advantages and disadvantages, as well as functions, in the attainment of the objectives of the conservation of the species' germplasm. In fact, propagation plays an important role in determining the successful outcome of any collecting activity. The application of either or both methods in a specific activity is regulated by various considerations. These include, principally, the objectives of a programme, the resources at its disposal in quantitative and qualitative terms, the condition of the trees encountered in the target population and the physical conditions that exist in the area.

Theoretically, the most efficient manner of sampling the diversity of a population is through the collecting of as many fruits as possible. It can be assumed that each fruit produced by a tree is the product of fertilization, either by its own pollen or by that of neighbouring trees. In this hypothetical situation a fruit of a heterozygous genotype produced by self-fertilization would result in a family that is heterogeneous for the genes it possesses. On the other hand, the fruits resulting from hybridization would contain samples of genes from the pollen parents. Consequently, they would provide a larger sample of the population's diversity. Thus, fewer numbers of individuals would suffice to provide a representative sample of the genes in the population.

The above theory rests on various assumptions. First, that there is sufficient heterogeneity in the population to provide an adequate gene pool. In this respect the effectiveness of sampling is governed by the population size. It is assumed that all, or most, of the individuals have fruits at a stage for harvesting. Another assumption is that inter-fertilization occurs freely and at random within the population.

In practice, these assumptions are never realized. Cacao trees in natural forest habitats produce few fruits and, even when ripe fruits should be available, they are frequently rendered useless on account of diseases and pests and the depredations of animals and birds. Fruits are lost in some rivers owing to flooding coinciding with the time of ripening. Where unaffected fruits are available these can be out of reach of normal methods of harvesting. Another aspect that is related to the ability to satisfactorily sample the populations' diversity is that, frequently, few seeds are contained in the fruits.

The post-collection treatment of the seeds regarding maximizing the utility of an activity is as important as the collecting itself. Cacao seeds are recalcitrant; that is, they do not possess a dormancy period and germination starts as soon as the seeds are exposed to moisture. Consequently, the life of a seed outside a germinating medium may be as short as a few days. Seeds are also sensitive to cold and lose their germinating ability at low temperatures, which sometimes happens in the case of material transported by air. Deterioration of the cotyledons occurs and this can be detected by an odour similar to iodine.

The above problems can be mitigated and eliminated by certain techniques. Seeds can remain viable in the fruits for a few weeks but may lose some of the ability to germinate as time progresses. Methods for packing and transporting the seeds out of the fruits in a condition that will result in a satisfactory establishment rate are available. These require that a certain amount of care is given to the seeds when they are unpacked on arrival at the destination, especially when a long period in transit is involved. Protection from cold is also a foregone essential.

When it is not possible to rely on the availability of fruits to obtain a representative and random sample of the diversity that a population possesses, the alternative is to use vegetative propagation. Even when ripe fruits are present vegetative propagation can be resorted to in order to reproduce and conserve certain genotypes that exhibit interesting or valuable characteristics, as an adjunct to the normal sampling process.

Several collecting programmes have been undertaken with the objectives of obtaining selected germplasm that appears to possess desired traits such as resistance to one or more pathogens. In such cases, involving specific individuals, vegetative propagation may be the most effective manner for conserving the desired traits. However, by their selective nature, these activities do not consider, and do not achieve, the conservation of the genetic composition of the population in its totality.

Although vegetative propagation would be an ideal tool for germplasm conservation its use is fraught with certain problems that seriously affect the contribution that it can make in this respect. The usual and economically most effective method of vegetative propagation utilized in genetic resources programmes is that of budding. This provides the easiest way of handling the material collected in the field and is cost-effective in terms of the quantity of material that can be transported to the propagation facilities.

The use of vegetative propagation in sampling the diversity present in a given population implies that the proportion of trees sampled needs to be fairly large. However, the usefulness of the method depends on the availability of material in a condition that would guarantee adequate success in budding. The problem is that this requirement is difficult to achieve in most populations that are growing under natural conditions. Branches that would be collected for budwood are sometimes beyond the reach of the collectors (as in the case of tall trees with single trunks, where the canopy is high above ground; see Plate 2c). The growth patterns of the trees may result in suitable material being unavailable at the time of collection. In addition, it is necessary to consider the presence of pathogens on the branches, such as thread-blight and algae. These factors demand that infected trees be rejected as sources of material that, in addition to sanitary considerations, by impairing the success of propagation, cause the loss of the genotypes and result in wastage of resources. On the other hand, good success has been obtained in budding material that would appear unsatisfactory according to the norms for budwood quality that are current on the basis of experience.

The vegetative material for propagation is subject to a short useful life. This means that, in addition to the problems of obtaining adequate material, it is imperative to organize a collecting programme so that transport of material is strictly controlled. On the receiving end the facilities for propagation have to be adequate for the handling of the quantities of material that are received in each shipment. This implies the availability of the rootstocks and the necessary manpower to undertake the propagation in the shortest possible time and with the greatest efficiency, in order to ensure the maximum benefits from the labour and costs expended in collecting and transportation.

The general experience that has been gained from various collecting activities is that the rate of success from budding is usually low. As a result its use for obtaining representative samples of the diversity existing in a population is often debatable. Vegetative propagation can make some contribution towards germplasm conservation when measures are taken to make the process efficient on the basis of mitigating the losses that can occur. Many factors need to be taken into consideration at all stages and decisions made regarding the use of the technique according to the prevailing circumstances, keeping in mind that the actual process of propagation is only one stage of the entire germplasm conservation programme.

Vegetative propagation would be most successful when the target area is located near a facility for propagation or a genetic resources centre. Such circumstances permit collecting to be carried out in a target area on various occasions during the periods when suitable material is available and can be utilized to advantage. The collecting activities that have been conducted on a casual basis with reliance entirely on vegetative propagation have in all cases been unsuccessful in fulfilling their expectations because of the losses suffered from natural causes. In addition to the limited numbers of trees that could be sampled during any activity the proportions of genotypes established relative to the original sample sizes have been low. The consequence is that the opportunity to conserve specific components of the germplasm in the target area is lost given that, in programmes conducted to date (with one exception), no attempts have been made to recover lost genotypes by repeating the collections.

The most effective manner of overcoming the restraints imposed by natural conditions in obtaining representative samples of the diversity of a population is that of collecting both fruits and budwood. This policy involves collecting fruits and budwood from the trees that have ripe fruits and budwood at random from trees without fruits. The concept was put into practice since the inception in 1976 of the Brazilian Amazon Genetic Resources Programme. In terms of the class of material collected the individual parent trees of a population would be represented in the genetic resources centre by progenies descended from seed alone, clones of the parent genotypes and other parents represented by both their clones and their progenies from seed.

The combination of vegetative and seed propagation has several advantages, not only with regard to the conservation aspect of the programme, but, principally, concerning the characterization and evaluation of the diversity. The procedure allows comparisons to be made between the phenotypes of the adult trees occurring in natural habitats and those of the same genotypes in their juvenile stage of development and to determine their response to other habitats. The observations made on the developing clonal plants would reveal important features related to plant habit, leaf characters and flowering behaviour. These observations would be applied in the characterization of the diversity contained in the populations in terms of the attributes exhibited by the adult trees growing under natural conditions.

Seed progenies also provide the same information about the juvenile growth characters. Among other advantages they are important in revealing the existence of characters controlled by recessive genes and the presence of mutations, an example of which is shown in Plate 31a. Some of these could be valuable in breeding programmes but are not identifiable in the cloned genotypes. In practical terms the scale and nature of the intra- and inter-family segregation in the progenies would be used to measure the homozygosity status of the parents and the diversity contained in the population. This segregation, considered together with the traits of the parents, contributes to the analysis of the inheritance of the individual traits, both qualitative and quantitative. Analysis of the quantitative traits may provide indications regarding such aspects as the individual's combining ability for yield components and their inheritance. The knowledge could then be applied in the selection of genotypes of potential use in the improvement process.

It has been the practice in cacao genetic resources programmes to organize and conduct collecting activities specifically for the species. These activities have been given the title of 'expeditions'. The expression covers a multitude of types of activities. These range from visits of a few days' duration to a single location in which a few items are collected to activities that last several months or cover large areas of territory with the collection of large amounts of material. However, it is not necessary that collecting of germplasm be undertaken solely in this manner. Programmes that are based on the accumulation of as large a sample of genotypes that will represent the diversity of one or more populations, or of a particular area, could be conducted using other methods. Collections can be made in association with botanical expeditions or similar scientific activities.

Material can also be obtained through the services of persons resident or working in the target area. Once the provider of these services understands the objective of the programme and details of the procedures involved material would be supplied at stages when conditions are suitable. Such methods obviate the need for expensive activities and can provide an adequate supply of material that, perhaps, would come closer to achieving the aims of sampling the diversity of the area than would be possible from a single casual visit.

The effective operation of large programmes of collecting would depend on the *a priori* availability of information about the target area and the cacao populations that exist in it. The Brazilian programme was based on the principle of undertaking a preliminary reconnaissance of a target area on which the planning of the principal activity would be based. Although the aim of a reconnaissance was to gather information about the occurrences of cacao and conditions for exploration it was an established practice to collect some specimens from trees in the area visited. This practice, in addition to contributing to the conservation objective, is useful in providing some preliminary information about the nature of the population concerned.

The Cultivated Populations: their Role in the Creation of Diversity and its Conservation

Cultivated populations are also repositories of the genetic diversity of the species. Some may be looked at from the point of view of constituting a controlled stage of the development of the diversity. In accordance with the propositions regarding the origin of the species' diversity on which this work is based, all cultivated populations, regardless of the time that they have existed, would ultimately be descended from the primitive diversity and, therefore, from the diversity of the Amazonian Region.

The cultivated populations are derived from specific components of this diversity and exhibit considerable variability in terms of the genetic material they contain and the circumstances of their evolution. The extent of the variability will become evident when the major regions of cultivation are described in Chapter 6.

However, it will be found from the histories of the creation and evolution of the populations cultivated in the individual areas that there are groups of such populations that are descended from genetic material from the same sources and are related through the exchange of varieties among them. From the point of view of conservation of their resources the related populations should be considered to belong to the same gene pools. Although conservation programmes may be conducted in local populations those with the same genetic background should be treated jointly as components of the same portion of the species' diversity.

With regard to the genetic composition of the cultivated populations there are perhaps three basic situations that need to be considered from the point of view of conservation. The simplest one is that in which a single homozygous genotype is involved. In such cases no variability will be expected except for the possibility of mutations. When these occur they will be expressed only in the succeeding generations, if reproduction occurs.

Next in order of increasing complexity is the situation where the source material concerns a single heterozygous genotype. The populations obtained from it will exhibit a degree of variability depending on the nature of the genotype. Subsequent generations will produce further evidence of variability but all individuals will be closely related in spite of exhibiting a wider range of phenotypic variability than that of the original source genotype.

The third situation refers to populations formed from introductions of diversity from more than one source. In these complex situations there will be enlarged variability resulting from hybridization among these varieties in the subsequent generations. Concomitantly, segregation will be generated in the descendants of the genotypes of the individual source varieties by self-fertilization. The result of these two lines of reproduction will be the creation of populations that are complex in terms of the genotypic combinations produced.

Cultivated populations have certain features that are advantageous for the processes of acquiring samples of the diversity and its characterization and conservation. The advantages of many cultivated populations include such conditions as easier access and uniformity of age, density and tree habit. The composition and appearance of cultivated populations are, of course, dependent on the particular situation and history of each field. Several factors operate in determining the phenotypic variability that is observed in individual situations.

The relative uniformity contained in some cultivated populations and the ease with which the comparative observations on the trees can be performed makes the task of collecting much more reliable. In these circumstances the process of collecting samples of diversity implies a certain degree of selection for specific characteristics displayed by the trees. At one end selection is performed on the basis of the identification of certain specific traits that are desirable in agronomic or commercial terms and, at the other end, for those traits that are considered to represent the range of variability for other characters that the population possesses.

The process of selection presupposes that there has been some form of characterization based on the indicators of variability. In the past, ignorance about the modes of inheritance and such aspects as the relationships between phenotype and genotype and genotype–environment interaction meant that selection would have been based on intuition. In more recent times attention has been given to the development of the bases for making selection more effective in terms of the results desired.

Although many different situations occur within the concept of 'cultivation' the establishment of cultivated populations that were formed up to the mid-20th century, and certainly the older enterprises, would have resulted from the introduction of small quantities of planting material. The material introduced would have originated either in spontaneous populations or from an intermediate stage of cultivation. Whatever the source of the planting material that composes a field, it is probable that some degree of consanguinity will occur in most of such assemblages of plants. However, it is not possible to generalize on this subject and examples, based on their historical backgrounds, of the genetic composition of cultivated fields will be presented in Chapter 6.

The different systems of sowing and cultivation practised in various countries have an important bearing on the composition of cultivated fields. The practice largely followed in Trinidad at the end of the 19th century, as described by Preuss (1901), is a good example of the manner in which fields would have been established in many growing regions. It would be presumptuous to attribute the origin of this practice to Trinidad. The method described involved using fruits harvested from selected trees. The seeds from each fruit were sown in the same area, such as within the same row. Where several fruits from an individual tree were used differences among the progenies from the various fruits may have occurred because of uncontrolled hybridization but they would all have been related. Hence, it would be expected that neighbouring trees would exhibit similar phenotypes. This situation can be observed in several of the fields formed in this manner when careful comparisons are made between neighbouring trees. An example of how the sowing of seeds from a single fruit in the same row produced plants having the same phenotype was that of a row of trees having fruits of strikingly similar appearance encountered in a field in Ecuador.

Mention was made previously of the practice of certain indigenous peoples to plant a few cacao trees in or near their villages. This could be considered to be the first natural stage of domestication of the species. Presumably, the seeds for these plantings derived from spontaneous populations.

Within a given area the trees that occur in plantings formed in this manner would probably be descended from previous plantings that were abandoned and passed to the sub-spontaneous state. These sub-spontaneous individuals and their predecessors would form a lineage from perhaps a spontaneous individual. In other words it may be possible to trace a link between the individuals at one location and those of other populations. The linkage of these populations is important in the context of the formation of new occurrences of the species through the migrations of the indigenous inhabitants of the regions concerned. The process of migration applies to both the Amazonian and Circum-Caribbean Regions.

The establishment of cacao cultivation in the Amazonian Region, which occurred after the species was discovered, involved, principally, the intentional colonization of new areas or, possibly, the expansion of the areas occupied by natural stands. The cultivated areas in the Region can be considered to be another stage in the process of enlargement of the habitat range of the species and its diversity.

As the seeds used in these plantings are descended from the spontaneous occurrences it should be possible to relate the cultivated populations to their parental varieties when the latter are identified. Even when the parental varieties cannot be identified, for example, on account of their having disappeared, the analysis of the diversity existing in the various cultivated populations would constitute an important contribution to the understanding of the primitive diversity of the Region. In this way collections of individual genotypes under cultivation and their progenies form an integral part of the conservation of the genetic resources of the Region.

The populations that have existed in the Circum-Caribbean Region and in the secondary areas where the species occurred in pre-Columbian times and those subsequently established through introductions are of similar importance in the context of germplasm conservation. The hypothesis of a common origin of the species implies that the cultivated populations outside the Amazonian Region, although they may have existed for many centuries, owe their diversity to descent from heterogeneous sources and mutations that occurred during the period of their existence.

In some cases no documentary evidence exists regarding the sources of the planting material used in establishing areas of cultivation. Where it is possible to undertake an analysis of the variability of a population the origins of its components may be conjectured on the basis of the observed characteristics and their relationships to those of other varieties. An analysis of the heterogeneity of a population would entail the identification of the primary or ancient germplasm, the introduced variety or varieties, the hybrids formed between them and the segregants produced by recombination in the succeeding generations. However, owing to the fact that our knowledge of the diversity of the species is far from complete it is not appropriate, at the present time, to attempt to reach definitive conclusions on the subject.

There are several populations, such as those formed in more recent times in the Old World, where some knowledge exists concerning the sources of the varieties that were introduced. Even in these cases the characteristics, as well as their origins, of the varieties grown in the source countries are often poorly defined. The introduction of varieties possessing unique characteristics into areas in which cultivation existed would be expected to result in contamination of the varieties that had been cultivated previously through hybridization and formation of intermediate types. The hybrid generations would then provide the planting material for subsequent cultivation. However, it is observed that during several generations selection pressure for certain characteristics may result in individual populations that do not exhibit the introduction's characteristics although its genes may be present in the hybrid progenies.

Practical Considerations in the Establishment and Conduct of Conservation Activities

The contributions to germplasm conservation made during the 20th century in the secondary areas of distribution have been principally by-products of programmes of selection. In general, these have been based on predetermined objectives in which attention was given to genotypes expressing certain criteria regarding the traits sought as ideal for the formation of cultivars. Since we are referring to selection programmes conducted in cultivated populations the details regarding these programmes and their outcomes are described in the introduction to Chapter 6.

Starting from the late 1940s germplasm from new sources was introduced into the producing countries. Principally these were derived from the collections made during the 1930s in the Amazon valley that are described in detail in Chapter 5. These introductions had two effects, one of which was the acceleration of the process of substitution of the 'traditional' populations in some countries. The other effect was the release of new combinations of genetic factors that governed the traits expressed in the Amazonian varieties through reproduction in the subsequent generations, either by commercial cultivation of the progenies or in the experimental improvement programmes. However, in most cases, since the genotypes introduced were selected members of the populations concerned, the range of germplasm involved was a fraction of that contained in the source populations.

The cloned cultivars obtained in selection programmes, notably those undertaken during the period from 1930 to 1950, have constituted the basis of the origins of the concepts of germplasm collections created through the accumulation of the genotypes from several countries. Much of the knowledge on which the management of germplasm collections and the characterization and evaluation of genotypes is based has developed from these collections of clones. The first collection specifically established as a genetic resources collection was that in Trinidad (Bartley and Chalmers, 1971). As will be explained below, this germplasm collection was based on concepts of the genetic resources that were far wider than those considered in the formation of collections of clones that had, up to then, been considered to represent the species' diversity.

In accordance with the definition of the genetic resources established in Chapter 1, the original Trinidad germplasm collection included genotypes from as many sources as were available, including elements of what could be considered to have been the primitive diversity. Also included were progenies selected from a wide range of families produced by controlled pollination. The motive for including the progenies was based on the concept that they would possess the same genes as the parent genotypes that had been transmitted to their progeny and would be equal representatives of the diversity.

Natural reproduction involves various degrees of inbreeding and hybridization and could proceed for various generations. Through these processes the genes contained in the primitive diversity can be expressed in the progeny of the individuals belonging to the original populations. The observations made on these progenies are the means for determining the genetic composition of the primitive and cultivated populations in which characterization of the trees is not feasible. Therefore, the production and maintenance of families of progenies is essential for describing the universe of diversity of the species, in addition to their role in its conservation.

Consequently, it is essential that a genetic resources programme be based on the principle of seeking for or producing the progenies that are derived from the primitive elements of the diversity and that are equal components of it. Hence, the collections of germplasm that are established within the genetic resources conservation programmes must represent all of the individual elements of the diversity regardless of their derivation.

The conservation of cacao germplasm, especially considering the threats of extinction of the natural populations (by natural and human factors) and the changes that occur in cultivated populations through the introduction of alien varieties, is indispensable for providing the material for future improvement of the crop. The most effective manner by which to guarantee its conservation is through the accumulation of the diversity in specified locations. The function of these collections goes far beyond the mere aim of conservation. Accumulation of variability from various populations at a single site permits comparative observations to be made from which to establish the distinguishing characters of the individual populations. The availability of several varieties at the same site enables the programme to use the individual genotypes for the production of the inbred and hybrid progenies that will form the basis for collections of genotypes that possess specific traits and characteristics. These progenies also supply the genetic material that can be utilized in future improvement programmes, the identification of the useful sources of breeding material being based on the due characterization and evaluation of the components of a germplasm collection.

In the formation and management of germplasm collections it is essential to consider certain factors related to the nature of the species. The variability shown by genotypes regarding their habits, tree architecture and growth rates must be taken into account in planning the layout and establishment of the accessions. In most cases, of course, characteristics of this nature would not be recognized until the plants have been established in the field and are in the development stages.

When the traits that are specific to the accessions have been identified alterations to the planting schemes need to be made to provide optimal conditions for the normal development of the accessions. Such alterations include the use of spacing distances related to the architecture of each accession and the arrangement of the accessions in order to mitigate the effects of competition resulting from large differences in tree sizes, whether within a heterogeneous family or whether between accessions. Excessive competition is a factor that inhibits expression of the specific characteristics of accessions, especially those with slower growth rates or smaller tree sizes. The inhibition of character expression caused by inadequate management and planning results in accessions that possess desirable properties being given indifferent treatment. When indifferent attention is paid to individual genotypes and families their important characteristics are frequently overlooked with regard to their utilization, and their future status in the gene banks may be put in jeopardy.

Since cacao populations can be found in a large range of habitats it may be inferred that specific components of the variability have become adapted to certain environments. The concept of centralizing the genetic resources in specific locations usually results in accessions being brought to and established in environments that differ substantially from those in which the populations evolved. Therefore, the curators of the genetic resources centres need to be aware that a specific variety may not be adapted to the conditions prevailing in the collection that they manage. The consequences of placing plants in environments to which they are not adapted are, again, the inhibition of character expression or, worse, the loss of the accessions concerned.

There is a general lack of appreciation of the fact that a genetic resource centre is not merely a farm on which cacao is cultivated. It is necessary to pay attention to the specific needs of the individual accessions as they become apparent from observations made on the plants and to take the measures necessary to ensure their conservation and their ability to provide the desired information.

Considering the various undertakings in the field of the conservation of the genetic resources of cacao it is a matter of regret that several of them, including those that ought to be the most important, are situated in areas where conditions are inappropriate for their intended purposes. Inadequate management combined with, or resulting from, the unsuitable growing conditions prohibit normal development and character expression. One alarming feature has been the high rate of loss of plants.

The history of collecting cacao genetic resources is a long account of losses of material at all stages from the transport of the material to the propagating site to the years following establishment in the field. Some of these losses may have been unavoidable; for example, those resulting from inferior quality of the material for propagation. However, losses sustained in the field owing to inadequate conditions, inexperienced management and lack of continuity of the effort required for the maintenance of the collections, are deplorable. The losses are regretted, not only on account of the failure to maintain material that may no longer be available, but because the material collected often represents considerable expenditure of the resources applied in its acquisition.

As regards the financial aspects of maintaining genetic resource collections inadequate conditions demand greater inputs to achieve the results expected from them. The resources for these inputs are inadequately provided. At the same time, their unsatisfactory field performance results in the inability of the accessions to attain satisfactory levels of productivity, naturally resulting in reduced income that would, under optimal growing conditions, be applied in the maintenance of the collection.

Another great constraint regarding the functioning of genetic resource centres is that they are usually limited in scope in relation to the extensive genetic diversity that the species contains and the activities that they should perform.

The most important factors concerned with the management of genetic resource centres are related to the fact that cacao is a perennial tree species with an indefinite longevity and a fairly long reproductive and generation cycle. These characteristics imply that space is essential for the natural development of each individual that would occupy the same site for an indefinite period of time. In addition, the determination of the characteristics of genotypes and their interrelationships will take several years before they are concluded. Also, as mentioned above, slow growing and unproductive trees would not contribute to the economic value of the collection. In this way, once a space is attributed to an individual it is lost for use for the expansion of a collection when additional accessions are acquired.

These factors demand that genetic resource centres be planned so that sufficient space is available for the quantities of germplasm that may be introduced. Space would also be required for establishing the families that would be produced for purposes of characterization and evaluation as well as those connected with the objectives of conservation. Obviously, it is necessary that the centre be provided with adequate resources that would be needed in order to function for an indefinite length of time, a matter that is frequently overlooked.

Various measures can be adopted to mitigate the effects of the inherent properties of the species and the needs of the genotypes, which influence the procedures to be adopted in sampling the diversity. In order to apply these measures effectively it is necessary to treat a germplasm collection as a dynamic enterprise. All too often the attitude is taken that, once the accessions are established in the field, that is the end of the matter and the trees are left to their own devices, sometimes, unfortunately, to perish. All of the measures require constant observation of the behaviour of the trees and the appropriate actions taken when necessary. This demands a degree of flexibility in the management of the collection's resources. One underlying principle is that optimal conditions of management and resources would enhance the development of the plants and minimize losses. In turn, it will become possible to maintain smaller numbers of plants and accelerate the other stages of the conservation process. The fundamental condition underlying the preceding principle is that any collection must be located in the environment that is appropriate for its purposes.

Among the measures that can be taken to achieve effective use of space and resources are: (i) using spacing in accordance with the growth habits of the individual genotypes; (ii) grouping together genotypes with similar characteristics; (iii) reducing the numbers of individuals in families of progenies by propagating those representing the variability expressed in the family – in the case of families of homozygous parents a single progeny would be sufficient; and (iv) re-propagating individuals at risk (e.g. those in unfavourable situations) and replanting them in situations where they would benefit from intensive care. The land made available would be used for establishing new additions to the collection.

The genetic content of the genotypes that prove to be less easy to grow in the prevailing circumstances, especially those that may be at risk of being lost, may be preserved by hybridization with another variety and conservation of the progenies produced. The most suitable genotypes to use as parents in this procedure would be those that are self-compatible and homozygous, especially those carrying recessive marker genes. Through the application of appropriate breeding techniques the intrinsic characteristics of a genotype or variety to be saved would be identified. The alleles responsible for the expression of the characters would be recovered for conservation and use.

The matter of the sizes of families of progenies that should be included in collections is a debatable question. What should be done depends on the parent genotypes and the variability existing in the families resulting from the collections. Accordingly, as mentioned above, a single inbred progeny of a homozygous parent would be a sufficient representative of its genotype. On the other hand, when extensive heterogeneity is concerned there may be no limit to the numbers of plants that would represent the variability of the population from which they may be derived. In Brazil it was observed that progenies descended from different fruits of a parent tree exhibited variability in their juvenile phase, which would indicate that the fruits derived from fertilization with different pollen donors. Consequently, it was considered advantageous to adopt the system of planting samples of the progenies of each fruit in order to facilitate the identification of the contributions of the male parents.

In one undertaking comprising what should have been one of the most important germplasm collections, the policy was adopted of planting a sample of the progenies, the size of which was determined on the basis of an estimate of the numbers of genes that would be represented. The calculations did not take into account the specific circumstances of each population sampled during collecting or the pollination dynamics of the species, as determined by the compatibility status of the individuals comprising each sample. This theoretical sampling may have produced the desired information about the populations if the appropriate studies had been carried out on the progenies. In view of the fact that the collection was established on land cleared from the forest with some of the forest trees left to provide shade, conditions developed such that, with the inadequate management that the collection received, significant mortality occurred. Consequently, some of the families were irreparably lost. In contrast with this story it should be emphasized that the most successful collections have been those in which all of the progenies were established.

Some of the activities that constitute the stages of a genetic resources conservation programme, as set out in Chapter 1, include the characterization and evaluation of the accessions, either as single genotypes or as families. The summary of the variability indicators in Chapter 3 gives an idea of the array of items that could be considered in the characterization of the accessions. The detailed observations of the plants' characters would constitute a formidable undertaking when all of the available germplasm is considered. For this reason, the quantity of such information available is limited and usually incomplete as regards any population. Although some extensive characterization has been undertaken and large amounts of data have been produced a great deal of it remains unpublished.

In various parts of the text reference has been made to the recognition, from the earliest times since variety collections were made, of the need for descriptions of the varieties. The selection programmes undertaken in the 1940s and the subsequent exchanges of clones underlined the usefulness of descriptions of these clones that would be available on a general basis especially for the purposes of identification. The large volume of information that subsequently became available from collecting and conservation activities and improvement programmes led to the adoption of methods for recording this information and making it universally available. The International Cocoa Germplasm Database (ICGD) was set up about 1990 with the objective of accomplishing this task. Various descriptions of the Database have been published such as End *et al.* (1994).

The Database is constantly developing and growing in terms of the number of genotypes included and the information about them, so that it has to be kept up to date. Two printed versions have been published and two CD-ROMs later produced as well as an online version which is updated. Several thousand entries are involved including clones and families (the latter erroneously included under the term 'clone'). Data are presented according to an established order of descriptors and it is possible to obtain lists of genotypes that possess certain characteristics. However, the present organization does not provide information on the distinguishing characteristics of genotypes and the populations to which they belong in a manner that is easy for most users to discern. Another drawback refers to the fact that no distinction is made between observations recorded on parent trees and those of their progenies, either clones or families. Another source of confusion relates to the entering of completely different descriptions from two or more sources which is the result of misidentification. The use of average values often hides the true characteristics and there is little information given regarding inheritance. One defect concerns the recording of information about the reaction of genotypes to diseases. Absence of infection under certain circumstances is inferred as 'resistance', while subsequent infection (or susceptibility) on a given genotype is omitted.

On the whole there is a misunderstanding about the purpose of characterization. The objective is not merely to gather data about all of the traits of a genotype but to arrive at a description of its distinguishing characteristics and those of the population to which it belongs.

Although much has been accomplished in a general manner of all stages of the genetic resources conservation of cacao, the above discussion illustrates the negative aspects or failures to conserve and adequately characterize the populations that are known to exist. On account of this situation the presentation of a comprehensive discussion of the species' genetic diversity at this time is impracticable. In the following chapters as much as is possible of the existing knowledge about the populations that comprise the diversity will be presented with the intention of arriving at an understanding about its nature. It is hoped that the views expressed above will encourage the establishment of adequate programmes that will guarantee the conservation of the diversity and the availability of the information that may be essential for the effective utilization of its components.

5 The Foundations of the Diversity

PART 1: THE AMAZONIAN REGION

In the introductory chapters the concept was advanced that the basic diversity of the species originated in the Amazonian Region. In the first section of this chapter the knowledge that has been acquired about the diversity that exists in the Region will be summarized according to its component populations. The evidence about the diversity will be interpreted on the basis of the observations that have been made on the morphological characteristics of what may be considered to be the primitive gene base and the variability that has been derived from it.

Although the volume of information at our disposal probably represents only a fraction of the total diversity, the magnitude of this complex subject is such that the presentation of an analysis of the data is an enormous task. The area drained by the R. Amazon alone covers over 7,000,000 km^2 to which another 10% perhaps has to be added to complete the region as defined in Chapter 1. Over 1000 watercourses of various magnitudes make up the hydrologic systems of the Region. On the basis of the hypothesis that each component stream could represent a distinct cacao population the description of each would be an impossible task within the confines of the present volume.

Although the Brazilian Genetic Resources Programme constitutes the major activity of this type that has as yet been undertaken it is far from covering the potential area of the distribution of the species. Further, the losses that have occurred since collecting commenced result in less than half of the parent genotypes sampled being available for study. Some of the diversity described herein is known only to the author and is not included in the material that can be found in the conserved state. The scope of the Brazilian Genetic Resources Conservation Programme can be comprehended from the inventory of the accessions established in the field up to 1985 (Almeida *et al.*, 1987). A considerable quantity of additional material was obtained in the succeeding years, which is included in the summary of the results of the programme provided by Almeida *et al.* (1995). A general picture of some of the populations sampled may be obtained from Barriga *et al.* (undated).

Topics Related to the Diversity Applicable to its Description and Analysis

To obtain a complete understanding of the global picture of cacao diversity and its distribution it is necessary to establish some facts and dispel several false concepts that have been adopted in the course of time by persons unfamiliar with the Region. The principal subjects to consider in this respect are:

- It was realized early in the history of the attempts to collect cacao diversity that, for the most part, the cacao populations on the tributary rivers contained varieties that are unique to each river as regards their phenotypic characteristics. The reasons for this geographical distribution would be consistent with the fact, as explained in Chapter 4, that cacao populations, whether spontaneous or cultivated, would be descended from limited amounts of reproductive material and having a restricted gene base as a consequence of differential adaptability and selection.

- Prior to the arrival of the Europeans the Region was not artificially partitioned into national divisions or subdivisions. In other words the evolution of the cacao diversity took place over the entire primary distribution area. For this reason the description of the individual cacao populations and varieties must be based on their distribution in the various drainage systems without reference to past or present political divisions. However, individual countries have conducted genetic resource programmes at national levels including the formation of national collections. This situation demands that attention be paid to these conservation efforts although the present treatment of the subject would involve the interpretation of the results from sources of information other than those relating to national programmes.

- Another aspect regarding the distribution of cacao populations that must be considered is the fact that, previous to European colonization, the areas of the occurrence of the species were occupied by indigenous peoples that, in most cases, are not the same as those actually found in these areas. Consequently, most of any human involvement in the dispersal of cacao populations would have occurred in prehistoric times and little can be attributed to the present distribution of Amerindian peoples.

- It is necessary to be clear about the fact that only part of the Amazon Basin and the adjacent areas on its periphery was occupied by cacao at the start of European colonization of the region. Since the 17th century large areas in which cacao previously did not occur have been brought into cultivation. Hence, it is important to recognize that there is a distinction between the areas where the primary germplasm occurred in a, presumably, spontaneous state and the secondary areas of cultivated germplasm.

- On the main R. Amazon the primary areas, that is, those existing prior to European occupation, would appear to have been located west of the mouths of the Tapajós and Trombetas rivers. It may be more correct to define this area as embracing the region west of the mouths of the Madeira and Negro rivers.

- There are, however, some few exceptions to this distribution pattern, such as the island of Gurupá near the mouth of the Amazon on which, according to early reports, a significantly large population of cacao existed during the 17th century. The origins of such populations will be discussed in the appropriate sections.

- The greater part of the primary cacao diversity would, therefore, be located in the western and southwestern quadrant of the Region. Some of the drainage basins to the east of the area defined above also appeared to contain occurrences of cacao in pre-European times and were also exploited during the extraction era but little is known about their varietal status or the extent of cultivation along these rivers. North of the Japurá-Caquetá drainage area the occurrences of cacao are relatively sparse. For example, in the valley of the R. Negro environmental conditions were found to be unfavourable for the development of cacao, except in isolated patches. These circumstances make it improbable that there were any significant natural occurrences of cacao in this valley. The exceptions to this would be the R. Branco, a tributary of the R. Negro, and perhaps a couple of other drainage systems that have not yet been explored and those tributary streams that form the beginnings of the R. Orinoco.

- All secondary areas of occurrence would have been derived from planting material obtained from the primary area. In view of this situation it is taken for granted that the cacao diversity represented in the cultivated populations would be part of that in the primary areas. The cultivated populations would, by being representative of the varieties contained in the primary populations, be repositories of the same genetic factors but, as a consequence of recombination and mutation, may exhibit a wider range of variability.

- It has become a custom to refer to Upper Amazon and Lower Amazon regions. This tendency is derived from the use of the term 'Upper Amazon' to group together the varieties collected in 1937–1938 in the vicinity of the town of Iquitos in Peru. The stretch of the river between its formation by the confluence of the Rivers Ucayali and Marañon and the frontier with Brazil was known as 'Alto Amazonas', from which the application of the term originated. Obviously, the Amazon Basin has upper and lower regions. However, the persons who attempt to divide its cacao diversity into two sections are unable to define the dividing line between them, either on the basis of geography or on specific distinctions in the distribution of the species' diversity. While the primary cacao diversity is distributed within the geographical area that embraces the upper regions of the river basin it occurs mostly in discreet areas as distinct genetic entities. Therefore, it is somewhat meaningless to group these distinct varieties under a single term that implies that they belong to a homogeneous area as regards their diversity. On the other hand, because the cultivated populations are usually located in the 'Lower Amazon' they would be descended from the primary populations of the 'Upper Amazon' and, therefore, would have the same genetic origins as some of the components of the primary populations.

- Up to fairly recent times all cacao known in the Amazonian Region and the varieties outside the region that are descended from the diversity in the region have immature fruits of a green colour. In this respect the Amazonian cacao appeared to be distinguished from that of the Circum-Caribbean Region where the presence of red fruits constitutes one of the main distinguishing features. However, as will be described below, the discoveries made during the past few decades show that this distinction no longer applies.
- During the course of history cacao varieties have been exported from the Amazonian Region to various destinations. The discrete nature of most of these movements resulted in the formation of cultivated populations that are composed of easily identifiable components of the diversity. In the same way that the cultivated and sub-spontaneous populations in the Amazonian Region provide indicators as to the characteristics of the primary populations, the genetic bases of the extra-Amazonian populations also contribute to the characterization of the primary diversity.

The Diversity in the Primary Distribution Area

Note: the river systems, on the basis of which the account of the Amazonian Region is organized in this chapter, are shown in detailed maps of sections of the region inserted in appropriate places in the text. The location and orientation of these sections can be determined by reference to Fig. 1.

To set the scene for the presentation of the diversity in the Amazonian Region, attention is drawn to the introduction of a new concept concerning what may have an important bearing on the manner in which the diversity is distributed. In the western and southwestern sectors of the Amazon Basin there are two geographical zones, each of which contains the headwaters of several rivers that radiate from them in various directions. These zones are shown as shaded areas in the map in Fig. 5 and an appreciation of the situation can be made from them. It is considered that these zones have an important bearing on the origins of the diversity and its dispersal.

The zone in the western sector is located in the eastern foothills of the Andes within Ecuador and is identified by the letter 'A' in Fig. 5. As described previously this zone will be known as the 'Equatorial Oriental Piedmont'.

The zone located in the southwestern sector embraces parts of southeastern Peru and the adjacent border areas of Brazil and Bolivia. It is in this zone that many of the major drainage systems commence. As will be described later these drainage systems are important sources of the diversity in view of the abundant presence of cacao in them and the scale of variability

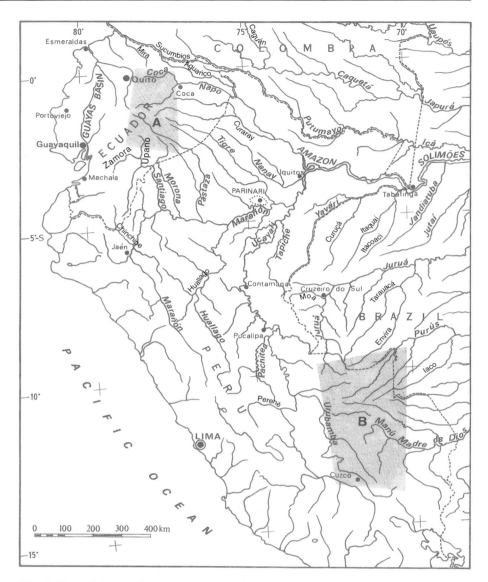

72 Chapter 5

Fig. 5. Map of the southwestern section of the Amazonian Region. The shaded areas indicate the locations of the focal zones of divergence: A, the Equatorial Oriental Piedmont; B, the Province of Caravaya.

occurring in these populations. The zone is defined by the letter 'B' and will be called the 'Province of Caravaya'. This is the name that was used in the past for the region and has its core in the range of hills with this name.

In accordance with the principles set out above it is intended to describe the cacao populations in the area defined as the primary area of the evolution and distribution of the species' diversity on the basis of the separate drainage systems of which the Region is composed. It should be emphasized that any attempt to describe the entire cacao diversity is naturally hampered by the absence or paucity of information concerning certain drainage systems in the areas. There is an additional problem to be considered with regard to the largely unknown movements of planting material between areas. The mixtures of varieties that have resulted from the introduction of foreign germplasm into the drainage systems that will be described produce difficulties in determining the natural status of each of the elements that are to be found in the areas comprising the Region.

For the purposes of facilitating the presentation of the distribution of the species according to the individual drainage systems a classification of these would be of considerable help. Such a classification would enable the reader who is not familiar with the geography of the Amazonian Region to understand the distributions of the river systems and their interrelationships. The drainage systems of the Amazon Basin and adjoining areas in Brazil were classified in this way and this classification was used to describe the spatial distribution of the individual populations and the descriptions of their characteristics. This classification is to be found in Almeida *et al.* (1987). A similar classification system for the entire Amazonian Region would be essential in obtaining a clear conception of the distribution of the species in the Region. The system of classification involves the attribution of hierarchical levels of the relationship of each watercourse in the main valley, of which it is a component.

In addition to the complexes of tributaries to the main river, the fact that some of the principal tributaries are major drainage systems in their own right both regarding their lengths, which can reach 2000 km or more, their individual properties must be taken into account. Along large rivers more than one variety could occur, in contrast to the smaller tributaries where single varieties would be more usual. This variability could be associated with the various minor tributaries in which specific components of the germplasm may exist. In some circumstances the varieties that are encountered on the second-order tributaries differ from those on the main tributaries. There is also the possibility of the occurrence of gradients in the genetic composition of the population along the length of a drainage basin. In addition, the germplasm existing at the confluence of rivers could represent a composite of the varieties of each river or a predominance of the varieties that occur in either river system.

If there is a natural tendency for the germplasm existing in the secondary drainage basins to flow towards the main drainage basins of which they are components the latter could function as receptors of this germplasm. Consequently, it would be expected that the diversity of a main drainage system might be larger than that of the individual tributaries, not only from the flow of their germplasm but resulting from hybridization and recombination.

The River Ucayali drainage system

It is appropriate to begin an interpretation of the Amazon cacao diversity with this system for two reasons. First is that the location of the headwaters of the system is considered to be the source of the R. Amazon. The second is that much of the upper regions drained by the system was occupied by 'savage' Indians who were, from the earliest European contacts, given the name of 'Chuncho'. It was in the land of the 'Chunchos' that the first recorded contact of Europeans with South American cacao occurred, as mentioned in Chapter 1. The relevant passage from Cieza de Leon's (1923) account of the exploration of the 'River of the Chunchos' by Pedro de Anzures (Peranzures) describing his return to Tacana states: 'Working their way with the usual hard labour, at last they found a quantity of *cacao* which was a great help. For three days they marched through forests thick with these *cacao* trees…'.

This leaves no doubt as to the extent and antiquity of cacao in this valley (as well as on the upper tributaries of the Madeira drainage system that are near to the stretches of the Ucayali system, at the same latitudes, which were also explored on this expedition). It is for this reason that the cacao that could be considered to be native to the Upper Ucayali is identified as 'Cacao Chuncho'. Apparently, the identification of true native cacao in its original habitat is hampered owing to the introduction and cultivation of other varieties that has proceeded during the past decades. The title of 'Criollo' in the sense of 'native' is also applied to the primitive variety.

One of the features of the Ucayali system is that individual stretches of the main river receive different names as it successively unites with its tributaries. The first (and most southerly) of the components of the system in which cacao apparently occurs is the R. Vilcanota that joins the R. Urubamba at about 12° 50′ S, both rivers being within the Province of Caravaya as shown in Fig. 5. Cacao plantations are found at altitudes above 1500 m. In addition to the extreme southern latitude the environmental conditions would be very different to those on the banks of the R. Amazon near the Equator.

The available descriptions of 'Cacao Chuncho' are somewhat confusing since the name is variously applied. It is possible that some variability exists in populations that can be considered to be of 'native' origin. Since the trees would be very old they can reach considerable heights. The fruits are described as having a short broad shape with a short acute apex, the surface being rough. The pulp surrounding the seeds has a pleasant sweet taste. The cotyledons are dark purple in colour. This variety appears to differ from another (presumably also native to the region) called 'Cacao Comun'. The descriptions of this variety indicate that the fruits are smaller than those of the 'Chuncho' variety, the shape being more rounded. The fruit surface is smooth or slightly rough and the husk thin (F. Coral, unpublished personal observations, 1987). The small fruit size would be consistent with the incidence of low temperatures and diurnal and seasonal fluctuations.

The R. Ucayali is formed by the uniting of the R. Urubamba with, successively, the Apurimac, Ene and Tambo rivers. Much of the Apurimac runs through a deep gorge and it would be unlikely that, even though some cacao may occur in the valley, such as that in San Francisco, the areas are probably restricted in size. As far as I know, no botanical specimens have been collected in the valley. The Indians on the R. Tambo (the last stretch of the valley), who belong to the tribe of its left-bank tributary, the Solipa, are reported as using cacao.

The R. Ucayali (Fig. 5) has several tributaries that flow into the left bank (coming from the Andes). One of the tributaries just below the confluence of the Urubamba and Tambo rivers is the Perené. The report of Clark (1891), which provides information of the natural resources of this valley for a proposed colonization scheme, contains the following details about the cacao of the area:

> Trees were found 50 feet high with a diameter of 18 inches to 3 feet above ground… So far as I could ascertain, there is found wild only one variety, the fruit of which when immature, is of a greenish colour, turning when fully ripe into an orange yellow. The pods, which are about 8 inches long, are deeply fluted or ribbed, and contain from 27 to 30 seeds of a triangular shape and of deep rose or purple colour.
>
> In no case did I find the plant cultivated. The natives, however, collect the beans and prepare a sort of cocoa from them.

The R. Ucayali was explored by F.J. Pound (1938) during his collecting mission to Peru in 1937 and 1938 and on his second visit (Pound, 1943). His reports give details of the cacao varieties seen from the R. Pachitea (Fig. 5) to the confluence of the R. Ucayali with the R. Marañon. Observations were made on scattered occurrences of cacao trees. The impression given from the statements in the report is that no attempt was made to penetrate areas further from the river banks or to explore the tributaries of the Ucayali except the R. Pachitea. The idea that the density of the cacao population seen on the river was low does not agree with information given in other sources.

Huber (1906b) records a visit to an area below the town of Contamana, situated on the right bank of the R. Ucayali, which he considered to be between this river and the R. Tapiche. The cacao population seen was reported to have been fairly extensive and considered to be truly wild. Unfortunately, Huber did not describe the variety or compare it with those he knew on the main R. Amazon but a herbarium specimen from this population exists. In the area of the R. Sarayacu, further downstream, the cacao population would have been equally large because, during the post-independence period, the mission was maintained solely on the sale to Brazil of the cacao harvested.

In the 1943 report Pound made it clear, however, that the varieties seen on the R. Ucayali differed from those on the Marañon and Alto Amazonas rivers. The conclusion was that the fruit colour of the Ucayali cacao was a darker green than that seen on the other rivers. He described a 'central' (i.e. more commonly occurring) type that would be dark green in colour, elongate in shape and having a slightly rough surface. These fruits possessed a short but blunt apex and were without the basal constriction. Evidently, there was a degree of variability, with trees having fruits with a rougher surface. Other fruits were described as having smooth surfaces and five furrows.

As Pound had a specific objective for his mission (the search for genotypes resistant to *Crinipellis perniciosa*) he did not collect samples of all of the types encountered. Only one collection was made at a location somewhere between the towns of Contamana and Pucallpa and consisted of progenies from the fruits harvested from one or more trees. The family of progenies was identified with the name 'Scavina'. During his stay in Peru in 1943 Pound revisited the area and collected material for propagation from a tree presumably belonging to the same group, the resulting clone being identified as 'P 31'.

If we can take Pound's word that this tree and its descendants represent the most common cacao type of the river, they provide a basis for analysing the diversity that has been described by other sources or from samples obtained during subsequent collections. The progenies that comprised the family were few in number, possibly because of the small quantities of seed contained in the fruits or because of a high mortality rate after introduction to Trinidad. At the present time the family contains some 13 genotypes. These are similar in most respects but differences among them are apparent in such phenotypic characters as leaf shape, some having broader leaves than the majority of the progenies. The trees grown from seed exhibit little variation with regard to their vegetative characters. They tend to have longish erect stems and relatively narrow canopies with dark green leaves, on which basis they are distinguishable from the other varieties. Although the canopies appear to be narrow and compact these genotypes tend to develop long terminal flushes and the compactness results from a tendency to produce many small branches from the nodes. There is also little variation in the flowers, the main feature of which is the lighter yellow ligule and lack of pigment in the stamens.

The fruit colour is a dark green. Under certain conditions some pigment may be expressed. The surface is slightly rough but the ridges are not noticeably pronounced. The apex is usually acute with a rounded end or nipple-like. The seeds are small, cylindrical in shape and very dark purple in colour. The variety is notable on account of the fact that all the progenies are self-incompatible but some are compatible with other members resulting from the existence of several self-incompatibility alleles. Some members of the 'Scavina' family are compatible with 'P 31' and others incompatible with it. It is concluded from this situation that 'P 31' is the parent of some of the 'Scavina' genotypes but not of all of them. A situation of this type would have arisen in a stand of several stems, derived from more than one seed, of which 'P 31' was one but that some of the fruits were collected from other trunks comprising the clump or from neighbouring trees. The fruits of 'P 31' are compared with those of one of the 'Scavina' progenies in Fig. 6.

The fact that the 'Scavina' family has received greater attention than most other cacao varieties provided an opportunity to study in depth some of the specific characteristics associated with the variety. One of these concerns the self-incompatibility status of all the genotypes belonging to the family. It was in this group of genotypes that the independent action of two or more alleles was identified. This situation has profound effects on the breeding behaviour of the genotypes but it also can be used to determine the relationships between the members of the family, confirming the conclusion, derived from the variations in phenotypic expressions, that they are descended from different parents.

Another intrinsic characteristic that was first revealed in these genotypes is a shorter developmental period for the fruits from fertilization to maturity than that occurring in other populations. This period was found to be about 130 days. The short development period may be related to the small seed size and

Fig. 6. Fruits of 'P 31' and 'SCA 6'.

one other important trait that has significant consequences concerning the commercial value of the products derived from the family. The trait in question is the higher moisture content of the fresh seeds, both in the pulp and in the cotyledons. As a result the ratio of dry seed weight to 'fresh' seed weight is significantly lower than that of other varieties, a factor that results in erroneous overestimation of the productivity of the genotypes and their progenies. The pulp is also notable for its sweet taste and the commercial product for its 'floral' aroma.

One of the more notable intrinsic characteristics of the 'Scavina' family is the appearance of dwarf segregants in inbred progenies of the clones, progenies from hybrids among them and sib-crosses within the F_1 generation (Bartley, 1967). The dwarf offspring are distinguishable in the juvenile stage of development as show in Fig. 7. The dwarfing character may persist when the plants reach the adult stage and are recognizably different to the normal plants.

The above characteristics of the 'central' type of the R. Ucayali valley are important measures for determining the scale of diversity in the valley and identifying the foreign germplasm that may have been introduced. The evidence that a greater degree of variability is to be found in the area of Contamana than that represented by the 'central' type is given by the existence of genotypes such as the clone 'U 10' whose fruits are shown in Fig. 8. The notable characteristics of this genotype are the large size, the smooth surface and the blunt apex.

More recent collecting activities involved populations between the towns of Contamana and Pucallpa on the R. Ucayali and some of its left-bank tributaries. Also explored was part of the R. Pachitea (Fig. 5), one of the more important of the tributaries of the Ucayali that joins the river a short distance south of Pucallpa. The general impression obtained from the available information (F. Coral, unpublished data, 1987) is that cacao is fairly common in the area and that the range of diversity is fairly large. However, since the

Fig. 7. Dwarf progenies from inter-SCA crosses compared with a normal plant at 6 months of age.

Fig. 8. Peru, R. Ucayali population fruits of the clone 'U 10' from the Contamana area.

details of the types collected are scanty, an analysis of the variability will not be attempted until further information becomes available.

The R. Pachitea has been the target of various colonization schemes. In the 19th century a German colony was established on one of the tributaries, the R. Pozuzo, and cacao was one of the crops exploited. Apparently, the colony still exists but details are lacking about its cacao involvement.

Collections have also been made between the towns of Contamana and Requena but like those referred to in the previous paragraph the details of the characteristics of the genotypes are insufficient to obtain a satisfactory picture of the diversity along this stretch of the valley. One of these genotypes, a clone named 'U 18' shown in Fig. 9, shows some affinity to the 'Scavina' type but the ridges of the fruits are more pronounced than those of the variants described by Pound (1938).

In 1973 a few genotypes were collected in the vicinity of the town of Requena and the lower reaches of the R. Tapiche. This river is virtually the only right-bank tributary of the Ucayali and is one of the longest. It has some importance in the story of the species' dispersal on account of its source being located near the Brazil–Peru boundary. The town of Requena is situated near the mouth of the R. Tapiche and it can be considered that the cacao population sampled near it has the same origin. Further collections were made in 1987 but these have yet to be analysed.

The 1973 collections were identified as 'UCA' (= Ucayali) and 'TAP' (= Tapiche) for the respective areas. On account of the adverse conditions existing at the time of the visit the details of the original collections were not recorded (W.S. Chalmers, unpublished report of a visit to Ecuador, Peru and Brazil,

Fig. 9. Peru, R. Ucayali varieties: fruits of 'U 18' from near the town of Requena.

1 March–14 April, 1973). It has been possible to record some observations on the few surviving clones and families. As occurred with most of the samples from the valley, trees were selected principally with regard to the absence of infection by *Crinipellis perniciosa* so that they are not expected to represent the diversity that occurs in the area.

The report on this mission (W.S. Chalmers, unpublished report, 1973) states that the fruits seen on the R. Tapiche had five pairs (each pair close) of pronounced ridges and the surfaces were very rugose and light green (*blanco*) in colour. The apex was very short and nipple-like, the husk being fairly soft. This description could apply to the fruits of the genotype depicted in Fig. 9.

The clones from some of the parent trees indicate that the 'Tapiche' types belonging to one location may represent two varieties, although some of the differences may be due to contamination. In general, the fruits have a narrow elongate shape with prominent, paired ridges but some differences in surface roughness and gloss are apparent. There was one distinct type with small broad elongate fruits with an acute apex and less pronounced ridges. One of the above genotypes appears to differ from the others in possessing very large seeds. Only one of the R. Tapiche seed progenies appears to have survived, this being descended from the parent identified as 'TAP 1'. The fruit of this tree, as shown in Plate 13a, is of small size with a short broad shape. The surface is somewhat rough. Although the tree habit and some fruit characters indicate an affinity to the 'Scavina' family the genotype differs in possessing rather large seeds and, most importantly, pigmented stamens.

On the same occasion that the first collections on the R. Tapiche were made, a search for cacao was undertaken in and around the town of Requena (W.S. Chalmers, unpublished report, 1973), where it appears that cacao was relatively rare. Although no trees were encountered in the forest near the town, evidently cultivated trees of relatively recent plantings were located in back gardens in the town. The three trees sampled in this area could, therefore, be considered to belong to cultivated and probably introduced varieties. The fruits of the few samples collected on one farm were described as having ridges in close pairs (fruit shape not described), the surface slightly rough or smooth, husk soft and the apex nipple-like. One of the parent trees was noted as possessing broad leaves. In general, these genotypes possessed some similarities to the R. Tapiche types but differed in the fruit surfaces being less rough and exhibiting some pigment.

The observations made on the few seed progenies from these samples indicate that, although the families showed some differences, in general the fruits were small with a rough surface. These families did not include the parent tree with smooth fruits. Some similarity with the 'Scavina' (or central) type described by Pound (1938) was noted. The similarity of the fruits of the 'UCA' progenies to those from the R. Juruá was also noted. However, the 'UCA' trees differ in having stamens with intense pigmentation. Although the report indicated the presence of trees in the town gardens that differed from those seen on the R. Tapiche, possessing fruits with a more pronounced basal constriction, the only tree sampled did not have fruits that would relate it to the other trees.

No descriptions of cacao varieties along the stretch of the R. Ucayali from just below Requena to its confluence with the R. Marañon have been found. It is probable that the area is fairly densely populated and that much of the cacao that occurs is cultivated from introduced varieties. The information contained in the report of the collections made in 1987 refers to the presence, in areas close to Requena, of cacao trees with fruits similar to the 'Nacional' variety of Ecuador. Although these observations can only be verified by personal observations and comparisons of the varieties concerned, the possibility of some introgression of varieties from the R. Marañon basin could be considered.

The River Marañon Drainage system (Fig. 5)

Historically the formation of the R. Amazon as known by this name has been considered to begin from the junction of the Ucayali and Marañon rivers. Accordingly, the latter is equally important as a component of cacao germplasm of the Amazon Basin. Cacao would have existed along the river from ancient times. La Condamine (1745) descended the R. Chinchipé at the start of his journey down the R. Amazon and, arriving at its confluence with the R. Marañon, described that both banks of the latter (and presumably those of the R. Chinchipé) were covered with wild cacao trees. He added that the native inhabitants valued cacao as much as gold and that its quality was not inferior to that of the cultivated type (i.e. the product of the Ecuadorian littoral). Hence, the commercial exploitation of these stands would have commenced and cultivation probably undertaken subsequently. The towns of Jaén and Bagúa are situated in the area described and constituted an important production entity. Unfortunately, the traditional varieties of this area have not been described in a way that can be used to relate them to other varieties.

The River Huallaga

There is only one important right-bank tributary of the Marañon river. This is the R. Huallaga (Fig. 5) on which cacao would probably have existed from prehistoric times but has also been the scene of cultivation, at least from the 18th century. The Jesuits had their regional headquarters at La Laguna near the mouth of the Huallaga. They were great promoters of trade in cacao. Initially, this would have involved the product obtained from exploitation of natural populations and later, from cultivated areas. The absence of sufficient information about the cacaos of the Huallaga does not permit a reliable analysis of the natural populations to be made.

Pound surveyed the lower part of the R. Huallaga from Yurimaguas to its mouth and the observations contained in his report (Pound, 1938) are virtually the only source of information available about the varieties present in the valley. However, the descriptions contain several vague terms that are unhelpful in identifying the characteristics of these varieties. For this reason and the fact that they contain references that are important with regard to other varieties it is pertinent to reproduce the following descriptions to which appropriate comments have been added.

After stating that the cacao around Yurimaguas 'was quite distinct from the cacao of the R. Marañon', Pound wrote:

> The bulk of the trees found were large and tall, some being well over 50 feet high. The pods belonged to two classes, one a normal green long warty type without pronounced bottleneck or conspicuously long point while the other was a shorter but broader super-green type.
>
> The first type recalled at once the 'Nacional' cacao of Ecuador and probably this population belongs to the same strain particularly since both the habit of the trees and the pod characteristics conform. The cotyledons of the beans too were much paler than those from the Rio Marañon 'lagarta' types although not so pale as the Arriba Nacional beans from Los Rios in Ecuador.

One of the cases of ambiguity in the description is that concerning the fruit colour. I suppose that, on the basis of the currently proposed system of colour descriptors, 'normal green' and 'super-green' would be translated as 'intermediate green' and 'dark green', respectively. The cacao of the R. Marañon referred to would be that which Pound encountered in the Parinari area, which will be described later. During the succeeding descriptions it will be necessary to refer to Pound's use of the term *lagarta*, which implies a very different meaning to that normally associated with the word. In the context of his usage the term appears to be applied to elongate fruits, regardless of the roughness of the fruit surface. The description 'long warty type' would conform to other authors' concepts of the term *lagarta*.

To my mind the first type described would refer to the criollo type of the Upper Ucayali system and the second to the Ecuador 'Nacional', a variety that is referred to later.

One of the principal tributaries of the R. Huallaga is the R. Moyobamba. This stream is one on which cacao appears to be present in significant quantities, but again the information about it is insufficient to obtain a concept of the variety situation. Pound (1938) stated that the reason for his visit to Yurimaguas 'was to examine the cacao there which was reputed to be similar to a high quality cacao growing near Moyobamba in a valley of the Cordilleras to the West and also similar to the cacao near Cuzco in the headwaters of this same river.'

The above account is confusing since it assumes that the variety referred to was growing in the headwaters of the R. Huallaga, while later on he refers to Urubamba, which, as we already know, is part of the Ucayali system. However, the statements made in this regard confirm that it should be the 'long warty type' that is present in the valley of the R. Moyobamba and 'the other' is the type similar to the Ecuador 'Nacional'. One herbarium specimen (R.S. Matthews, No. 1653) dating from 1835 (i.e. probably before any change in the population on the river had become effective) has small leaves.

When he surveyed the flora of Peru in the late 18th century Hípólito Ruíz (Lopez and Cuatrecasas, 1953) referred to the cacao near the town of Cuchero (perhaps the present day Tingo Maria or in its vicinity) on the R. Huallaga, indicating the possibilities of cultivation of the crop. Ruiz recognized that the cacao he encountered near Guayaquil on a subsequent visit to Ecuador differed from that seen in Peru by giving a specific name to his herbarium specimen of the Ecuador variety. The two types described by Pound (1938) are therefore distinct. If they had been present in the same area for a long period of time it would be expected that some hybrids between these types would have been produced.

The establishment of relationships of these types is important in assigning the varieties selected in the Huallaga valley in recent years to their respective origins.

The Upper River Marañon

Although the upper R. Marañon flows in a northerly direction more or less parallel to the Huallaga and Ucayali rivers it is probable that much of its course is through mountainous terrain unsuitable for cacao. However, towards the end of the course before it flows eastwards, the cacao of the Jaén and Bagúa region occupies both banks of the river as described by La Condamine. As mentioned above, virtually no information about the cacao in this region is available. The upper Marañon valley may have had some importance in prehistoric times in view of the location in it of more advanced indigenous cultures.

After beginning its easterly course the R. Marañon is successively augmented by several tributaries that join it on its left bank. Nearly all of these tributaries have their sources in the Equatorial Oriental Piedmont but in the lower courses they flow through Peruvian territory. Herein lies a problem with regard to the description of the cacao populations of these rivers. Although a satisfactory basis for describing the varieties that exist in Ecuador is at our disposal there is no supplementary data from Peru that will enable us to obtain a complete picture of the population structure and sequence of varieties of each river. Similarly, the population situation on the lower course of the R. Marañon and its relationship to the varieties on the tributaries is unknown at the present time. The same applies to the area where the border between the two countries has been in dispute, a situation that has prevented exploration in what may be a key area for the dissemination of the diversity in the region. Consequently, it will only be possible to describe the diversity existing in those areas of Ecuador from which collections of living and herbarium material have been made and descriptive information is available.

The cacao varieties that occur within the radius of the Equatorial Oriental Piedmont possess one common feature. This is the occurrence of trees that have flush leaves lacking anthocyanin pigment, a trait that is genetically associated with seeds whose cotyledons also lack pigment. A citation in Patiño (1963) suggested that the presence of unpigmented cotyledons was known in the past. It was for this reason that the early documents referring to cacao in the region equated the variety encountered with that of the colonial Kingdom of New Spain. This latter country is present day Mexico, where, as will be described in the discussion of the Circum-Caribbean Region, the varieties belonging to the *Criollo* group were exclusively cultivated. Several travellers to the Equatorial Oriental Piedmont have referred to the combination of unpigmented cotyledons and flush leaves that distinguishes the cacao of this area. The first living specimens from this region were collected in 1948 from near the headwaters of the R. Pastaza. These specimens still exist and have had some important applications in the development of new cultivars.

The first of the tributaries that enters the R. Marañon at the point where it changes its general direction to the east is the R. Chinchipé. Although this river has its source in Ecuador it flows through Peru for most of its course. It is a shorter river than the other rivers that will be described but virtually nothing is known about the presence of cacao in the valley, except for the reference in La Condamine (1745).

The Zamora–Santiago valley

The next principal tributary entering the R. Marañon on the left bank is the R. Santiago. Within Ecuador this river is formed by the junction of the R. Zamora and the drainage system that flows from the north, which starts from the source of the R. Upano. From the map of the region in Fig. 5 it will be appreciated that the Zamora–Santiago drainage system is complicated and not easy to describe. It consists of innumerable small tributaries with different names for each river course. Various specimens have been collected from some of these streams and, owing to the relief and separation of the valleys, it is possible that local strains of the general population are present. The result would be a complex population structure for the drainage system in which appreciable genetic diversity may be encountered.

The complexity of the drainage system in terms of the distribution of cacao and the fact that the specimens that have been collected have not been adequately characterized and analysed renders it impossible to describe the apparently ample diversity in this system in its entirety. The phenotypic variation observable is depicted by some of the examples given in Plate 10. On the whole this variability is extremely interesting and deserves to be studied in greater detail than the superficial observations made so far.

The northernmost section of the Zamora–Santiago drainage system mainly involves the R. Upano. The river commences by flowing in a northerly direction then turns back to flow towards the south. The source and upper course of the R. Upano lie within the Piedmont area and at the point of the change in direction are close to the source of the R. Pastaza.

The R. Upano and other rivers of the same system that flow through the foothills of the Andes are usually confined to deep valleys surrounded by land at fairly high elevations. On account of this topographical feature cacao is found at various elevations from the floors of the valleys to the high plateaus of the surrounding hills at elevations ranging from 200 m to over 1200 m.

In 1961 (K.D. Doak, unpublished, details given in Soria, 1970), a collecting traverse was made along the Upano river from north to south. Unfortunately, the material collected was subjected to neglect with the result that no analysis was made of the variability in the population of this river. The collection established has not been in a condition to provide the information that should be obtained from it. The most frequent characteristic of the genotypes from the upper course of the R. Upano is the combination of unpigmented cotyledons and flush leaves with fruits that are short and broad although having fairly rough surfaces. The few surviving specimens from the southern end of the traverse near the confluence with the R. Zamora exhibit an evident shift in phenotypic expressions.

Subsequent collections in this drainage system confirm the existence of the gradient in variety characteristics, the populations in the southern sector of the Zamora valley differing in many respects from those of the R. Upano. Although there is the tendency for the combination of white cotyledons and unpigmented flush leaves to occur this is less frequent. The Zamora valley populations show a greater degree of diversity, with some trees being homozygous for pigmented cotyledons, trees of greater vigour being present in some localities and, in rare instances, anthocyanin pigmentation occurring in the fruits. The range of fruit shapes and other features is also greater than found elsewhere. The presence of trees with fruits bearing a strong resemblance to the traditional types of the Pacific littoral is a notable feature of the populations of the drainage system. Several trees have fruits with almost smooth surfaces and apices that are more rounded than those of the varieties that occur in the northern segment of the system. What is evident from the progenies is that the populations may contain a much larger degree of diversity than can be gauged from the characters of the parent trees.

The genotypes collected on the R. Santiago between its origin from the confluence of the Zamora and Namangoza (continuation of the Upano valley) rivers and the Peruvian frontier possess traits that are very distinct. The plants from these collections that the author was able to examine were still small and too poorly developed for any reliable observations to be made. Few fruits were acquired at the time of collection and these provide insufficient data on which to base a reliable evaluation of the diversity of this population. However, some of the specimens, such as that in Plate 10d, produced large fruits with seeds that were as large as those of any other variety known. It would be interesting to determine whether the cacao in the Peruvian section of the river possesses these same outstanding traits.

The River Morona area

The next tributary system to the east is that of the R. Morona (Fig. 5). The river and some of its tributaries have their sources in Ecuadorian territory but it is difficult to ascertain the exact situation on account of the area falling in the zone of conflict and perhaps not very well known. The only specimens known from the Ecuadorian segment of the drainage system were a few from Taisha on the R. Panguientza, one of the streams that form the R. Cangaime, the major source of the R. Morona. The cacao population appears to be sparse. Establishment of the progenies was poor and virtually no information has been available about its characteristics. There is one preserved fruit belonging to a herbarium specimen (Cazalet and Pennington, no. 4653, 1962) from the same area, which is green in colour, small in size (perhaps partially developed) with a rough surface. The cotyledon colour appears to have been uniformly white. Thus, the population of the area could be considered to be part of the general population of the rivers that radiate from the Equatorial Oriental Piedmont.

Pound (1938) stopped on the R. Morona on his journey to the R. Huallaga. The location visited was not specified but it is assumed that it was not far from the mouth of the river. In fact, although the material derived from his collections was identified as 'Morona' (acronym 'MO'), Pound wrote: '...in the region of the Rio Morona'. This implies that the material may not have been collected on the R. Morona. It is also significant that the list of the collections sent to Trinidad from the expedition does not include any reference to R. Morona but the following entry can be found 'Ecuador Nacional – introduced from Peru 1938' that is not mentioned elsewhere in the report. From his account it can be inferred that the population seen was fairly homogeneous with regard to the phenotypic characters, as may be inferred by the brief description of the fruit shape as being 'short oval'.

The lack of specific descriptions of the other tree characters and details about the material collected renders the evaluation of the surviving progenies very difficult. The progenies established in Trinidad, which bear the acronym 'MO', have numerical identifiers up to 130. A quantity of this magnitude would have been obtained from at least three fruits and these collected from more than one tree. Whether more than one parent tree was concerned is unknown. The resulting situation is that we do not know if all of the progenies are descended from the trees at one location or at several.

The author undertook a study of the 37 'MO' progenies that survived in the 1960s. The data are incomplete with regard to several of these genotypes. Thus, the basis for a complete evaluation of the group is lacking. Most of the genotypes possess short broad fruits with length:diameter ratios under 2.0. Five genotypes had ratios in the range of 1.5–1.7. On the other hand, a few genotypes were notable in possessing elongate fruits with acute apices. There was also a degree of variation with regard to the fruit surface, which was generally rough with prominent ridges. Several genotypes are noted as having traces of surface pigment that could be more evident in some of the fruits. One feature that characterizes this group is that of a fruit surface with an appearance that has been described as 'mealy' (or 'granular' as stated in Chapter 3). Although this trait was first described in the 'MO' progenies it also appears in other populations, as will be described later. On the whole this group of genotypes exhibits a significant degree of variability that indicates that the base population was not as homogeneous as the description implied. The short broad fruits have phenotypes that are similar to those of the Ecuador 'Nacional', hence the confusion in terms. Some of the genotypes are weak, as regards their growth, and the combination of the light cotyledon colour and light green flush leaves also has been observed. The group of genotypes segregates for an incompatibility reaction, the determination of the degrees of fertilization among the genotypes possibly offering a means of allocating them to the families to which they belong according to descent from the respective parents.

The River Pastaza drainage system

This is the next most important system in the sequence of left-bank tributaries of the R. Marañon. This system also has its sources in the Equatorial Oriental Piedmont. Little is known about the cacao populations throughout the whole course of the R. Pastaza. Although this river was one of the main lines of communication between Quito and the Marañon it was not an easy artery for travel and possibly little attention was paid to the flora along the river banks. However, a volcanic eruption in 1698 sent a lahar down the river, destroying the vegetation in its path. Consequently, the vegetation along the system would have been formed during the succeeding decades, a situation which, in addition to the nature of the terrain, may have resulted in a sparse cacao population and of recent formation.

The only specimens representing the Ecuadorian segment of the system resulted from two collecting activities confined to the vicinity of Montalvo on the R. Bobonaza, a left-bank tributary of the Pastaza. It appears that cacao was not common in the area and the specimens collected do not provide sufficient data for evaluating this population. It would be assumed that the fruit characteristics do not differ from those of the nearby R. Napo tributaries to the north. However, the specimens from the first group were descended from parent trees described as being very large, the fruits of some of them having a degree of red pigment. The few fruits collected suggest that they may attain a large size when they contain a larger quantity of seeds. The seed size is outstandingly large but those described from the original fruits had pigmented cotyledons. The pigment perhaps is associated with the pigment on the fruit surface.

The River Tigre drainage system

The most easterly of the major tributaries of the R. Marañon and the one nearest the junction with the R. Ucayali is that comprising the R. Tigre drainage system. Within Ecuador the system is formed by two larger streams, the R. Corrientes and the R. Conambo, which join to form the R. Tigre, the entire course of which is within Peruvian territory. The only specimens from this valley comprise ten trees sampled in a restricted area on the R. Conambo near its junction with the R. Corrientes (Allen, 1987). Although the original description of the fruits from these trees lists them as having a green colour, in fact all of the cloned progenies that were established have red fruits, as evident in Plate 12b. This is an important discovery. The place where these specimens were collected is not far from the Montalvo location on the R. Bobonaza, suggesting a possible link regarding the distribution of red pigment.

From the observations made on the samples obtained in the valleys of the tributaries of the R. Marañon discussed up to this point it appears that the frequency of occurrence of trees with fruits that have smooth or nearly smooth surfaces is higher than that of the R. Napo system.

The Parinari area

At this stage it would be appropriate to insert the information regarding the cacao population examined by Pound (1938) in the area on the left bank of the Marañon River, lying between the mouths of the R. Pastaza and the R. Tigre. This area was also affected by the lahar (flow of volcanic debris mixed with water) that destroyed the forests along the R. Pastaza in 1698 and it could be assumed that the vegetation in the area that we are concerned with had evolved following the disaster. Consequently, it is probable that the age of the cacao population examined by Pound in 1937 would have been a little more than 200 years at the most. Herndon (Herndon and Gibbon, 1853) travelled down the R. Marañon and included in his report a quotation of Castelnau to the effect that 'Cocoa, recently introduced into these two provinces, gives its fruit at the end of three or four years, at most.' We may infer that the presence of cultivated cacao, as opposed to the populations that we may suppose to have existed in the natural state, may date from as recently as the beginning of the 19th century.

The traverse of the R. Marañon that Pound made was from east to west; that is, in the opposite direction to this description of the cacao populations of the river. The descriptions given in his report (Pound, 1938) need to be considered in the light of the terminology used for the descriptors, which differs from that in current usage or from the terms used in this book. An additional problem regarding the descriptions is that those applied to the R. Marañon cacaos are based on definitions established in previous sections of the report. Hence, it is necessary to interpret the descriptions of the R. Marañon varieties in more specific terms. The specific descriptions can be verified in comparison with those of the progenies descended from the parent trees sampled during the traverse.

In no case in the account of the R. Marañon is there an indication as to whether any of the genotypes seen could be considered as belonging to a 'native' regional variety or how representative they were of the regional diversity. Certainly, the material that survives from the collections differs from the cacao varieties seen elsewhere in the drainage system. It would be logical to suppose that the trees seen belonged to cultivated populations descended from imported planting material, probably introduced from Brazil. Hence, we have to look at these populations with reference to others that will be encountered during our survey of the diversity of the Amazonian Region.

As in much of the other information in his report Pound (1938) does not provide enough detail to give the reader a sufficiently accurate picture of the situation on the river or about the material he collected. The only material collected (as fruits) on the river is identified as being from a location called Parinari, the progenies of which are usually identified by the acronym 'PA'. With regard to defining the area there is a degree of uncertainty concerning the exact location of the parent population. The report suggests that the area of Parinari was located below the mouth of the R. Tigre, when in fact all places bearing this name are above this mouth. Neither are we informed about the other locations he visited on the R. Marañon. In relation to the Parinari specimens we are in the dark as to whether the parent trees that provided the material occurred at a single site or whether they were at various sites. Details of this nature are basic to the understanding of the relationships between the parent trees and, therefore, of their progenies.

In common with all of the material resulting from this expedition the progenies attributable to a particular area were mixed at the time of planting. This means that their individual identifications do not provide any clues regarding the families that are descended from the individual parents and, therefore, how an individual progeny is related to the others. Consequently, this unsatisfactory identification situation and the scanty details about the original population hamper the genetic analysis of the group of progenies and the understanding of what may be considered to have been the representative cacao population of the lower course of the R. Marañon. It is important to understand this point because Pound (1938) referred to the cacao type, of which the Parinari population is our standard, on the basis of possessing 'lagarta'-shaped fruits, as being less frequent as he proceeded upstream. The type with short and broad fruits of the upper course became progressively more frequent until it was the only type present. A sample of fruits from the R. Marañon with shorter and broader shapes is illustrated in Fig. 10.

On the assumption that Pound's word 'lagarta' specifically implies an elongate fruit shape with an acute apex the various fruit types described by him can be translated as follows:

(i)   Shape elongate, basal constriction, apex acute, surface smooth, colour light green;
(ii)   Shape elongate, basal constriction, apex acute, surface smooth, colour dark green;
(iii)   Shape elongate, basal constriction, apex acute, surface rough, colour light green;
(iv)   Shape elongate, basal constriction, apex acute, surface rough, colour dark green;
(v)   Shape elongate, basal constriction absent, apex blunt, surface smooth, colour light green;
(vi)   Shape elongate, basal constriction absent, apex blunt, surface smooth, colour dark green;
(vii)   Shape elongate, basal constriction absent, apex blunt, surface rough, colour light green;
(viii) Shape elongate, basal constriction absent, apex blunt, surface rough, colour dark green;
(ix)   Shape short and broad, basal constriction absent, apex blunt, surface smooth, colour variable.

The interpretation of the descriptions indicates that the trees having elongate-shaped fruits were the most frequent. However, it is apparent that, if it could be considered to be a single entity, the population concerned possessed a significant degree of variability. The progenies of these trees in general conform to the above categories of fruit types. On this basis the cloned 'PA' progenies could be divided into fruit type classes and those within each class considered to be descendants of the same parent. For example, 'PA 285' could be assigned to fruit type class (ii).

Fig. 10. Peru, fruits of a variety from the R. Marañon.

The clones 'PA 81' and 'PA 150' may belong to class (viii) and their relationship is augmented by common features, such as their greater vigour, which sets them apart from the other clones. However, there are other traits besides those of the fruits that are basic features of the population or contribute to the variability among its components. It must be taken into account that the attributes that are observed in the progenies are the products of the actions of the genes that determine the expressions of these attributes.

On the basis of the observations made on some of the progenies it could be concluded that among the common features are the white flowers with no stamen pigment or this being very slight. Cotyledon colours appear to be uniformly dark purple. One feature that is common among the progenies is the tendency to produce a higher proportion of aborted seeds than found in other varieties.

The scale of variation in the fruit characters is exemplified by the segregation that occurred in the hybrid of 'PA 148' and 'PA 16'. Although the fruit of these clones appears to be similar, those of the progenies exhibit significant variability as shown in the specimens in Fig. 11.

The progeny clones exhibited a considerable range of expressions for other characters. Owing to the size of the sample represented by the progeny clones it is not possible to determine and describe the entire range of diversity present in the group. References to some of the more outstanding genotypes and the contrasts among them would serve to illustrate the scale of variability that is known.

For example, 'PA 13' has an elongate rough fruit corresponding to class (viii) (see Fig. 12) but is distinguished by having a low branching habit and

Fig. 11. A sample of fruit types segregating in the cross of 'PA 148' and 'PA 16'. The fruits of the parents are shown in the bottom row: left, 'PA 148'; right, 'PA 16'.

dense canopy with long, dark green leaves. The clone 'PA 107' shares some of these traits. On the other hand, 'PA 15' forms a vigorous tree with small leaves and has a short, dark green fruit. The main feature of 'PA 285', besides being distinguished by the oddly shaped elongate fruit with an acute apex having a tendency to curve, concerns the horizontal branching system. The trait appears to be dominant and, perhaps, sets this genotype in a category distinct from the other known descendants of the 'Parinari' parents.

One procedure that has been adopted to determine the genetic relationships among the cloned progenies is to identify the incompatibility alleles that they possess. The 26 'PA' clones that exist in Brazil have been classified in this way. All have been shown to be self-incompatible. Through the inter-fertilizations among the clones (e.g. Yamada *et al.*, 1982, 1996) it has been possible to place them in four groups on the basis of the possession of what is presumed to be a common incompatibility allele. In this way the similarities between such pairs of clones as 'PA 81' and 'PA 150' and between 'PA 30' and the genotype identified as 'PA 121' that are suspected on the basis of other observations have been confirmed.

Fig. 12. Peru, R. Marañon: fruits of 'PA 13' (×0.31)

During the 1960s a programme of inter-crosses among the 'PA' clones existing at that time was carried out in the original planting in Trinidad. The programme involved 101 genotypes among the 143 that existed at the time to which the group acronym 'PA' had been assigned. The analysis of the results would be valuable in order to complete the study of the inter-genotype relationships but the data and conditions to analyse them have not been made available. The only data from this study that are at our disposal are provided by the lists of compatible crosses obtained and established in the field. A few of these combinations confirm the results obtained in Brazil and the remainder indicate which of the clones may not be related by the absence of common incompatibility alleles.

One result from the inbreeding of some of the 'PA' clones in Brazil and their F_1 progenies was the discovery of segregation in the inbred families of plants with yellow leaves markedly different to the normal leaves, as shown in Plate 31c. The gene producing the abnormal leaves has a lethal effect resulting in mortality of the individuals carrying the gene after a few weeks. The condition was present in progenies of 'PA 30' and the genotype identified as 'PA 121'. The possession of the same gene in both genotypes indicated that they are related, a fact that confirmed their common descent postulated on the basis of the possession of the same incompatibility allele. This finding provides a tool for determining which other clones are related to them.

The River Amazon from the Ucayali to the mouth of the River Javarí

For various reasons it is convenient to divide the course of the R. Amazon into segments. One obvious reason being its length, which involves the traverse of several ecological zones, as shown in Fig. 1.

The first segment, see Fig. 5, that will be discussed in the section concerns the course of the river that flows through Peruvian territory. The segment is also important in terms of features that are peculiar to it.

From the confluence of the Ucayali and Marañon rivers the occurrence of cacao continues in what appears to be a more or less continuous sequence until the mouth of the R. Napo. This segment has received a certain amount of attention with several collections of living material obtained during various activities. One of the most important was that carried out by Pound (1938, 1943) in the vicinity of the city of Iquitos. In the first report Pound described the existence of a variety he called 'Criollo de la montagne', an unfortunate mixture of languages for what should have been written in Spanish as 'Criollo de la montaña'. In spite of the fact that Pound attempted to dispel any suggestion of a connection of this 'variety' with others in the *Criollo* group, readers of the report who are unfamiliar with the sense of the name have interpreted the term in ways beyond what it represents. In simple words the term is translated as 'native of the forest'. The term, therefore, refers to a variety (or varieties) that existed before cacao was cultivated in the area.

The 'Criollo de la montaña' was described as having fruits that were light green in colour on which some red pigmentation was superimposed, and an elongate shape with a rough surface, the cotyledon colour being dark purple. In presenting the above description Pound made it perfectly clear that this variety was not related to any of the other varieties he encountered along the banks of the river. It represented a type that was common on the R. Napo and areas to the north. These observations on the 'Criollo de la montaña' have two significant implications.

First is that the variety belongs to the complex of populations that occur in the area bounded by the Marañon and Amazon rivers and including those of the R. Napo valley. The knowledge at our disposal about the cacao populations in this area would permit us to place the 'Criollo de la montaña' in its proper perspective.

The second implication is that the cacao types that are found along the banks of the R. Amazon in its initial course are cultivated and are not native to the region. They would, therefore, have been introduced from another region. They are not related in any way to the native varieties of the Ucayali and Marañon rivers. The most probable source of the original planting material would be some area in Brazil with which the original (such as the missionaries) and later settlers had trading and other relationships. Soria (1965) mentioned that farmers in the area north of Iquitos had imported seed from Brazil. It would seem that much of the native cacao within reach of the rivers had disappeared by 1964 and the trees shown to him that were called 'native' were remnants of former cultivated varieties, most of which had similar origins.

Three examples of fruit types from a location just below the junction of the R. Ucayali and the R. Marañon are shown in Fig. 13 demonstrating the range

Fig. 13. Peru, R. Amazonas: three distinct fruit types from a location below the confluence of the R. Marañon and R. Ucayali, demonstrating the scale of variability present in the cacao populations of the region.

of variability encountered in the area. The characteristics of the fruit on the left indicate a possible origin in the R. Napo.

The above conclusion complemented by the knowledge gathered during various breeding activities, which indicate its validity, can be applied to dismiss the supposition that the genotypes derived from the various collections made in the Iquitos area belong to what have come to be known as 'Upper Amazon' varieties. The suggestion that these varieties differ from those that occur in Brazil is unacceptable on this basis because no attempt has been made to compare them with the diversity present on the lower course of the R. Amazon.

The collecting activities undertaken were restricted to specific locations in the area. Therefore, they would represent the variety introduced to the specific farm and should never be construed as being representative of the diversity of the region.

As referred to above, the first collections of cacao in the area were made by Pound, first in 1937 and 1938 and later in 1943. The material collected derives from two areas. One was the strip of land between the final course of the R. Nanay and the banks of the R. Amazon. Here the trees at the time differed greatly in size on account of differences in age. It was inferred that the very large trees belonged to one variety, which was considered to represent the earliest planting established. These trees produced fruits that are distinctive and fairly uniform in their characteristics. In 1937 fruits were collected from several trees of this type and the progenies were grouped under the denomination 'Nanay' or the acronym 'NA'. Material for vegetative propagation was obtained from several trees belonging to the type in 1943, as well as from other types of the area. The resulting clones, although possessing some of the characteristics described by Pound (1943), exhibit a certain degree of variability, some individuals having distinctive characters.

The characters of the fruits of the type (referred to as the 'central' – more or less – type) as described in the report are: fruit colour 'half blanco' (i.e. lighter green), without pigmentation; long oval in shape without basal constriction and having a rounded apex and a slightly rough surface. Pound proposed naming the common variety 'Nanay' and apparently 14 trees of this type were the parents of some of the progenies that are identified with the acronym 'NA'. The majority probably conform to this description. Probably the clones resulting from the selections made in 1943, identified as 'P', would include representatives of the type. It would be expected that among these clones, which derive from trees in the Nanay location, the majority would belong to the type designated 'Nanay'. However, among the 22 clones from this location few actually correspond to the description. The clones 'P 11', 'P 16' and 'P 19' may be taken to represent this 'central' type. However, as mentioned above and from observations made on a few of the progenies, enough variability exists to conclude that the original population was not as uniform as the report suggests.

The report of the 1937–1938 collecting activity (Pound, 1938) is clear about the fact that trees exhibiting variation for other characters also occurred at the Nanay location, some of these being attributed to subsequent introductions. The characters of the fruits of these trees ranged from 'short, oval, smooth half blanco unpigmented' to 'elongate, more warty, more blanco' and it was stated that these occurred 'more frequently'. Although the fact is not specified in the report the 'NA' progenies would also have been descended from trees with 'long, generally warty pods with a pronounced bottle neck and a conspicuous point'. Pound stated that this type of fruit would be called 'Lagarta' or 'Parinari' as it was the 'central' type in the Parinari region.

In summary, the genotypes descended from the trees at the Nanay location possess a large range of diversity. The large number of individual genotypes involved has precluded their complete evaluation and characterization, the knowledge about the group being confined to a small proportion of genotypes that have by chance or their outstanding traits been used in breeding. The 22 'P' clones from the 1943 collection may be considered to represent the basic variability that is presumed to have occurred at the location. These include the almost spherical fruits of 'P 30' to elongate fruits with a rough surface, such as 'P 4'. The latter may be a hybrid descended from the 'Criollo de la montaña'. It is not possible to confirm that any of these clones resemble the progenies descended from the trees at the Parinari location on the R. Marañon, except for a similarity of the 'P 4' and 'PA 13' fruits.

Although there may be a preponderance of genotypes that are descended from the 'P' clones with characteristics of the 'central' type, there is significant variability in the aggregate of progenies from the collecting activity of 1937 and 1938, identified as 'NA'. The range includes outstanding genotypes such as 'NA 277', which is a vigorous tree possessing several distinct traits. It is possible that a few progenies are descended from 'P 4', obviously one of the parents, which would have transmitted its elongate rough fruits to its descendants. The

varieties attributable to the Nanay location mostly have small seeds with dark
purple cotyledons. The most constant feature of the group is the possession of
stamen filaments that are intensely pigmented. The progenies identified as 'NA'
could be attributed to the families descended from their respective parents,
including the 'P' clones, on the basis of their incompatibility relationships, if the
data from the studies carried out in Trinidad were suitably analysed.

The other location visited by Pound on both occasions is that of the island
in the R. Amazon in front of the city of Iquitos. He reported that the varieties
grown on the island differed from those occurring at the Nanay location. Two
types were singled out for attention and from which fruits were collected in
1937–1938. Presumably, the parent trees were collected as material for
vegetative propagation in 1943. The resulting clones are included in the 'P'
series and named 'Island'. The seeds derived from the original collection were
mixed and established in Trinidad with the identification of 'IMC'. As a result, it
is not possible to tell exactly which 'IMC' clones are progenies of the respective
'P' parents.

On the basis of similarities of their fruit and flower characters to those of
the parent trees some of the 'IMC' clones can be ascribed to the respective
families. The fruit of the parent, 'P 18', was described as being the main type
seen on the island, 'a very large lagarta calabacillo which was completely oval

Fig. 14. Peru, R. Nanay: fruits of a genotype from this river.

in shape and very smooth.' The use of the word *calabacillo* is puzzling since there is nothing in the description to relate it to the meaning generally applied to the word. It is possible that the connotation in the description refers to the smooth surface but this would be a redundancy. When *Lagarta* and *Calabacillo* are translated as elongate and smooth, respectively, the description is better understood. The majority of the 'IMC' clones conform to this description of the fruits.

The parent and its progenies are notable for the flower characters. They are of a fairly large size compared with the other types in the region and are readily distinguished by their coloration. This is due principally to the pigmentation of the stamens, the appearance of which is specific to the variety, and the coloration of the surrounding organs.

Another significant feature concerns the high ovule numbers. The ovule number determined in the parent 'P 18' is 74 and most of the progenies with similar fruits have numbers above 65. The highest seed number that has been found in fruits of an 'IMC' clone is 77. The 'IMC' family was the first such variety known to have ovule and seed numbers of this magnitude. This attribute is probably responsible for the predominance of the 'P 18' progenies in the 'IMC' group. Within the family, variability is exhibited with regard to other fruit characters such as the thickness of the husk. In addition, a significant degree of variability is observed in some of the progenies of intra-family crosses with segregants having fruits with distinctive shapes and flowers with intense pigmentation. The few studies of the compatibility system in the family indicate that a mechanism that is specific to the family may be involved. The intra-family diversity that is encountered would be due either to the heterozygous status of the parent tree or resulting from fertilization from other trees in the source population.

The other parent tree from the island, 'P 21', could have belonged to another type described in the report. The more probable candidate is the type described as having 'equally large pods, somewhat warty at the stem (peduncle) end but smooth and round like Calabacillo at the apex.' (In this case the term '*Calabacillo*' may have been correctly applied.) A few 'IMC' clones conform to this description being conspicuous by their smaller fruits with a short and broad shape. Seed numbers are lower than those of the genotypes that appear to belong to the 'P 18' family, the flowers being less pigmented and the fruit husks generally harder. If the two types were growing in proximity the possibility may exist of some of the 'IMC' clones being hybrids between them.

As with the other locations on the R. Amazon and the R. Marañon, from which living material was collected, the 'Island' population could have developed from more than one introduction.

There have been two subsequent collecting activities in this stretch of the R. Amazon and the results of these should be examined in comparison with those of the 1937 activity. Chalmers visited three sites above Iquitos, from two of which some living plants survived but with such small samples that determination of the populations' diversity is unreliable (W.S. Chalmers, unpublished report, 1973). The general description of the cacao in the area indicates that the plantings were recent and the variability substantial for the small population sizes. Mention was made of the presence of trees with fruits having a 'characteristic minute slightly rough, minutely pimpled surface'. The fruit shape is not described nor whether any of the material collected had fruits of this type. The surface described recalls that of the fruits from the R. Morona collections.

The population at a location situated higher up the river appears to have had a measure of variability and some trees possess fruits approaching the 'Parinari' genotypes in appearance as well as flowers with unpigmented stamens. However, their fruits differed in possessing hard husks. The progenies of one parent sampled possess fruits that are elongate in shape with a basal constriction, a description that may correspond to Pound's *lagarta* type. Only one specimen came from another location lower down the river. The fruits of its progenies appear to have phenotypes similar to the 'Nanay' type and the flowers also have intensely pigmented stamens.

Chalmers also visited the island and collected from four trees (W.S. Chalmers, unpublished report, 1973). Although the collector reported 'The general pod type was almost identical to 'IMC 67'' and described the stamens as being pigmented, only one specimen with these characteristics appears to have been collected and this is not represented in the surviving collection. The parent trees of the other three specimens from the location had unpigmented stamens. The conclusions about the characteristics of these trees are hampered by doubtful identification of the clones and/or their progenies. The fruits of the parent tree of one specimen were described as having hard husks but the progenies' fruits are rather different from those described. The observations on the few extant progenies indicate that they resemble the 'central' type on the R. Ucayali. Presumably they derive from an introduction made after 1937, from the upper reaches of the river.

The banks of the R. Amazon between the mouths of the R. Nanay and the R. Napo and along a road from Iquitos to the banks of the R. Nanay were subjected to a superficial visit (Soria, 1965). The locations visited on the R. Amazon include the one formed from seeds introduced from Brazil. The relevant report describes seeing 'wild' cacao trees but, since none appears to have been related to the 'Criollo de la montaña', it can be assumed that such trees, from which material was collected, were remnants of former cultivation. The vague descriptions do not allow a definitive assessment of the population structure to be made but the general fruit characters described and the pigmented stamens of the flowers indicate that some of the trees were related to Pound's 'Nanay' type. The collector reported seeing large fruits similar to the Island type, 'P 18'. It is not known if any of the progenies of these trees survive. This collecting activity included the first mention of cacao exclusively within the R. Nanay drainage system although the location may not have been very distant from the banks of the R. Amazon. Although the forests of the R. Nanay have been explored by botanists no herbarium specimens of cacao have been located.

The segment of the R. Amazon between the mouths of the R. Napo and the R. Javarí is virtually unknown with regard to the existence of cacao. The section is remarkable for the absence of major tributaries of the Amazon. The paucity of information and absence of specimens is expected not to be an indication that this section marks a gap in the distribution of the species. During the colonial period missionaries were active and they would have exploited the produce as they did in other regions. However, in more recent times logging and the cultivation of sugarcane became the primary activities and it is probable that the forests along the river banks in which the species occurred had been destroyed. The only botanical reference to cacao in Peruvian territory in the lower part of the segment is the collection of specimens from Caballo Cocha, near the Colombian 'Trapezium'. This is the name is given to the area of Colombia that was created to give the country a port on the R. Amazon. Also at this level there is an island called 'Cacao', where the species could have existed as is found on other islands with this name. Collections of herbarium specimens and reports of cacao in the Colombian territory perhaps could provide a greater amount of information than is available in the Peruvian territory in this section.

The extent of deforestation on the banks of the R. Amazon in Colombia, which probably resulted in the extinction of bygone cacao populations, was referred to by Soria (1970) when he reported on his visit to the area in 1969. The descriptions of the few trees seen are too vague to serve to obtain a notion of the basic features. Apparently, some variability was observed but, since the material collected did not survive, we have no means of confirming the observations reported.

The River Napo drainage system (see Fig. 5)

The R. Napo is the major tributary of the R. Amazon in this section of the river, joining it on its left bank. In common with the other left-bank tributaries of the R. Marañon and R. Amazon, the course of the R. Napo is divided between Ecuador (occupying the upper course) and Peru, the greater length of the valley belonging to the latter. Since 1949 when the first variety collections were made in the R. Napo valley a significant amount of effort has been expended in collecting and conserving the germplasm contained in the Ecuadorian section. On the other hand, no effort has been made in Peru, with the exception of some oblique information concerning varieties said to have originated in the Peruvian section.

In Ecuador two major conservation programmes have been undertaken. The first was that under the Trinidad–Ecuador Cooperative Programme (TECP), which was conducted from 1967 to 1973. The second programme was funded by the London Cocoa Terminal Market Association and referred to as the London Cocoa Trade Amazon Project (LCTAP), which operated effectively between 1979 and 1986. The information acquired during and after the collecting activities could be analysed to obtain a comprehensive record of the diversity in the Napo valley. Although many of the basic observations made during the LCTAP have been described (Allen, 1987) no similar analysis has been produced with regard to the TECP. In view of this lack, the absence of studies on the products of the collections and the fact that the collections have been subjected to considerable losses it is not possible to obtain a picture that adequately describes the range and composition of the diversity. The situation is regrettable as the population contains some extremely important components in addition to the contribution that they can make to areas of research, such as inheritance.

Like all of the tributaries described above, the R. Napo drainage system originates in the Equatorial Oriental Piedmont. From its source the R. Napo first flows in a generally northerly direction and then towards the southeast. As a consequence of the location of its sources the R. Napo cacao population shares the same basic distinguishing characteristics as the tributaries of the R. Marañon system that have been described. The most notable characteristic, of course, is the presence of white cotyledons associated with light green flush leaves. It appears that this trait is predominant at least as far as below the junction with the R. Coca, a left-bank tributary of the R. Napo. At the end of the 19th century Tyler (1894) reported that on the R. Curaray, a lower tributary of the R. Napo (the course of which is divided between Ecuador and Peru), there were 'immense plantations of wild white cacao'. Hence, it is probable that cacao with this gene was amply distributed throughout the eastern foothills of the Andes.

Most of the genotypes possessing the 'white' gene are plants of relatively inferior vigour, low survival ability and adaptability in other environments. They are also more difficult to propagate. Consequently, the losses sustained in collections of this type, being higher than those of more vigorous genotypes, result in a bias in the sense that the collections from the region that have survived contain a lower representation of white cotyledon genotypes than may actually be the case.

The presence of trees with fruits containing purple cotyledons has been recorded from most areas of the valley but it is difficult to estimate the distribution of such trees and to make inferences as to their status and origins. The published data, such as those reported from the LCTAP programme (Allen, 1987), contain means of observations of various fruits. Thus, in cases where the fruits contained seeds with pigmented cotyledons, it is impossible to judge the genotype of a given parent tree in order to estimate the extent of homozygosity for either colour phenotype.

The situation on the R. Curaray is an example. If, at the time of Tyler's (1894) journey down the R. Napo, the cacao population consisted entirely of trees with white cotyledons, the presence of purple cotyledons on the river at the end of the 20th century would probably (excluding mutation) be the result of introduction of other varieties. Presumably, these introductions and their hybrid progenies could be distinguished from the native variety. The presence of trees with red fruits could be ascribed, for the most part, as resulting from the introduction of other varieties to the region, many of which carry alleles producing purple cotyledons. However, if native red-fruited varieties exist in the region these would constitute a complicating factor. The TECP reports on the R. Curaray population (the river is shown in Figs 5 and 21) drew attention to the existence in the area of introduced varieties but this appears to have been ignored in the LCTAP sampling of this population. The existence of such situations inhibits the proper analysis of the distribution of the factors responsible for the absence of pigmentation. However, observations on some collections, such as those made sequentially downstream along the R. Napo up to the frontier with Peru, indicate that the expression of purple cotyledons or mixtures of degrees of pigmentation increases along the course of the river.

Although genotypes with white cotyledons may comprise a large proportion of the population, a significant range of variability is expressed for other characters. While some trees have a bushy habit, which is noticeable when they are young, this is infrequent. In some valleys, such as that of the R. Payamino, the trees possess a distinctive habit that consists of a tall stem with a relatively small canopy (Plate 11a). On account of this habit one of the features of the R. Napo varieties is the tendency for the flowers and fruits to be produced entirely, or nearly always, on the plagiotropic branches in the canopy. This situation makes the fruits inaccessible for harvesting and, therefore, collecting. Plate 2c shows a tree that appeared to belong to this type, on the trunk of which there were no signs of flowers and fruits having been produced.

Besides this singular, essentially ramiflorous, flowering habit the Napo population tends to produce flowers that are predominantly without pigment except for the staminodes and guide-lines. Generally, the peduncles are green in colour and the buds are white and sometimes green on the outer surfaces of the calyx. Also, the shape of the buds tends to be elongate, ending in an attenuate apex on account of the sepals being narrowly elongate. Some of the flowers have guide-lines that are pale in colour but this trait is not necessarily associated with the pale flush leaves. Some uncertainty exists regarding the extent to which pigmentation of the stamens occurs. Some descendants of Napo genotypes have pigmented stamens although the trait has not generally been recorded. Flower characters were not recorded in the LCTAP.

The variability of fruit types is fairly expressive. The occurrence of fruits with a high degree of surface roughness (as shown by the fruit type depicted in Fig. 15) is one of the principal features of cacao genotypes in the Napo valley. The most outstanding of these genotypes, which seems to be unique to the region, is that with the broad fruits depicted in Plate 11b. The broader fruits with length:diameter ratios of 1.5 or less are often associated with low seed number contents. Some genotypes have fruits that are more elongate and similarly distinguished by rough surfaces (Plate 11c). The fruit depicted in Plate 11d may be mistaken for another type but this particular fruit is distinguished by having a very thick husk. However, elongate fruits with less rough surfaces are perhaps more common. It has not been possible to relate the specific fruit types to any particular region, although it would be expected that, if the populations of specific areas are composed of related genotypes, the trees in them would possess similar fruits.

The Napo valley populations exhibit an ample range of seed sizes, varying from very small to some of the largest seeds in the species. Some of the large seeds could be attributed to the fact that the fruits in which they occur under natural conditions contain few seeds. The low seed numbers would result from the combined effects of low ovule numbers and self- and cross-incompatibility

Fig. 15. Ecuador, R. Napo system: fruits of the accession 'COCA 3370'.

that accounts for the low fertilization rates observed. The fact that the seed numbers reported do not exceed 59 indicates that ovule numbers are not likely to be much above this figure. However, it is necessary to conduct detailed investigations of the progeny of the collections to verify the observations from the naturally occurring individuals.

The R. Aguarico (see Figs 5 and 21) is the principal left-bank tributary of the R. Napo and one that has seen the greatest human activity since colonial times, as it was the most important transportation route from the cordillera to the R. Amazon. Much of the course of the R. Aguarico runs close to that of the R. Putumayo drainage system and observations on cacao varieties of both rivers indicate that some overlap of genes has occurred. For this reason it is possible to find genotypes from the R. Aguarico that possess characters that resemble those of genotypes occurring in the Putumayo system and, consequently, differing from those of the remainder of the Napo system. There were few specimens from the Aguarico valley attributable to LCTAP. Several specimens were collected in the watershed between the Napo and Aguarico rivers but the locations of these with regard to each valley are not well defined. More detailed information about the cacao populations of the Aguarico valley is obtainable from the TECP reports but considerable effort is required to collate the data and observations in order to obtain a coherent picture of the cacao population of the valley. The area sampled was relatively small and the results, therefore, cannot be used to make inferences about the characteristics of the population of the entire valley. It would seem that the white cotyledon character almost exclusively occurs in the native genotypes of the sample area, except for the intrusion of a slight pigmentation that occurs to a small degree.

Allen (1988) undertook an analysis of the observations on the genotypes collected during the LCTAP programme in the populations derived from the

Equatorial Oriental Piedmont. In this analysis the genotypes attributable to the R. Napo drainage system were compared to those from the tributaries of the R. Marañon. On the basis of the descriptors used it was concluded that no differences could be detected. This inference is not corroborated by the author's own observations. The comparisons reported in Allen (1988) were not based on all of the characteristics that differentiate the populations in the two regions. It is possible that a more detailed analysis of the data may reveal the true differences among the various populations.

Since the principal feature that distinguishes the populations that contain descendants of the Equatorial Oriental Piedmont is the association of unpigmented flush leaves and unpigmented cotyledons, the knowledge of the distribution of these characters is basic to the determination of the differences among the local population in the region. The acquisition of this knowledge depends on an understanding of the mode of inheritance of these linked character expressions. On this basis, the genotypes can be attributed to the individuals in the populations and the frequency of the occurrence of each of the three genotypes in the local populations may be calculated. These frequencies will then be combined to obtain a picture of how the character combination is spatially distributed in each region.

In the R. Napo valley, the homozygous recessive genotype is predominant. In some local populations the frequency of its occurrence is almost 100%. It is possible that in other populations where there is introgression from the R. Putumayo germplasm, the heterozygous genotype occurs with a higher frequency than in the adjacent R. Napo populations.

With regard to the populations within the R. Marañon drainage system, it is probable that the homozygous recessive genotype occurs with high frequency (also around 100%) in the extreme tributaries, such as the R. Upano. The frequency of this genotype decreases in the populations nearer the R. Marañon where the homozygous dominant genotype would predominate.

Fig. 16. Ecuador, R. Napo system: fruits of two genotypes from the R. Aguarico.

As stated previously, no information is available regarding the cacao population along the R. Napo in its passage through Peruvian territory. Tyler's description of the cacao variety of the R. Curaray probably relates to the area along the final course of the river that is within Peru. Pound (1938) travelled up part of the lower course of the R. Napo and observed that cacao occurred sparsely in the area visited. He concluded the account of his contact with the R. Napo as follows:

> In the montagne (forest) around the Napo Experimental Station (at Santa Clotilde) quite a number of small stunted cacao trees were found, most of which were flushing crimson showing that the population of the montagne belongs to the variety 'criollo de la montagne'...

The only conclusion that can be drawn from this statement is that the light green flush leaf character was absent or occurred at low frequency. The absence of fruits inhibited the determination of the variety so that Pound's conclusion would be a mere supposition.

Our only knowledge about the Peruvian R. Napo varieties is derived from descendants of material obtained through two chance contacts, in both cases the material resulting from Pound's activities. The first case concerns a planting of trees in the Nanay area descended from seeds said to have been brought from an unspecified locality on the R. Napo (Pound, 1943). He described the 'Napo variety' trees as distinguishable from those of the 'Nanay' type by being more vigorous and by possessing what is called a 'weeping habit' of the branches; that is, the tendency for the branches to grow horizontally and curve towards the ground. Unfortunately, the account gives no information about the fruit and seed characters of the trees or about such traits as the flush leaf colour. From 27 trees of the 'Napo variety' seen at the site seven individuals were selected and budwood from them sent to Trinidad for reproduction together with the other clones identified as 'P'.

The author carried out some studies on the resulting clones in Trinidad but the lack of access to the data obtained together with the paucity of data from other sources about their characteristics does not allow more than a superficial description of the variety to be produced. This situation prevents the characters of these 'Napo' progenies from being compared to those of the populations sampled since 1948. All of the 'P' clones said to belong to the 'Napo variety' are self-incompatible. They produce fruits that are of small size, with an elongate shape having an apex that is inclined to be attenuate, the surface being somewhat rough. One of the distinguishing features of the progenies is the fairly intense coloration on the petal ligule.

The other group of progenies apparently attributable to Pound's journey along the lower R. Napo in 1938 (Pound, 1938) resulted from a fruit he gave to the Experiment Station at Palmira, Colombia, according to information in Patiño (1963). The clones of the progenies from this fruit were assigned the group denomination 'SPA', in Colombia, and several of them have been introduced into various germplasm collections. Data of various degrees of usefulness have been produced about the clones in the collections. The accumulation of these observations (excluding genotypes identified with the same acronym but alien to the population) permits a fairly complete description of the family's composition to be obtained.

In spite of the stated origin of the fruit, its source is a mystery since Pound stated that he did not see fruits on the 'native' trees he encountered at Santa Clotilde on the R. Napo. The possibility is that the fruit was obtained from a tree in the experimental planting. Consequently, the determination of the relationship of the parent tree to the R. Napo populations is a subject that is involved in the characterization of the family.

Most members of the family possess fruits of a similar shape and other characters with some variation. Fruits of two members of the family are depicted in Plate 18d, showing the broad and elongate shape, the slightly rough surface and apex that is bluntly acute. The basic fruit colour is green but this varies in intensity from intermediate to dark. The flowers of the known progenies have a basic white colour but segregation is observed with regard to the stamen pigment, which is absent in some progenies but intense in others. The intensity of stamen pigment may be linked to the fruit colour.

The most distinctive characteristics of this family reside in the habit consisting of a large branch angle, resulting in spreading canopies in orthotropic plants, and distinctive leaves. On the basis of this characteristic they may be related to the 'Napo variety' from the Nanay peninsula. The leaf blades are broad with a rough appearance that makes these genotypes readily recognizable. The typical form of the leaves of the family may be appreciated in Fig. 17. Inbred progenies obtained from selfing or sib-crossing segregate for flush leaf colour from white to red. When these clones are combined with genotypes carrying the anthocyanin inhibitor gene, seeds with white cotyledons are obtained indicating that these genotypes are heterozygous for white cotyledons. These two characters provide a basis for attributing the family to a R. Napo origin.

Fig. 17. Peru, R. Napo: the broad thick-textured leaves that are characteristic of the 'SPA' family.

However, members of the family appear to possess an incompatibility allele that acts in the same manner as the allele found in the 'P 18' family from the Island at Iquitos. On this basis it may be suspected that these families are related. In fact, Pound (1943) must have come to this conclusion when he wrote, 'It is not improbable that the original group of disease-free trees located in 1938 in Nanay has some connection with a chance introduction from the Napo.'

The Solimões drainage system

The name 'Solimões' is applied particularly in Brazil to identify the stretch of the R. Amazon from the place where it enters Brazilian territory at the mouth of the right-bank tributary, the R. Javarí, to its confluences with the R. Negro and R. Madeira (Fig. 18). The exact definition of the stretch of the R. Amazon to which the name could be applied is uncertain since it could also be considered as starting from the mouth of the R. Napo where the R. Amazon begins its eastward direction.

For our purposes the separation of this section of the R. Amazon from the remainder of the system has a particular significance. The Solimões is the region of the river basin that receives the most important tributaries, not only on account of the large drainage basins that some of them occupy but that some of the most extensive populations occur on them. Accordingly, it is of fundamental importance to our survey of the diversity of the species. In this discussion it is intended to demonstrate the fact that the Solimões itself is the repository of perhaps the most extensive genetic variability that is amplified many times when all of its tributaries are included in the system. The importance of the region resides not only in the amplitude of the diversity but also in the fact that some of the most valuable genetic material for purposes of improvement of the species is to be found in this basin.

The Solimões region was probably the most important source, in quantitative terms, of the cacao harvested during the long period of exploitation of the Amazon's cacao production by the extractive process. It is probable, also, that the region supplied a greater part of the planting material from which cultivated populations were formed. The analysis of the existing variability in the Solimões basin, therefore, is important in understanding the origin and composition of the cultivated populations in the Amazon Basin and those in other countries that have imported varieties from the region.

The documentation concerning the region that dates from the extractive period provides much information about the distribution of the cacao populations and the quantities produced, leaving no doubt as to the expressive occurrence of the species along the Solimões. Martius visited part of the Solimões during the exploration of the Amazon in 1816 (Spix and Martius, 1828). He wrote enthusiastically about the abundance of cacao on the Solimões and its major tributary, the Japurá. Among other matters Martius refers to the custom of the Indians in pre-colonial times to plant cacao in the vicinity of their settlements.

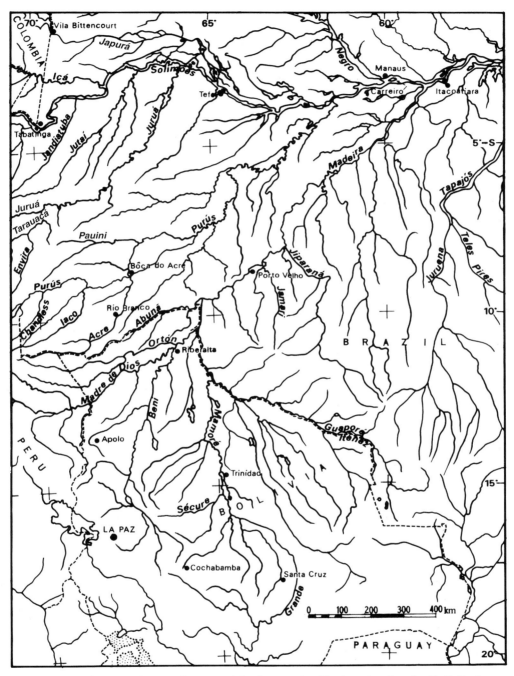

Fig. 18. Map of the south central sector of the Amazonian Region showing the R. Solimões and the drainage systems of its southern tributaries and the drainage system of the R. Madeira.

Besides this information virtually no attention has been paid to the cacao populations of the Solimões drainage system until fairly recently. Pound (1938) referred to the river when he travelled up it on his way to Peru but he gave an exceedingly inaccurate description of the variety situation that he was able to observe. However, some years later, in Pound (1945), he corrected the inadequate description given in the 1938 report by stating:

> In the middle Amazon which stretches from the Rio Negro to the Rio Ucayali the central type changes perceptibly. The pods are still completely unpigmented but longer and more corrugated and often definitely warty. Sometimes a pronounced bottle neck appears. The shell may be thicker and the pod approaches the cundeamor type in places. The seeds though still relatively small and uniformly purple in colour are plumper than those of the lower Amazon.

The word 'shell' used here corresponds to the term 'fruit husk' used in this text.

Although a few specimens derive from a visit to the middle Solimões in 1971 and a reconnaissance of the same area in 1981 the knowledge of the scale of occurrence and the diversity of the varieties along the river only became a reality in 1982. The collecting activities accomplished in 1982 and 1983 resulted in the acquisition of a large quantity of very variable material. However, the entire length of the river has not yet been explored for the purpose of conserving its genetic resources.

One collecting activity concentrated on the stretch of the river from the frontier with Peru up to a point just below the mouth of the R. Putumayo/Içá. The material collected in this zone exhibits an ample degree of variability, some of the unusual types observed in the progeny being depicted in Plate 15. The characteristics of much of the cacao in the area include fruits of an elongate shape with high ridges and thick husks. Also present are varieties with short and broad fruits of the shape that is considered to represent the traditional variety of Ecuador.

An important feature of the cacao populations along the course of the river from Tabatinga to the mouth of the R. Putumayo is the occurrence of trees with pigmented fruits of various types (Plate 12d, e). Included in the range of diversity are fruits with phenotypes similar to those encountered on the upper R. Putumayo basin that have been sampled by the programmes carried out in Ecuador.

Lower down the Solimões sampling of the diversity was conducted between the lake of Tefé to below the mouths of the R. Caquetá/Japurá. On the north bank of the Solimões this river and the Japurá (as it is named in Brazil) run close together and are connected by a network of canals, which breaks the land into numerous islands, as is apparent in Fig. 18. The area thus takes the form of a delta of the R. Japurá, in which it is impossible to define the course of the river. This situation results in the development of isolated occurrences of cacao, which individually may comprise specific varieties with their own characteristics depending on the mode of their formation and evolution.

Since the area linking the valleys concerned is undefined it is necessary to treat the cacao populations in the 'delta' region as belonging to an integral geographical entity and they will be described in the respective part of the discussion of the R. Japurá.

The stretch of the Solimões, the populations of which have been sampled, contains genotypes with diverse characteristics and Plates 16, 17 and 18 present a few examples of them. Variability is expressed in terms of tree habit ranging from plants that produce the first whorl of plagiotropic branches at a low height (Plate 18a) to those reaching heights that are usual for the species. Leaf characters vary considerably, the most notable variation being encountered with regard to leaf shapes that range from narrow elongate to broad and rough in texture.

The variability of the fruits is notable. Although the collection report gives the impression that most of the parent trees had fruits that were elongate in shape, it was found that the aggregate of genotypes covers a multitude of combinations of the various component characteristics. It could be concluded, from the observations on the clones established, that the occurrence of identical genotypes is rare, even within the same locality. At one extreme of the range of shapes are several parent trees with spherical fruits. The samples from the vicinity of Tefé have short fruits with almost parallel sides that are similar to the Ecuador 'Nacional' fruits. Fruits of a type similar to that of the 'SPA' family (Plate 18d) appear to be common (see Estampa 7D in Barriga *et al.*, undated).

Among the most interesting types of fruits from the area is one that is found in segregants in the progenies of one variety that has the typical *Pentagona* characteristics. The fruit is shown in Plate 17a, where it is compared to the large heavy fruits that are more typical of those of the family. The fruit of the parent tree of the family (Plate 17b) is also large with ridges arranged in close pairs. The close pairing of the ridges is more pronounced in the developing stage, the cross sections of which approach those of the pentagonal character. In addition to this family another parent tree was recorded as having fruits with the *Pentagona* characteristics. The progeny, the fruit of one of which is shown in Plate 17c, did not segregate for fruits of the pentagonal type.

Another feature of these populations is the presence of genotypes carrying the anthocyanin inhibitor gene. One of these genotypes is notable for its poor growth and very small fruits. The genotype has the additional abnormality in that, in both the parent clone and its obviously selfed progeny, seed numbers are very low, perhaps not exceeding one-third of the ovule number. In common with other cacao genotypes that express the same phenomenon the low seed numbers can be attributed to a sterility mechanism that has to be determined.

One of the features of the varieties of the Solimões is the occurrence of naturally produced fruits with high seed numbers, indicating that ovule numbers could be as high as 65. Although there was a wide range of seed sizes and numbers, in general the latter tended to be considerably higher than those obtained in the R. Napo populations, even admitting that the largest fruits would have been selected during collection. Some of the Solimões fruits weighed over 1200 g.

The conclusion that is reached from the analysis of the variability expressed in the cacao population of the sample area is that it is of such magnitude that it may be composed of a mixture of varieties from different origins. It could be apposite to conceive of the Solimões basin as the recipient of genes from the populations that inhabit its numerous and extensive tributaries. The concept will be better appreciated when the respective cacao populations are described. The range of variability precludes the presentation of a detailed account in this work. The diversity present in the Solimões sector merits a special publication after its whole length has been surveyed and when it has been comprehensively analysed.

The River Javarí drainage system (Fig. 5)

This is the first drainage system that can be considered to be part of the Solimões basin. The R. Javarí joins the R. Amazon on the right bank and forms the border between Peru and Brazil. In common with most of the other systems in Peru little is known about the cacao germplasm that occurs in the side of the valley in that country, except for a few uninformative herbarium specimens that probably were collected from near the mouth of the river. It would, however, be expected that the germplasm occurring on the Peruvian side would be the same as that found in Brazil.

The extent of the occurrence of cacao in the valley is a matter that has been the subject of conflicting reports and it is uncertain whether the species is distributed throughout the valley. The Brazilian Genetic Resources Programme collected on the R. Javarí, up to the mouth of the right-bank tributary, the R. Curuçá, and on the latter river. Also explored were the right-bank tributary, the R. Itaquai and its tributary, the R. Quixito.

The human population of the region has been sparse and its remoteness perhaps resulted in the cacao population escaping exploitation during the extractive period. Consequently, the population of the Javarí valley can be considered to be of spontaneous origin and not likely to have been subjected to human interference. From its particular characteristics this population appears to have arisen independently of the other populations of the Solimões basin. From all accounts the cacao population density is rather high.

The accessions from the Itaquai/Quixito valley possess several features that indicate that the cacao variety is unique in its composition. The plants are generally slow-growing in the juvenile phase. The flowers are very distinct on account of their delicate, open, appearance and an almost complete absence of pigment. The delicate appearance is produced by a combination of long ligule straps and very pale yellow (almost white) ligules that are long and narrow. The fruits are also unusual. They were described (P.F.R. Machado, personal communication) as being notable for a bluish tinge on the surface as shown in the photographs in 'Estampa' (Plate 6) in Barriga *et al.* (undated). In fact, none of the progenies has exhibited this fruit colour. The fruit colour is really a very light green enhanced by the almost smooth but slightly granular surface; the shape approaches spherical with inconspicuous, very close-paired ridges. An example of a typical fruit is shown in Plate 16a.

One of the other features of the cacao of the Javarí valley is the occurrence of flush leaves without anthocyanin pigment (Plate 3e). The shade of whiteness of these leaves appears to differ from that of the unpigmented leaves of the R.

Napo population. In the area of the confluence of the R. Curuçá with the R. Javarí more elongate fruits are encountered. The most interesting finding connected with the population in the area is that several of the families produce fruits with notable red pigment on the surface. It has not been possible to ascertain whether the pigment results from the action of a specific allele that is related to the allele that controls the anthocyanin pigment in the *Criollo* group. Since some of the families are composed entirely of trees with red fruits that are elongate in shape, such as those of the tree shown in Plate 12a, it may be supposed that a specific allele is present in the population that results in the expression of fruit pigmentation. In contrast to the pleiotropic effect of the pigment allele in the *Criollo* group genotypes and their descendants the Javarí allele appears not to produce a noticeable pigmentation in the leaf pedicels. The absence of pedicel pigment may be associated with the action of the specific allele for absence of pigment in the flush leaves.

This important finding has a precedent in the discovery by R.L. Froes (personal communication) of trees with red fruits on the R. Jandiatuba, another right-bank tributary of the Solimões, this being the next valley to the east of the R. Quixito (see map in Fig. 5). The region involved, perhaps, is the birthplace of types that are in several respects similar to the genotypes of the *Criollo* group. However, it should be emphasized that there is no intention to suggest that there is a relationship between them.

One of the genotypes from the same location as a family with red-fruited trees had flowers that were deeply pigmented. Thus, the presence of anthocyanin pigment may be fairly common in the area that is at the junction of the R. Javarí and R. Curuçá.

Unfortunately, the unsatisfactory conditions under which the R. Javarí genotypes were established and the lack of continuity of the observations that the population deserves have prevented the material from being adequately investigated.

The subsequent tributaries of the Solimões, the R. Jandiatuba (as mentioned above) and the R. Jutaí are known to be valleys where the species occurs. Although some scarce information regarding the properties of the populations is available, owing to the lack of specific collecting activities and exploration of these rivers having been undertaken their cacao populations will not be described.

The River Juruá drainage system

The R. Juruá and its tributaries form probably the second most important drainage system on the south of the Solimões. The situation and upper and lower courses of the drainage system are shown in the maps in Figs 5 and 18, respectively. The valley has been important as an area of exploitation of its natural rubber resources since the late 19th century but, although cacao is mentioned in some accounts, it is not known for the existence of the species. The long period of rubber exploitation may have had some effect on the distribution and genetic diversity of cacao in the valley.

Like most of the southern tributaries of the Solimões the R. Juruá has its sources in Peru in the Province of Caravaya but nothing is known about the cacao existing in that country. Equally, since the course of the river in the State of Amazonas was not considered to be a priority area for conservation the cacao population of most of the drainage system is unknown. The only surviving specimens from the R. Juruá itself were collected in 1967 from the vicinity of the town of Cruzeiro do Sul, situated near the northern boundary of the State of Acre.

The fruits of some of these specimens are shown in Plate 13b; all of them derive from a small area along the R. Juruá below Cruzeiro do Sul. It is an area where the principal activity is related to rubber collecting and where there has been evident alteration in the course of the river, through erosion and simultaneous rebuilding of the banks. The specimens concerned belong, basically, to a single variety although differences among the genotypes are evident. Although most of the parent trees were extremely large (having some of the largest trunks seen by the author) the resulting clones are relatively slow-growing, developing into trees that are small in size. The difference between the sizes of the naturally occurring trees and their cloned progeny indicates the longevity of the former or the high fertility of the soil.

At the time of collection it was very noticeable that the fruits and vegetative characteristics of the cloned progeny resembled in certain respects those of the 'Scavina' variety from the R. Ucayali. Unfortunately, the material collected from the trees that most resembled the 'Scavina' type was lost in propagation. Among the comparative traits is the tendency to produce narrow crowns when grown from orthotropic shoots and the propensity to reproduce in this way. The location of the area of collection on the R. Juruá is relatively close to the corresponding area on the R. Ucayali. The Juruá location is also connected to the R. Ucayali system through its tributary, the R. Moa, whose headwaters are close to the source of the R. Tapiche. This situation would indicate a possible means of interchange between the two drainage systems by Indian tribes or during the exploitation of rubber on these rivers.

Chandless explored the R. Juruá in the mid-19th century and mentioned the presence of cacao on the river. In Chandless (1869) he referred to the fact that the Indians on the upper R. Purús had contacts with the R. Ucayali and the Sarayacu; they even knew Moyobamba. He also mentioned that the connection with the R. Ucayali was through the R. Juruá. If these contacts included exchange of cacao this would explain the presence of similar varieties on the Juruá and Ucayali rivers and their respective tributaries.

The most uniform characters of the genotypes collected are found in the flowers. In particular, the petal ligules are distinguished by their shape in which the edges are curved upwards to form a 'spoon-shaped' organ. These ligules are also light yellow in colour with traces of pigment along the veins.

The fact that the genotypes resulting from the 1967 collections appear to be related and distinctive should not be interpreted that they are representative of all the native cacao varieties of the R. Juruá. The fruits illustrated in Plate 13c and Fig. 19 indicate that the diversity is somewhat larger with other varieties being present. The fruits in these illustrations were from trees in

Fig. 19. Brazil, R. Juruá system: a sample of the fruit types from the upper River Juruá drainage system, from genotypes other than those represented in living collections.

locations on the R. Moa and upstream from Cruzeiro do Sul, the samples of which were unsuccessfully established.

The River Tarauacá sub-drainage system

The R. Tarauacá (Fig. 5) is a major right-bank tributary of the R. Juruá, and its cacao population, as well as that of its tributary the R. Envira, deserves special consideration on account of its peculiar and potentially useful characteristics. The area of the valley that was sampled lies entirely within the State of Acre and includes the R. Envira, a tributary of the R. Tarauacá that joins the latter on its right bank. In contrast to the conception of cacao being distributed along the river banks the report of the collecting activity (P.F.R. Machado, unpublished data, 1981, Relatório da Expedição Botânica para Coleta de Cacau Silvestre nos Rios Tarauacá e Envira, no Estado do Acre) refers to stands of cacao occurring on the banks of small streams and in locations some distance from the rivers.

One of the outstanding features of the germplasm encountered is the development from orthotropic shoots of branches at a small angle from the trunk producing narrow canopies with sparse foliage (Plate 1a and 1b). The trees of this variety are readily distinguished by possessing leaves that are fairly thick in texture but have a peculiar yellowish tinge that is difficult to define.

Segregation for flush leaf pigment occurs with leaves without pigment being frequent. So far, it has not been possible to identify any individual parent genotypes that are homozygous for this characteristic, which would be inherited as a recessive trait. As no special characteristics of the flowers have been observed it could be possible that the non-pigment character is inherited in the same way as that of the R. Napo population. The occurrence of non-pigmented cotyledons has not been recorded, precluding the determination of the relationship of cotyledon colours with flush leaf pigmentation.

Although the trees apparently possess poor growth traits they are capable of producing large crops of fruits in relation to their size. In spite of the apparent uniformity of the vegetative traits the population exhibits a significant degree of variability in relation to fruit characters, examples of the extreme fruit shapes being shown in Plate 14. The common distinguishing fruit characters are the expression of surface pigment and the granular appearance of the surfaces owing to their being covered with slight protuberances. The fruits are small in size but, since the husks are thin, the ratio of seed weight:total fruit weight tends to be considerably higher than that found in other varieties. This characteristic compensates for the small fruit size and confers on the genotypes a potential importance in economic terms. The small fruits are the consequence of the very small seeds they contain.

Few of the parent trees sampled survived transportation and propagation, a situation that precludes a definitive assessment of the properties of the population. From the surviving genotypes it appears that on the R. Tarauacá the majority of the trees have fruits that are short, broad in shape with rounded or slightly pointed apices. On the other hand most of the R. Envira genotypes produce fruits that are elongate in shape, although a few genotypes have fruits that are short and broad in shape. Several fruits contained more than 55 seeds. From this result it could be supposed that ovule numbers could be as high as 60, or even more. The only characteristic that the R. Tarauacá variety seems to share with the genotypes from the main R. Juruá is the extremely soft husk of the fruits, due probably to the absence of the woody mesocarp.

The observations on the flowers did not reveal any distinctive characteristics although pigmented ligules occur. There is segregation of ligule shape and size and also with regard to the position of the sepals, in which horizontal and reflexed sepals are found.

The River Purús drainage system

The R. Purús is the most easterly of the tributaries that join the Solimões on its right bank. With a length of 3148 km the R. Purús' drainage system covers a large land area. The drainage system is delineated in Fig. 5 and Fig. 18 and it can be seen that, in common with the other tributaries of the Solimões that have been described, the sources of the Purús and its major tributaries lie within Peruvian territory in the Province of Caravaya. Although several accounts provide details on the occurrence of the species in the headwaters region of the valley, one of which refers to a forest of cacao just south of the frontier with Brazil, no information is available about the variety situation in Peru. Accounts of the Amazon valley during the early period of colonization provide the knowledge that the tribes that inhabited the valley at that time had the custom of eating cacao seeds. Exactly what was involved was not described but it is evident that cacao played a role in the lives of these people. This role may have had an important influence on the distribution of the species in the valley.

The upper reaches of the Purús valley have received significant interest in terms of the development of its natural resources, the initial phase being related to the exploitation of rubber. For this reason, the region has been the subject of the attention of botanists and ecologists. The upper Purús valley lies in an area of alluvial soils that are considered to be among the most fertile in the Amazon Basin. This fact makes the region particularly vulnerable to agricultural development and consequent erosion of the natural vegetable resources contained in the once primeval forest.

The R. Purús is noted for its meandering course and this feature has resulted in its having undergone incessant changes through the concomitant effects of erosion of its banks and formation of new land areas. The consequences of this situation regarding the development and constitution of the cacao populations were explained in Chapter 4.

The first noteworthy account of cacao on the R. Purús was given by Huber (1906a), who described it as existing as mainly spontaneous populations and indicated that some variability was to be encountered among the stands of cacao. One of the noteworthy features of the cacao population of the upper Purús valley is that much of the cacao is to be found in large areas in which an occurrence may comprise several thousand trees. The formation of such extensive populations could be attributed to their establishment on new land where competition from other vegetation would be lower than that in established forest. Alternatively, such populations may have resulted directly from, or been influenced by, human activity either by the indigenous inhabitants or the new settlers who began to arrive during the 19th century.

On account of the pressure on the cacao habitats due to development efforts the upper Purús valley was considered to be a priority region for conservation of its resources. Several collecting activities have been undertaken over a period of years. The first activity, which took place in 1965 in the vicinity of the Acre State capital of Rio Branco, demonstrated the potentially enormous genetic variability that occurs in the region.

The magnitude of the diversity has been confirmed by the subsequent conservation efforts undertaken in the valley. The R. Purús and its tributaries have been explored in an area bounded by the mouth of the R. Chandless, the headwaters of the R. Acre on the east and as far north (downstream) to and including the left-bank tributary of the R. Pauini. Several hundreds of parent trees were sampled during these activities and are represented by large numbers of progenies. The specimens obtained perhaps represent only a fraction of the diversity the region contains since the Purús population appears to be the numerically largest that can be encountered even though the exact magnitude of the variability has yet to be determined.

Some idea of the range of variability of fruit types may be obtained from Plate 19a and b, the former being a composite of some genotypes, which derive from the parent trees collected on the R. Acre in 1965. Although the 18

Fig. 20. Brazil, R. Purús system: fruits of 'RB 34' or 'RB 35' showing the equidistant ridges.

surviving genotypes were obtained from only six sites the scale of variation within and between sites is obvious. The extent of diversity, as expressed in all parts of the plants, is so large that it is impossible at the present time to identify specific characteristics of the populations that inhabit the various stretches of the main river and its tributaries.

On the R. Iaco, another tributary of the R. Purús, the range of fruit characters is equally large, examples of the extreme types being shown in Plate 19c and d. Although most of the fruits that were collected on this river were short and broad, those of the cloned parent trees that lacked fruits at the time of sampling and the progenies of the seed collections are more frequently elongate with rough surfaces. Some of the segregants in the families produce fruits that possess characteristics not far removed from those used to define the fruits of the genotypes of the *Criollo* group, especially the specimens from near the headwaters of the valley.

Several characteristics are found in the flowers of members of the Purús population that can be considered to be specific to the cacao of the drainage system. The flowers are fairly large in size and present an appearance of thickness, or compactness, in most of the floral parts. The ligules are broad, held in a horizontal position and are darker yellow in colour with some pigment. The notable feature is the intense, bright pigmentation of the stamen filaments, which, qualitatively, differs from those of other known varieties. Although this combination of floral characters appears to be specific to the upper Purús region it is by no means generalized in the area. Some trees possess flowers that are pigmented only in the staminodes and guide-lines. The observations on the R. Pauini specimens indicate that the flowers of these trees are less pigmented than the flowers of genotypes belonging to populations upstream. On the other hand, the specimens obtained from the R. Purús between the mouths of the R. Acre and the R. Pauini are distinguished by the greater pigmentation of the petal ligules that are often entirely red in colour. The degree of ligule pigment is perhaps the most intense in the species. However, the occurrence of other genotypes on the same sector of the R. Purús having the same pattern of flower pigmentation as those of the R. Pauini suggests that such pigmentation may be expressed to a lesser degree as the river is descended.

Certain genotypes are notable for their tree architecture and habit. Although 'RB 29' possesses very small fruits it is capable of developing into very large trees. Another genotype produces very short flushes but has a tendency for abundant flower production, the flowers being very red owing to the intense pigmentation of the ligules. The genotypes from the vicinity of Brasileia, near the headwaters of the R. Acre, appear to produce weaker plants than most of the other specimens from the valley.

One feature of the population that was in some ways unexpected was the appearance of families segregating for trees with unpigmented flush leaves. The trait was observed to occur in only a few families. However, in contrast with the populations on other rivers, where genotypes having unpigmented flush leaves are usually less vigorous than their sibs with pigmented flush leaves, the R. Purús genotypes with unpigmented leaves stand out as being taller than their normal relatives.

Leaf types also vary considerably from large, broad, thick blades to those that are elongate with a fine papery texture. Another feature of the population is the sweet taste of the pulp and the strong aroma of the dried seeds, even when unfermented. Some specimens have fruits with phenotypes superficially similar to the R. Juruá specimens in Plate 13b but they differ considerably in regard to the flower characters.

In summary, the variability contained in the cacao population of the R. Purús is extremely large. The limitations that circumstances have imposed on the characterization of the numerous samples obtained in the valley have made it impossible to obtain an adequate profile of the ample variability revealed by the observations performed so far.

In terms of its genetic variation the cacao population of the R. Purús from the mouth of the R. Pauini to the several mouths by which it flows into the Solimões is unknown either from herbarium specimens or from collections of living material. From all accounts the occurrence of cacao near the mouths of the R. Purús is extensive. Above its mouths the river meanders through a large flood plain where conditions may be inappropriate for the existence of large cacao populations and, for this reason, the river was virtually unknown during the era when extraction of cacao from natural populations was practised.

The River Putumayo drainage system

We can now turn our attention to the tributaries that join the Solimões on its northern banks. The R. Putumayo, most of the course of which is shown in Fig. 21, is an important river that has its source in the Colombian Andes and

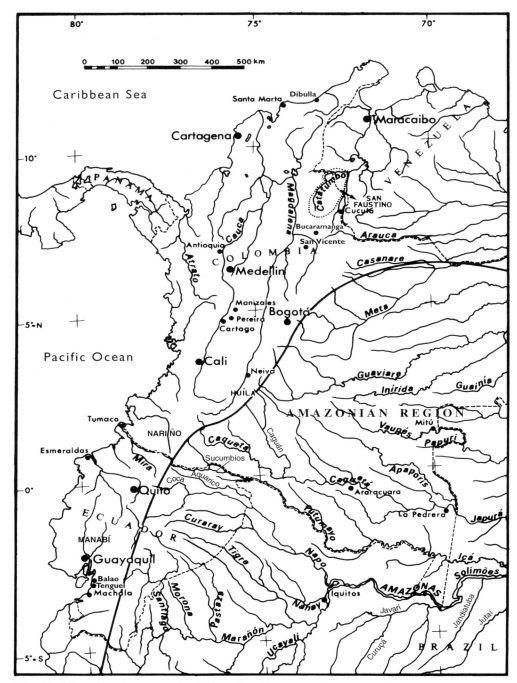

Fig. 21. Map of northwestern South America. The approximate limit of the Amazonian Region is shown, including the northwestern tributaries of the R. Amazon and western tributaries of the R. Orinoco. To the west of the line marking the limit of the Amazonian Region the areas of original cultivation of *Criollo* varieties and the major zones of cultivation in Colombia and Ecuador are shown.

flows in a southeasterly direction towards the R. Amazon. In geographical terms the origins of the river are derived from several sub-Andean streams that join to form the main river when it enters the Amazonian plain. Several of these tributaries are adjacent to the upper streams of the R. Caquetá system and the possibility of some common origin of the cacao populations near the foothills of the Andes may be considered.

According to Patiño (1963) documentary sources attest to the occurrence of natural cacao populations in the upper regions of the Putumayo valley as late as the 19th century. More recent explorations of the region indicate that there has been a significant degree of introduction of planting material of other varieties used to establish cultivated fields. For this reason it is difficult to determine to what extent the cacao trees may be derived from the original populations.

In spite of the size of the area of the system and the potential importance that the cacao population may have regarding the species' diversity, the knowledge about the cacao varieties contained in it is extremely sparse. The areas that have been sampled and the numbers of specimens that are available for study constitute a small fraction of the total area of the valley.

Although he would have travelled along part of the R. Putumayo on his journey to Colombia in 1938, Pound (1938) does not describe the cacao of the river in the report. He would have made some observations since, on other occasions, he does refer to the existence of 'cacao de monte' on the river with the implication that the Putumayo cacao was the same as that on the Peruvian stretch of the R. Napo. Subsequent collecting activities have shown that this conclusion is without foundation.

This may be an opportunity to explain the meaning and application of the term 'Criollo de monte', which Pound previously translated as 'Criollo de la montagne', which was mentioned earlier. Although from his statements, we may be led to believe that the entire region between, and including, the R. Napo and R. Caquetá systems was populated by a single variety, his other descriptions made it clear that an expressive degree of variability occurred. The fact is that the term 'Criollo de monte' would be employed by the inhabitants of a locality to indicate cacao trees that existed prior to their settlement of the site. Therefore, the term would apply to the particular local genotypes irrespective of their relationships to the genotypes that occur in other locations.

For the purposes of germplasm conservation the first survey of the population of the R. Putumayo was carried out by the Anglo-Colombian Cacao Collecting Expedition in 1953 (Baker et al., 1954). It was evident at that time that intrusions of varieties from the zones of cultivation had taken place in the headwaters region of the valley and that it was impossible to determine what native varieties, if any, existed. For this reason it would be appropriate to ignore the area of the Andean foothills for the purpose of determining the place of the sub-Andean cacao populations in the range of variability specific to the valley.

Because of the absence of undamaged fruits no living specimens were collected in the valley by the 1953 expedition, with the result that no progenies were available to study the genetics of the populations encountered. Lack of access to the original records and photographs hampers the adequate interpretation of the observations made during the survey and to relate them to the genotypes sampled during subsequent collecting activities.

Cacao was seen in apparently small populations near the town of Puerto Asis in the upper reaches of the valley. The common type was described as having fruits of small size, oval in shape with a very rough surface, the colour being very light green with traces of surface pigment, and the cotyledons purple. Soria (1970) reported, on the basis of a short visit, that cacao was scarce in the area.

The LCTAP collected a few specimens in stands of cacao below Puerto Asis on the R. Putumayo, the area sampled being between the mouth of the tributary, the R. Piñuna Negra, and the mouth of the R. Sucumbios. It was recorded (Allen, 1987) that some of the cacao trees seen on this stretch differed from those lower down the river. The former possessed fruits that were small in size, with smooth, slightly pigmented surfaces with small seeds having dark purple cotyledons. The small fruits would have been the consequence of low seed numbers, a single specimen from closer to Puerto Asis having elongate fruits with many seeds. However, the progenies of this parent tree have small fruits similar to those belonging to the general sample. The combined features of the LCTAP specimens correspond to those described by the 1953 expedition, except for the roughness of the fruit surface, this perhaps being a case of a local variation.

The main conservation activities have taken place along the major tributary of the R. Putumayo, the R. Sucumbios or San Miguel (see map in Fig. 21), and on the adjoining stretch of the main river as far downstream as Puerto Leguizamo. This activity relates to the fact that the lower course of the R. Sucumbios flows entirely through Ecuadorian territory and, therefore, within the area of action of the TECP and the LCTAP. There are two significant features to be considered regarding this area that may have an important bearing on the composition of the cacao populations. One is that the valley of the R. Sucumbios is situated a short distance to the north of the R. Aguarico, a part of the R. Napo system. The other feature is that further downstream from the mouth of the R. Sucumbios, in the area of Puerto Leguizamo, the R. Putumayo is separated from the R. Caquetá by a distance of only 25 km. Since, in both instances, the primitive distribution of cacao was not strictly confined to the banks of the rivers concerned it is probable that similar populations occur in both cases. For this reason, instead of treating the R. Putumayo cacao varieties as separate entities, they must be considered as components of the diversity contained in a larger area.

The 1953 expedition explored part of the R. Sucumbios as well as on the main river. The conclusion was reached that the cacao in the area was genuinely spontaneous, considering that, at the time of the expedition, the forest would have been relatively undisturbed and not subject to the development activities that made a large impact on the region after 1960. For this reason, there would be a greater guarantee that the observations made describe trees that represented the original population.

The main features of the trees encountered in the vicinity of Puerto Nuevo included the leaf shape being narrowly elongate with long slender petioles.

The flush leaves were pinkish in colour. The flowers were normally white in colour except for the guide-lines. Two types of fruits were seen, apparently occurring together. One class of fruit had an oval shape with an almost smooth surface, no basal constriction and a blunt apex. There were indications of surface pigmentation. The fruits belonging to the second class were larger in size with a length of about 220 mm, the surface being very rough, with a blunt but tapering apex. Several of the fruits of both types had large seeds with pale purple and white cotyledons with the former apparently being more frequent. Presumably, the photograph of a fruit from the R. Sucumbios (San Miguel) in Baker *et al.* (1954) depicts a genotype belonging to the second class. This fruit would be similar to that of one of the progenies of the TECP collections shown in Plate 11b. Similar fruits occur in parts of the R. Napo system indicating a general distribution of the combination of features throughout a wider area. The cacao type found upstream from Puerto Nuevo differed slightly from that of the trees at this site but the surface pigmentation appeared to be absent.

The R. Sucumbios population was sampled by the TECP on various occasions at several locations, including the area near Puerto Nuevo. These locations were also sampled by LCTAP. Although the combined programmes collected from a large number of trees the information provided about the nature of the populations is rather scanty. The number of available clones and progenies of the parent trees is very small. In addition, the sample sizes of the progenies are insignificant. The combined efforts do not permit us to arrive at satisfactory conclusions about the nature of the population on the river or regarding its relationships to other populations. At least, it appears that some of the trees sampled possess characteristics similar to those described by the 1953 expedition.

On the other hand, a detailed study of the genotypes from the R. Sucumbios may reveal the existence of a significant range of variability. The tree depicted in Plate 12c differs substantially from the description of the parent tree. One of the genotypes from the TECP attributed to the valley possesses smallish fruits that are distinguished by having ridges that occur in very close pairs, albeit with a distinct division between them, and a wide flat area between the pairs. This gives the fruits an appearance similar to that of a pentagonal shape. There is also a very wide range of seed sizes, some being exceptionally large.

As planted trees the collections from the R. Sucumbios contrast significantly with the trees of most of the R. Napo genotypes in terms of tree habit and leaf characters. The R. Sucumbios plants are shorter with thicker trunks and their canopies are denser and wider than the R. Napo trees, which have tall straight trunks and narrow canopies. The contrast is also noticeable with regard to the normally pigmented flush leaves of the R. Sucumbios genotypes compared with the green flush leaves of the Napo and other populations from the Equatorial Oriental Piedmont.

The LCTAP also carried out a survey of the stretch of the R. Putumayo below the confluence with the R. Sucumbios up to Puerto Rodriguez, a distance that involves the joint boundary between Ecuador and Colombia.

Some idea of the population of the area had been obtained previously by the TECP through observations of trees encountered on the R. Sucumbios that were derived from seed introduced from Puerto Rodriguez. The fruits of these types are normally green in colour and elongate in shape. Segregation for cotyledon pigment is an additional feature, accompanied by pigmented flush leaves. Allen (1988) found that the R. Putumayo genotypes differed from those of the R. Napo population.

The populations of the stretch of the R. Putumayo and the R. Sucumbios between longitudes 75° W and 77° W contain genotypes whose fruits are elongate with rough and pigmented surfaces and possessing seeds that, while having white or slightly tinged cotyledons, produce plants whose flush leaves are pigmented. Such combinations of characters are reminiscent of those considered to distinguish the genotypes ascribed to the *Criollo* group. However, no connection between the two groups is suggested, until the relationship has been adequately investigated.

Soria (1970) explored the part of the R. Putumayo near the mouth of the R. Cuembi and found that the original (the primitive forest has probably disappeared) cacao population was extensive and wild. Of the trees seen, the fruits were described as globose in shape (near spherical) with 'deep whitish green ridges and a slight bottleneck'.

The Anglo-Colombian Cacao Collecting Expedition stopped lower downstream at Puerto Leguizamo, where environmental conditions apparently differed from those of the region around the R. Sucumbios. For this reason cacao was not found in Colombian territory, although other *Theobroma* were encountered. An excursion was made on the right bank of the R. Putumayo in Peru. This yielded several trees, which, although spontaneous, were of small size. The fruits of these trees differed considerably from those seen at the previous locations. The fruits were small, apparently of a light green colour and had a tendency to be pentagonal in shape. The apex was blunt and there was a marked basal constriction. The surface was fairly smooth and the husk soft with a thin mesocarp. The cotyledon colour was pale. Hence, at this point of the river there was a significant shift in both vegetative and fruit characters from the upriver populations.

Between this location and the frontier with Brazil nothing is known about the cacao populations. It may be expected that considerable alterations occurred to the composition of the forest during the period of exploitation of the rubber resources since it was on the Peruvian side that the 'Putumayo Scandal'[1] took place, with the resulting damage to the environment.

[1] The 'Scandal' was related to the activities of a Peruvian company which, at the beginning of the 20th century, exploited the rubber on a large concession between the R. Napo and the R. Putumayo. The principal concerns were the atrocities committed against the native inhabitants who were forced to work for the company. The other concern was the policy of the company to fell the rubber trees in order to obtain the highest yield of latex in the short term. In this way, the capacity for future production of rubber was destroyed. The policy would have resulted in alterations to the forest flora.

When the R. Putumayo enters Brazil it receives the name 'Içá' (but for purposes of uniformity the name Putumayo will be used for the entire system). The entire course of the river in Brazil (Figs 18 and 21) was worked during the Brazilian Genetic Resources Programme. In the course of this activity several representative samples were obtained of the existing trees at 28 sites, the progenies of which were established in the field.

At the location nearest the frontier with Colombia the genotypes were distinguished by possessing unpigmented flush leaves. On account of this character being common in the areas already described, it may be presumed that this is the principal feature of the entire system. However, among the Brazilian specimens variability for other traits has been observed. Some of these specimens are remarkable for possessing significantly less pigmentation in the staminodes combined with pale guide-lines. In general, the flowers of these specimens are delicate in appearance, similar to those of the flowers from the Javarí system. In some genotypes cotyledons are white. However, details of fruits are lacking in the available information about these genotypes except the indication that the apices of the fruits tend to be mamillate.

The specimens obtained near the mouth of the river possess thick leaves with relatively sparse canopies. The flush leaves are uniformly pigmented. The flowers are white in colour with reflexed sepals and short staminodes. The fruits of these genotypes are elongate but ovate in shape and the surface essentially smooth in texture. Such genotypes perhaps belong to the diversity present on the Solimões.

In the intermediate stretch of the R. Içá, a greater degree of variability is observed. The trees in this area are possibly descendants of hybridization between the genotypes that are found at the extreme ends of the valley. All of the surviving genotypes possess pigmented flush leaves, although the intensity of pigment varies between them. Variability is expressed with regard to tree habit with some genotypes having a 'weeping' canopy. Here the fruits are small, dark green in colour, with an elongate shape and a fairly rough surface, having the closely paired ridges somewhat reminiscent of the fruit types from the middle stretch of the R. Putumayo. The relationship between these types and some of the specimens collected from higher up the river is manifested in the sharp-edged, close-paired ridges and extremely thin husks of the fruits. Plate 16d shows a progeny of one of the families from the intermediate stretch of the river that has fruits with a rectangular shape. Fruits with similar shapes are occasionally found in specimens from the Solimões, otherwise there is a suggestion of characters associated with some members of the *Criollo* group. Another member of the same family has characteristics of the fruits and other parts of the tree that are typically associated with those of the *Criollo* group, including pigmentation on the fruit surface. With regard to the flowers some of these genotypes are unique in that the guide-lines tend to be more prominently pigmented especially with regard to the greater visibility of the central guide-lines.

It should be noted that Barriga *et al.* (undated) refer to specimens from this river possessing fruits with a bluish surface, similar to those of the R. Curuçá. The specimens concerned are not specified and the observations recorded do not confirm the existence of such a characteristic.

The River Caquetá/Japurá drainage system

This is another river the course of which is shared between Colombia and Brazil with different names being used in each country, Caquetá in Colombia and Japurá in Brazil, for the upper and lower courses of the river, respectively. The drainage system is shown in Figs 21 and 18. On account of the fact that this river flows almost parallel to and not far from the main course of the R. Putumayo, there are no major tributaries joining the Caquetá on its right bank. Similar to the main R. Caquetá its two major left-bank tributaries, the R. Orteguaza and the R. Caguán, arise on the eastern flanks of the Andes. While there is no record of cacao on the main R. Caquetá in the piedmont the species appears to have been abundantly distributed on its tributaries. The historical reports on the initiatives that were taken for the exploitation of the cacao resources and references concerning planting of the species for cultivation indicate that some planting material was introduced into the region, especially on the R. Caguán. The native, locally occurring, germplasm may also have been utilized in the formation of new cultivations.

No information has been found regarding the existence of cacao on the R. Caquetá above its confluence with the R. Orteguaza. In the region of its headwaters the R. Caquetá is close to the headwaters of the tributaries of the upper R. Putumayo. Details about the cacao population of the R. Orteguaza are also scanty and not possible to confirm for lack of representative specimens and illustrations. The river was visited by Pound, and his report (Pound, 1938) contains a brief description of his observations on the cacao that was seen to occur fairly frequently in the forests. Unfortunately, the account is impaired by mistakes in the place names and the use of the term 'Criollo de la montagne'. The inappropriate terminology used requires that the descriptions of the cacao seen be translated into a form that makes them intelligible to most readers. On this basis it may be concluded that the common fruit type had an elongate shape with an obtuse apex, the surface being somewhat rough with indications of pigment. Near Florencia (which he called Venezia) the trees of a sub-spontaneous population had elongate fruits with a basal constriction and an attenuate apex. The surface was rough and apparently distinctly pigmented. The cotyledon colour was pale purple. The description suggests that the cross-section of the fruit was pentagonal.

The stretch of the R. Caquetá between the mouths of the R. Orteguaza and R. Caguán was briefly explored by the Anglo-Colombian Cacao Collecting Expedition in 1953 (Baker et al., 1954). Although several native stands of cacao were seen, no material was collected for lack of ripe fruits. The basic colour of the fruits seen was a very light green with a dull crimson flush on the upper part exposed to light. The surface had five ridges in very close pairs, which were prominent, producing a rough appearance. Presumably the fruit shape was elongate with a blunt acute apex. Basically, this type compares with the fruit described by Pound on the R. Orteguaza and it may be concluded that the same variety is distributed throughout the upper valley of the R. Caquetá.

The 1953 expedition had previously explored the R. Caguán from San Vicente to its mouth. Cacao was common in the forest, usually found as slender trees located near the river bank in the upper stretch of the river but, further downstream where the low lying lands are subject to flooding, cacao occurred only on higher ground. It appeared that the cacao along the R. Caguán belonged to a single variety. The fruit of this type (shown in Fig. 22) was described as being relatively small, the shape being elongate and elliptic, with a tendency for the shoulder to be broad. Although some fruits with more pronounced ridges were encountered, more commonly, the surfaces were fairly smooth with shallow furrows and thin husks. The cotyledons of the seeds were dark purple in colour. The distinguishing feature of the R. Caguán type was the intensity of red pigment on the fruit surface although the presence of pigment was determined by exposure.

It would appear that a single cacao variety (with some unimportant variation) inhabits the R. Caquetá valley above the confluence with the R. Caguán.

Virtually nothing is known about the existence of cacao along the R. Caquetá from the mouth of the R. Caguán to the Falls of Araracuara except for indications that the indigenous tribes prepared a drink from the roasted seeds.

On the other hand, the course of the R. Caquetá/Japurá below Araracuara has received significant notice both from travellers' accounts and by germplasm conservation activities. Martius (Spix and Martius, 1828) journeyed along the river and gave the following picture of the situation: 'In certain places,

Fig. 22. Colombia, R. Caquetá drainage system: fruits of a tree on the R. Caguán.

particularly in the humid, low-lying and sultry river banks, cacao and salsaparilla abound in an extraordinary manner… On all sides one passes with pleasure beneath the shady canopies of the first, where the soil is not too swampy…'. He also referred to the existence in the upper reaches of the river of the rich woods of cacao left by the Indians; a suggestion that the species was established by them.

Most of our knowledge about the distribution of cacao below the Falls of Araracuara (which form a natural break in the navigation of the river and, therefore, influencing the characteristics of the cacao populations) comes from a collecting activity (known as the Expedición Botánica Caquetá) carried out in 1984 (Ocampo, 1985; Allen, 1987). This comprised a survey of the banks of the river from Araracuara to the frontier with Brazil. Although data of fruit and seed characters of the specimens collected are given in Allen (1987) little attention was paid to descriptions of the individual parent genotypes, especially with regard to the other parts of the plants, thus leaving an incomplete picture of the variability encountered. In 1992 the author carried out a survey of the material established at the Napo experimental station in Ecuador and the analysis of the observations made provides the basis for the following descriptions.

The fruits of a few representatives from some of the more distinctive families (which are illustrated in Plates 20 and 21) provide a conception of the constitution of the local populations. Since the examples come from individual progenies, possibly resulting from recombination and segregation, they do not necessarily reproduce the characteristics of the parent genotypes. Only a few of the parent trees sampled were reproduced as cloned progenies and the minimal size of most samples of seed progenies rendered their evaluation unreliable. The insignificant quantity of surviving progenies from this collecting activity, although providing some indication of the scale of variability, does not permit a definitive assessment of it to be made. This situation applies particularly to the determination of the spatial distribution of the genotypes and their differences and relationships throughout the stretch of the river that was surveyed.

The specimens from the locations immediately below the Falls of Araracuara appear to belong to a single variety. The usual fruit type is shown in Plate 20a and c. The common features of the specimens are the small fruits with small to medium-sized seeds. The fruits are distinctive in their narrowly elongate shape and the long acute, essentially mamillate, apex that ends in a blunt point. The basic fruit colour is a light green on which a fairly intense red pigmentation may be superimposed. (In this regard this observation differs from that of Allen (1987), who recorded that the fruits from the same trees were not pigmented.) The ridges occur in close pairs that are relatively prominent, the effect producing a pentagonal shape in cross-section. The progenies of the parent trees from these locations vary in respect of their vegetative characters, probably as a result of the conditions under which they are grown. Some of the more outstanding trees that, probably, have developed normally have produced thick trunks. The height of the first whorl of plagiotropic branches ranges from very low to high and spreading branches are common. Some of the progenies produce flowers that are distinctly pigmented in contrast to the white colour that is more frequent. The author is of the opinion that the variety of cacao that occurs in the area below Araracuara is related to that on the parallel stretch of the R. Putumayo.

Similar fruits were encountered at locations further downstream except for a tendency for the appearance of smoother surfaces. The fruits in Plate 20d resemble basically those of the variety encountered on the R. Caguán as described above. However, with regard to the other plant characteristics a marked shift begins to appear.

A gradation continues as the river is descended with trees being larger in size and a corresponding increase, although small, in fruit and seed sizes. There is a tendency for the fruits to be broader in shape than those from locations higher up the river. However, there is one common feature throughout in the stretch of the river above La Pedrera, in that the basic fruit colour is light green and the flush leaves are lightly pigmented.

Between the extremes of the sites situated between Araracuara and La Pedrera there was one where a number of trees were found with larger fruits, for which reason several specimens were collected at this location. Examples of the general type of fruits produced by their progeny are shown in Plate 20c. These fruits were significantly smaller than those harvested from the parent trees but the seed size tended to be larger than that of the other specimens from the same area. The variability observed in the progenies indicates that the parent trees probably were hybrids between the extreme types occurring in the area explored. Possibly, they derive from seed introduced from the lower part of the river and planted at the location.

Most of the specimens from the stretch of the river above La Pedrera have fruits with long acute or attenuate apices or a shape tapering from the middle of the fruits. These apices are frequently pentagonal in cross-section.

Specimens were obtained from six trees on an island a short distance above La Pedrera. The records of the collections indicate that the fruits of these trees were of a large size. However, none of the few surviving progenies of these fruits produces fruits that are anywhere as large as suggested by the collection data from the parent trees. The common features that were possible to observe on the progenies that survived in 1992 were the large flowers and the narrow elongate fruits, the surfaces of which were smooth or slightly rough. It was assumed that the parents of the progenies belonged to a single variety but that this was of a hybrid nature judging from the variability of the other plant traits. From the progeny specimen with the highest number of existing plants one tree stood out because its fruits had a decidedly pentagonal form (Plate 21a), the shape being very narrow elongate and the surface essentially smooth. In the same family it was observed that the mature leaves of some trees had a whitish appearance, the denomination 'albino' being applicable to this character. The presence of a character of this nature may be responsible for the low establishment rate of the progenies and the poor growth of the plants in the field. The 'albino' effect may be associated with the seeds with light purple cotyledons that were recorded at the time of the collection but the proportion of these is unknown.

The Anglo-Colombian Cacao Collecting Expedition (Baker *et al.*, 1954) also explored the R. Caquetá in the vicinity of La Pedrera and the tributary, the R. Apaporis, which joins the R. Caquetá just above the frontier. As described in

Baker *et al.* (1954), the species was found to occur at low frequency along the R. Apaporis and it was considered that the trees were not of spontaneous origin. The population around the La Pedrera location comprised very large and probably old trees. The impression given by the report is that, on the basis of the fruit characters, it was composed of a single variety. The fruits were described as being of a 'pronounced Amelonado type, with a smooth white-base surface'. The husks were very thick and many fruits contained few seeds. The cotyledon colour of these was uniformly dark purple.

Only two trees were sampled for their fruits. Their progeny, of which the fruits of one individual of each are depicted in Plate 21b and c, exhibit characteristics that differ from the description and illustration of the type in Baker *et al.* (1954). The family to which the genotype in Plate 21c belongs is distinguished by having a yellow pulp (like that of some of the *Theobroma* species) and ovule numbers about 65. The specimen in Plate 21b is somewhat different. Both of these types appeared to be fairly homozygous. The fruit illustrated in Plate 21d belonged to a genotype collected in 1984 and some resemblance to the fruit in Plate 21c is evident. The specimens from the later collecting activity indicate that the population nearer the frontier is more heterogeneous with individual genotypes differing from these two families.

The cacao populations along the entire course of the river from La Pedrera to the Japurá 'Delta' and the adjacent stretch of the Solimões have been sampled in one way or another. The river and, especially, the Delta area is very wide and occupied by numerous islands. This situation is propitious for the formation of isolated populations derived from successive plantings, as indicated by Martius (Spix and Martius, 1828). An attempt was made to determine whether varietal differences in relation to sites did exist. At some sites the cacao populations definitely appeared to belong to specific varieties. However, the limited sizes of the samples that represented most locations prevented definitive conclusions from being obtained. As a result any generalizations on the subject could be merely speculative. One associated problem resulted from the manner in which the progenies obtained during the Brazilian programme were established. Although the quantity of samples obtained was large and the diversity indicated by the descriptions of the parent trees was apparently extensive the establishment of the progenies in scattered plots hampered their adequate characterization, rendering the studies incomplete.

The specimens from the stretch of the river immediately below La Pedrera manifest the same attributes of habit that are found in those previously described. These characteristics combine to form trees with thick trunks, the first whorl of plagiotropic branches produced at a low height and these developed in a horizontal position. In some families of progenies the combination of these characters produces a situation in which the planting can be almost impenetrable.

The populations on the Caquetá/Japurá system possess a number of unique features in terms of their vegetative characters. Some of these result from the variability in leaf forms. Plate 4c shows an example of a leaf with a distinctly serrated edge, produced in plants of specimens from below the mouth of the R. Apaporis. In some of these plants the leaves are indented to a greater degree than that shown, the result being a leaf blade with a shape approaching that of the species of *Quercus*. The progenies obtained from self-fertilized fruits of one genotype from the river segregated into four distinct leaf shapes. At the other end of the river several different plant types possessing, among other attributes, very particular leaf types as illustrated in Plates 5a and 5b, occur among the progenies. Variation in the shapes of normal leaf blades is also very evident (Plate 6b).

In common with the varieties on the higher stretches of the river, marked variability is observed in the pigmentation of the flush leaves. The range of colourless to definitely pigmented is exemplified by the family of progenies illustrated in Plate 4a. One notable feature of some of the juvenile plants was the occurrence of strong anthocyanin pigment in the stems, petioles (Plate 4b) and, sometimes, leaf blades in genotypes with green fruits. Previously, pigmentation of this nature had been associated with descendants of red-fruited trees of the *Criollo* group as a pleiotropic effect, as will be discussed in a subsequent section.

The mature leaves present ample variability in terms of shape and size. Reference will be made below to the similarity of the fruits of some specimens to those of the 'SPA' family, described under the R. Napo system. In the same way specimens from the R. Japurá have leaves with the same broad shape and undulating surface as the leaves of the 'SPA' family.

Flower forms, as applied to the entire stretch of the river below Araracuara, vary considerably in virtually all aspects. Most evident is the extreme variation in colours. In some genotypes the flowers are almost colourless with very light yellow ligules while, at the other extreme, intense pigment occurs in all parts.

It was a pity that more specimens were not collected in 1952 since the subsequent collecting programme carried out in 1984 in the area between La Pedrera and the border with Brazil yielded some interesting specimens, the exact characters of which are unfortunately unknown. Several of the fruits collected were of small size. However, the more outstanding genotypes have large fruits of a broad elongate shape with the surface being predominately rough. The similarity with the families collected in 1952 extends to the high seed numbers. The fruits of a few genotypes resemble those of the trees with fruits having outstandingly rough surfaces that occur in the R. Napo system. On the other hand, fruits of small size with smooth surfaces also occur. Some of the progenies produce fruits that are notable on account of having very large seeds. It would seem that a common characteristic of the fruits of the area is the thick and hard husks.

A review of the photographs of fruits of the parent trees, from which specimens were obtained by the Brazilian programme, called attention to the occurrence of fruits with shapes and coloration similar to those of the 'SPA' family described under the R. Napo system (Plate 18d). However, the Caquetá/Japurá fruits with this phenotype illustrated in *Estampa* (Plate) 7D in Barriga *et al.* (undated) may differ in having a combination of an almost cylindrical shape and a slight surface roughness. In the whole of the region concerned fruit shapes otherwise vary considerably. The range of shapes includes fruits that are almost spherical, some of these being of a size larger than the near spherical or short and broad fruits from other populations. At the other extreme the shape is very narrow elongate. The description of at least one of the genotypes with narrow elongate fruits corresponds to that of the fruits from the area below Araracuara (as shown in Plate 20c).

Taking all of the variability indicators together the area included in the Caquetá/Japurá and Solimões contains probably the greatest diversity in the species, at least in phenotypic terms. In this region practically all the known variations can be found or produced through breeding.

So far no collecting or survey has been undertaken of the cacao populations existing on the stretch of the Solimões below the 'Delta' of the R. Japurá. Historical records and travellers' accounts indicate that large cacao populations occurred in some locations, such as on an island named 'Cacao'.

The River Negro Drainage System

In the geographical sense and for the purposes of the survey of the cacao diversity the R. Negro marks the end of the Solimões system. Figure 23 gives an idea of the extent and disposition of the R. Negro drainage system and the connections of its tributaries to other drainage systems. The situation in the river itself is completely different to that of the rivers described so far. The R. Negro, as indicated by the name, is a black water river on account of its course running through geological systems whose ecological conditions are generally unsuitable for cacao. Although the R. Negro system flows almost parallel to that of the R. Caquetá/Japurá system there is nothing to compare the two systems.

A brief visit was paid in 1952, during the Anglo-Colombian Cacao Collecting Expedition, to the R. Guainía, the stream considered to be the headwaters of the R. Negro, and it was reported that the species was absent from the area. In some ways this absence is surprising on account of the proximity to the upper R. Orinoco where the presence of cacao has been known since the end of the 18th century as will be mentioned in the discussion of the Orinoco system.

The R. Negro is considered to start where the R. Guainía is joined by the Casiquiare Canal. This is the waterway that connects the R. Negro and R. Orinoco systems. The 1952 visit to the area around the mouth of the Casiquiare Canal did not produce any positive results regarding the characteristics of the cacao in the area. Sanchez and Jaffé (1989) reported on a few collections made during a Venezuelan collecting initiative in the area of the confluence of the rivers. With regard to these it is convenient to consider the specimens from the Casiquiare Canal as belonging to the R. Negro population.

Only data on fruits and seeds are given in the paper although it is suggested that the characteristics of the leaves and flowers do not differ from those of the specimens from the Upper R. Orinoco collected during the same initiative, which will be described in the appropriate section. The fruits are of a small or intermediate size and have a broad elongate shape, the length:diameter ratio being approximately 1.8–2.0. The fruit surface is slightly rough. The seeds of the specimens from the R. Negro are very small in size, but the specimens from the Casiquiare Canal appear to possess larger seeds. Although no conclusive evidence is provided by the small sample sizes on which the data are based, the apparently large differences in seed size suggest that two genotypes may be present in the area even though they come from

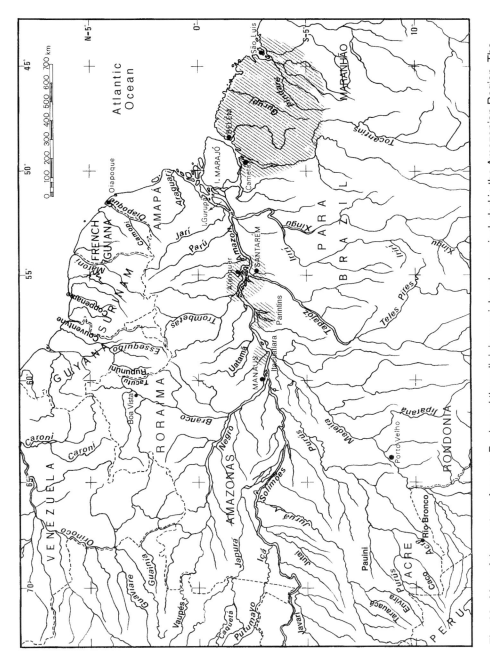

Fig. 23. Map of the eastern Amazon Basin and the adjoining peripheral areas included in the Amazonian Region. The major areas of cultivation are shown in grey, which includes the area defined as the Amazon Extension Zone.

nearby locations. In all specimens the cotyledon colours are described as having a uniform intermediate purple hue.

The R. Vaupés is the major right-bank tributary of the R. Negro. On this river (Figs 21 and 23), also with black waters and flowing through rocky terrain, cacao occurs infrequently as a few trees normally associated with existing or former settlements. Hence, it would be considered that such occurrences are derived from rudimentary cultivation, the trees perhaps descended from the same, unknown, source. The impression was gained that they belonged to a single variety reproduced in successive generations as new plantings were established. No details of the features of this variety are available but memory indicates that the fruits were dark green in colour, with a small to medium size, broader elongate in shape with slightly rough surfaces. This type was also found in a sub-spontaneous form or semi-cultivated in the vicinity of the settlement of Yavareté, having small seeds with dark purple cotyledons.

The R. Papurí, which is the largest right-bank tributary of the Rio Vaupés, provided material of a very different character. A few occurrences of cacao were found in isolated locations and apparently established from seed introduced from unknown sources. One occurrence consisted of many trees whose origins may have been ancient. The separate trunks seen may have been products of seeds germinating in the same locality or through natural vegetative propagation by suckers that had rooted.

Three trees were sampled on the R. Papurí, one of which, cultivated, had larger fruits with large seeds. This would be the specimen attributed to the river as illustrated in Baker *et al.* (1954) but which was not established later. Another specimen turned out to be homozygous and has played a role in research. The fruit (Plate 25c) is exceptional in some respects with the marked basal constriction and darker green colour on an almost smooth surface. The third specimen is also homozygous and proved to be unique in that the first whorl of plagiotropic branches was produced uniformly at about 60 cm from ground level. The cotyledons of the seeds have an intermediate purple pigmentation, for which reason the type was utilized in studies of the inheritance of cotyledon pigmentation.

The occurrence of such unique, homozygous and self-compatible genotypes in such a small area suggests that that they originated from parent populations in diverse locations and possessing the same genotypes. These populations have not been encountered so far and it is interesting to speculate on the locations in which they may be found. A connection between the R. Papurí and the R. Apaporis is known to exist and the possibility that these cacao varieties may occur on the upper reaches of the latter should be considered. It should be emphasized that these populations existed on the Colombian side of the river; there is no information on any occurrences on the Brazilian side.

From the mouth of the R. Papurí to its confluence with the R. Amazon, the R. Negro, as previously stated, is not considered to be a region where cacao occurs spontaneously. During the phase of colonization of the river from about the mid-18th century, which had the objective of discouraging possible Spanish invasion and to secure Portuguese sovereignty, cacao was one of the crops planted near the settlements that were established. These plantations, in general, were not successful. It is possible that some remnants of them may survive but the river has not been explored with regard to the existence of cacao and the genotypes that may be represented by the survivors. No information is available concerning the source of the planting materials used but it may be surmised that they had a narrow genetic variability, probably of a type with fruits of a short and broad shape.

The proximity of the R. Negro and R. Caquetá/Japurá systems would suggest the possibility of transfer of cacao germplasm from the latter along the connecting tributaries of these rivers, which are known to have been transit routes in the migrations of the indigenous tribes. However, the lack of any investigations on these tributaries has prevented this possibility from being determined.

The River Branco Drainage System

The R. Branco is the largest left-bank tributary of the R. Negro and, as the name suggests, is a white-water river. Consequently, the cacao situation on the R. Branco (the position of which may be ascertained from the map in Fig. 23) represents a marked contrast to that presumed to apply to the R. Negro. Since the earliest times of exploration and colonization the magnitude of the cacao populations along the river, especially on the lower reaches above its mouths, has been the subject of much admiration. Unfortunately, the nature of the genetic diversity along the river is scarcely known. The only investigation carried out on the river took place in 1976 during a reconnaissance of the valley. The period of the survey did not coincide with the period of fruit production and this prevented the composition of the population from being determined. A few specimens survive from those collected during the survey and all possess sufficiently unique characteristics to conclude that the genotypes belong to populations that are distinct from those encountered in other regions.

The specimens involved are depicted in Plate 22. The fruits in Plate 22a and e were obtained from the only two trees growing in the same location, both probably descendants of seed from two distinct sources. A large savannah separates the location from the main occurrences of cacao in the valley to the south from where, presumably, the seed originated. The fruit shown in Plate 22e, although having a short and broad shape, was remarkable on account of the suggestion of an almost pentagonal appearance. The flowers of this genotype are small and white with partially reflexed sepals. The tree whose fruit is represented in Plate 22a was isolated from that described, and well developed with a thick trunk and an extensive spreading canopy. Although some cacao specialists may consider the type to be a 'common Amelonado' and, therefore, homozygous, the constitution of the tree is heterozygous. One factor that it carries in the heterozygous state is the anthocyanin inhibitor gene. Some of its progenies (Plate 22b) are similar to the parent and also possess the anthocyanin inhibitor gene.

A third specimen consists of progenies from a young, planted and isolated tree that was reported to be a descendant of a fruit from the naturally occurring cacao populations along the river. The fruit of the tree was outstanding in the fact that its shape was roughly pentagonal. The progenies (two of which are illustrated in Plate 22c and d) exhibit a certain degree of variability, suggesting that their parent was heterozygous. They generally have fairly close-paired ridges and surfaces of intermediate roughness. The flowers of the parent tree and those of the progenies are distinctive, being of small size and well pigmented.

Several other occurrences of cacao were sampled but specimens of these did not survive. These occurrences include stands of more than 1000 trees. The few representatives of the valley's population that were seen can hardly be considered to provide a comprehensive view of the variability. The observations recorded on the few trees examined frequently refer to anthocyanin pigmentation at the base of the ligule that may be considered to be a unique character of the population. The tendency for the fruits to have close-paired ridges and a short acute apex with a blunt point also appears to be typical.

Cacao is reported to occur on the R. Jauaperi, a tributary of the R. Negro, closely adjacent to and southeast of the R. Branco, but no details are available. As a consequence of the presumed absence of spontaneous occurrences of cacao on the R. Negro, the valley of the R. Branco and, possibly, the other tributaries that join it on its left bank are isolated from the mainstream cacao variability. Therefore, it may be expected that distinct varieties have evolved in them or that these are related to the R. Orinoco populations through the connections by the headwaters of the two systems. The R. Branco, through its tributary, the R. Takutú, may be considered to be the gateway to the Guianas.

The River Madeira drainage system

The system, shown in its entirety in Fig. 18, is one of the largest and most important of the tributaries of the R. Amazon in terms of cacao variability. The name, R. Madeira, is given to the stretch of the river that starts approximately at the present-day frontier between Bolivia and Brazil. Accordingly, the name is applied only within Brazil, where the river flows in a northerly direction until it joins the R. Amazon on its right bank through several mouths. The Madeira is formed by a complex of several rivers that converge to form the R. Mamoré, before the start of the R. Madeira. Two of these rivers form part of the division between Bolivia and Brazil. In order to obtain an understanding of its cacao diversity it is important to perceive the pre-Madeira section of the system as being a vast inverted funnel as the various tributaries converge in the flood plain of the lower part of the system. The longest of the rivers that form the R. Madeira is the R. Bení (Fig. 18). The R. Bení has its source at about 15° S and until it joins the R. Mamoré has a length of 1600 km. The length of the R. Madeira and part of the R. Mamoré is estimated at 3500 km; thus the length of the system is over 5000 km.

The existence of cacao along the upper R. Bení has been known for a long time, especially in the valleys in the 'Yungas', as the region of the foothills of the Andes is called. However, it is not known if these cacao populations are native to the region. Since the species occurs in isolated valleys it is possible that variations in types may exist. Although these populations have been subjected to superficial collecting nothing is known about the specific characteristics of the trees. The only description of the cacao of the region may be obtained from Soria (1965), who visited the area near the town of Santa Ana. The location was stated to be east of the town and, therefore, on the right bank of the river further from the Andean foothills. This area is situated at about 15° 31' S, perhaps the southern extremity of the distribution range of the species in the Amazon Basin.

The description provides some significant information and is worth reproducing at this point (with some adjustment to the original translation in order to conform to the terminology used in this volume):

> Some farmers of the region stated that they find native cacao plants there, but it is more likely that these plants are survivors of old plantations and not wild cacao…
> However, the information given by the agronomists who work in the lower, typically tropical region of 'Bajo (lower) Bení', near Trinidad, and on the banks of the Mamoré River, indicates that in these areas there are forests of native cacao. Apparently, the cacao cultivated in the 'Alto (upper) Bení' is exactly like that of the native cacao found in the 'Bajo Bení'.
>   On the basis of the (few) fruits observed, the local variety can be classified as an *Amazonian amelonado* with very small fruits…of a dark green colour when young, elongate, with a short apex, slightly rough, (possessing) 10 superficial ridges and a thin husk. The shape of the flower bud varies from slightly round to elongate, but the petal ligule is always wide, orangey-red in colour, the base of the style being red and the stamen filaments unpigmented. These characteristics are general in all the populations. The cotyledon colour is purple and the seed size very small.

Towards the east of the 'Yungas' the plains known as the 'Llanos de los Moxos' were a significant provider of cacao for consumption in other parts of Bolivia and elsewhere. The Jesuit missionaries are considered to have been responsible for creating the plantations, using seed introduced from near Brazil. The plantations were abandoned when the Jesuits were expelled and the trees that are found in the area in what may appear to be virgin forest would be the survivors of the cultivated fields and not native to the region, as confirmed by Soria's (1965) account above. No descriptions of these trees have been encountered.

Cacao has been found in most of the valleys of tributaries of the R. Bení as the river is descended. The most cited area is that near the town of Apolo (Apolobamba), from which herbarium specimens were collected in the early 20th century (Otto Buchtien, No. 187, 1907). The distance between the R. Bení and the foothills of the Andes increases as the river's course progresses. The most northerly of the tributaries of the R. Bení is the R. Madre de Dios (Fig. 18) and this also seems to be the largest and most important stream in the western segment of the drainage system.

The R. Madre de Dios and some of its principal tributaries have their sources in Peru within the Province of Caravaya (Fig. 5). Therefore, their headwaters are located near those of the drainage systems of the R. Purús, R. Juruá and Urubamba/Vilcanota. From the point of view of the distribution of cacao diversity this fact is important since the area involved could be the genesis of common variability factors in the valleys through gene flow by one means or another.

According to the descriptions of the area the cultivation, or exploitation of the wild stands, of cacao started soon after the conquest of Peru, centred on the town of Paucartambo. This was the only area of the land of the 'Chunchos' from which ancient descriptions are available. However, constant attacks by the Indians resulted in the abandonment of the early settlements. For this reason, a similar situation arises as that described for the Upper R. Bení, in which it is now impossible to determine whether occurrences of cacao are native to these valleys or whether they are sub-spontaneous remnants of former cultivation. However, it is assumed that the planting material used to establish cultivated fields would have been obtained from the local natural varieties. Taking into consideration the small area of the Province of Caravaya the prospect arises of human factors being involved in the transfer of genetic material between the valleys.

The upper half of the valley of the R. Madre de Dios lies within the borders of Peru. Extensive botanical collecting has taken place in this area during recent years and, judging from the number of cacao specimens available, the cacao population appears to be extensive on this river and its tributary, the R. Tambopata. Without the opportunity to examine the specimens no descriptions can be made since the information accompanying the specimens provides few clues as to the nature of the individual trees. With regard to the course of the river in Bolivia the situation is quite different since no information is available from that country.

The other principal waterway forming the R. Madeira is the R. Mamoré, which drains the flood plain to the east of the R. Bení. The R. Mamoré basin probably drains a larger area and differs in character from the latter. Since the R. Mamoré drains the Llanos de los Moxos the cacao encountered along its middle course probably belongs to the variety cultivated by the Jesuits. The R. Chaparé is a left-bank tributary of the R. Mamoré and is perhaps the southernmost, at 16° S, of the rivers in which cacao occurs in the valley. The cultivation of cacao on the R. Chaparé may be due to the fact that this river was the principal access of the commerce of cacao from the Mamoré through Cochabamba. Further north is the town of Trinidad (due east of Apolo), a centre for cacao, as referred to in Soria (1965) and it may be assumed that the cacao varieties in the region conform to the description given in this report. However, an evaluation of the population would be necessary to determine if other variability exists that is not generally noticed.

Further north, prior to the confluence of the R. Mamoré and the R. Bení to form the R. Madeira, the former is joined by a major tributary that forms the frontier between Bolivia and Brazil. This has the name R. Guaporé in Brazil and R. Iténez in Bolivia (see map in Fig. 18). One tributary of the R. Guaporé in Brazil, called the R. Branco, according to personal information, has extensive stands of cacao in the valley and a significant amount of cocoa is obtained for sale in Bolivia. At approximately 61° W this valley would be the most easterly point of the distribution of the species in the Madeira valley. Most reports indicate that cacao is very common in the R. Guaporé valley and was cultivated in the missions during the 17th century.

On the whole, the R. Mamoré appeared to be one of the areas of greatest abundance of cacao as reported by Herndon and Gibbon (1854) and was the basis of a considerable trade. One of the features of the herbarium specimens of cacao from the upper Madeira basin that have been examined seems to be the occurrence of large narrow leaves (Williams, R.S., No. 806, 1901). Herndon and Gibbon (1854) wrote 'The largest cacao leaf that I could find measured one foot six and a half inches in length and five inches and three-quarters in breadth.' (This is a very narrow elongate shape.)

After the mouth of the R. Guaporé, the R. Mamoré continues to form the Brazil–Bolivia frontier until its meeting with the R. Bení. Cacao is also reported to be abundant in this area. Specimens were obtained from a small number of trees in an unspecified locality in Brazil. No information accompanied the specimens regarding the circumstances in the locality or about the characteristics of the parent trees. Hence, it is not possible to determine the status of the population sampled but it is assumed that the trees belonged to a variety that could be considered to be native to the area.

The few observations that could be made on the four families of progenies that were established indicate that there were large differences in growth rates and vegetative characters, which would be more as a response to unsatisfactory growing conditions and management than to intrinsic variability. As a result an unbalanced picture of their characteristics is available, which inhibits a true account of their nature to be obtained. The trees themselves differed from the neighbouring varieties with regard to their general appearance; normally canopies were relatively thin, even in the vigorous progenies. There appears to be little variability with regard to fruit characters, the fruits being of small size. Figure 24 shows the fruits typical of this sample. They have a darker green colour, a broad elongate shape with parallel sides and blunt apices. The furrows are shallow in close pairs with an almost smooth surface. The fruits are also similar with regard to the very thin husks and small seeds. The most common feature of all progenies concerns the flowers, which are basically white with the petals having ligules of a deep yellow colour with a broad shape, the edges being curved upwards into a spoon-shape. In some respects these progenies do not differ from the types described by Soria (1965).

To the west of the course of the R. Madeira, as formed by the meeting of the Bení and Mamoré rivers, there are several tributaries flowing from west to east almost parallel with and close to each other. It is reported that in the valley of at least one of these rivers, the R. Ortón, cacao was very abundant. The most northerly of these streams is the R. Abuñá, which also forms the limits of Bolivia and Brazil. Cacao does not appear to be very common along this river but a few specimens were obtained in Brazil. Because of the proximity to the R. Acre tributary of the R. Purús there is a tendency for the R. Abuñá types to

Fig. 24. Brazil, R. Madeira drainage system: fruit type of the small sample from a population on the right bank of the R. Guaporé.

possess some of the characters of the Purús system diversity, particularly the specific type of stamen pigment. The three specimens collected in 1967 along the watershed of the R. Acre and two tributaries of the R. Madeira, the R. Abunã and R. Iquiri, from three apparently isolated sites, differ significantly in almost all respects. The parent trees were some of the largest ever encountered, indicating that their establishment (natural or artificial) probably occurred at a very early date.

With regard to the R. Madeira itself, the upper course is characterized by several rapids and falls that have discouraged exploration although information is available concerning the presence of cacao along this stretch of the river.

The River Jí-Paraná and River Jamarí Drainage Systems

The most important of the tributaries of the R. Madeira, with regard to cacao is the R. Jí-Paraná, which joins the main river on its right bank at about 8° S latitude. As shown in Fig. 18, the R. Jamarí is another, shorter tributary of the R. Madeira, the valley of which runs almost parallel to the middle course of the R. Jí-Paraná and is located a short distance to the south. Because of the circumstances in which cacao is distributed in the area and the conservation activities that have been undertaken, it is convenient to deal with both valleys as a single entity. The cacao populations in these drainage systems were known from the period of extraction of the produce of the natural stands. At this time the cacao that occurred in the R. Jamarí valley was considered to be the most appreciated in terms of its quality, a fact that gives the population a special significance with regard to its place in the diversity of the species.

The sources of the R. Jí-Paraná lie in the open lands of the Chapada de Parecís, located to the southeast, which separate it from the R. Guaporé. There is a distinct divide between the open plains of the Parecís and the forests of the R. Jí-Paraná. The divide is located at about 13° S, 62° W and, therefore, marks the extremity of the distribution range of cacao. The other extremity of the principal range of natural cacao distribution in the valley may be taken to occur at about 10° S.

The area of cacao considered in this case possesses some important and distinguishing characteristics. One is that the area is somewhat isolated from the mainstream cacao diversity, except for the connection with the R. Madeira and the sources of other rivers that flow northwards, also into the R. Madeira or into the R. Tapajós, about which nothing is known in this regard. The other feature of the relatively small area concerned is that the topography is somewhat hilly. This is in sharp contrast to the other main drainage systems that have been described previously. The area contains patches of soils of higher fertility and, consequently, there are no limitations as to the type of terrain on which cacao can be established. As a result, cacao can be found at distances from watercourses and even near the summits of the hills.

Consequently, the distribution of cacao is almost continuous. The species can also occur in agglomerations of trees of high density in areas occupying several hectares. The general conditions prevailing in the drainage system are propitious for an even spread of genetic materials. Owing to the pressure on the land of the valley for colonization and the threat to the perpetuation of the cacao population considerable effort has been expended over several years to conserve as much as possible of the diversity that circumstances have permitted. Coverage has been given to collecting specimens of trees in a wide range of localities and environmental conditions.

The sample obtained so far includes some 500 individual parent trees conserved as their clones or seed progenies. This large sample cannot be described succinctly in a suitable way. In addition to the size and diversity of the population, the task of description is hampered by the fact that the specimens are located in two distinct stations with no recent results of equivalent characterization that embrace the entire range of specimens being available. In addition, no studies of inheritance or utilization of the germplasm have been carried out. One of the questions that would arise concerns the possibility of the existence of local variability. Some evidence of location differences has been observed in terms of agronomic traits but there appear to be no fundamental divisions in the range of fruit characters.

Within a certain range of traits the cacao population of the river systems involved exhibits an appreciable degree of variability. The fruit types shown in Plate 24 are only a few examples of what may be considered to represent the diversity of the region. On first contact the Jí-Paraná/Jamarí population appears to be distinct on account of the elongate leaves that are dark green with a smooth and glossy appearance. How well this description fits the whole population is as yet unclear. Flower types show variability; among the frequent characters that occur in the population are deep yellow ligules with significant anthocyanin pigment producing a bronze-like effect.

Although fruits of near-spherical shape occur, sometimes of small size, these are uncommon, but often inherited as a dominant character. The most usual shape is narrower elongate with or without surface roughness. The most outstanding fruit character is that of a gloss on the fruit surface, a character that also occurs in some parts of the R. Purús system. Surface pigment does occur as in the example in Plate 24c. The lack of adequate characterization of the specimens does not allow a definitive judgement to be made on the subject but, considering the ranges of seed numbers obtained in the naturally fertilized fruits sampled, the ovule numbers do not surpass 50. If higher numbers occur these are likely to be in rare instances. If low ovule numbers are more common this character would distinguish the cacao of the region from the surrounding populations in which high numbers occur.

Although cacao was, and perhaps still is, common along the length of the R. Madeira proper from the rapids to its mouths, the populations have not been subjected to collecting and characterization. The main river was a very important source of the product during the extraction era and details are available of the exact locations of the various places where the species occurred. The Jesuits established one of their most important missions just above the mouths of the R. Madeira and the cacao harvested in the area was an important source of revenue for them. It may be considered that this area was one of the sources of planting material for the cacao farms cultivated by the Order nearer the mouths of Amazon and Tocantins rivers. Like the main river the lower tributaries of the R. Madeira, such as the R. Aripuaná system, have not yet been sampled for the diversity of any cacao that exists on them. The author has not found any information about the presence of cacao on these rivers.

The River Tapajós drainage system (Fig. 23)

The Tapajós River is the most easterly of the great tributaries of the R. Amazon that flow more or less parallel to each other on the south side of the basin. Although various accounts of the earliest travellers along the river in the 18th century refer to the occurrence of cacao along the tributaries of the upper Tapajós, these have not been confirmed with regard to the present situation. The river itself is mainly a black water river broken by several natural barriers, because of which the existence of substantial amounts of natural stands of cacao would not be expected. Various attempts at cultivation were made in the past so some sub-spontaneous occurrences may be encountered. The scarcity of cacao in the valley would be evident from the information given by Herndon and Gibbon (1854) to the effect that the Mundurucú tribe, who inhabited parts of the valley, made an annual journey to the R. Madeira to collect cacao during the harvest season on that river. It is more than likely that any natural stands of cacao along some of the tributaries have vanished, as the forest was destroyed in recent decades as a result of gold prospecting.

The only specimens of what may be considered to be naturally occurring cacao in the system are derived from a few trees on the R. Teles Pires in the vicinity of the town of Alta Floresta (State of Mato Grosso) which were sampled when the area was being opened up to agricultural development. From the size of the cacao trees it would seem that they had existed for a long time. The four specimens that represent the population show little variability and appear to belong to the same genotype. The importance of this population is that it is located at what may be the extremity of the distribution range of the Tapajós system. However, nothing is known about the Juruena-Arinos valley, the other stream that joins the R. Teles Pires to form the R. Tapajós. In addition, since the entire length of the R. Teles Pires has not been explored, it is improper to suggest that this isolated variety is representative of the diversity in the valley.

The principal distinguishing feature of this variety concerns the leaf, which is of a type not yet found elsewhere, the nearest populations with leaf types having similar features are those of the R. Jí-Paraná system. The leaves of the upper Tapajós variety are narrow elongate, very dark green in colour with the leaf blades being thick and hard with a smooth, glossy surface. These are shown in Fig. 25. Although the parent trees were large and tall their progenies are slow-growing and the first whorl of plagiotropic branches is formed at a low height. The comparative insignificance of the plants has resulted in little attention being given to them. The fruit is short and broad in shape with a rounded apex, the surface being fairly smooth with low ridges.

In the middle Tapajós valley a few specimens derive from a plantation, probably established in recent years, at some distance from the right bank of the river. These specimens, perhaps, are descendants of a single fruit, as they possess some similar traits and may not represent any native population connected with the system. The trees are similar although possessing some variability. The fruit is normally broad and slightly elongate in shape with the

Fig. 25. Brazil, R. Tapajós drainage system: the distinctive leaves of the population on the R. Teles Pires.

short acute apex that is common in the lands along the stretch of the R. Amazon in the vicinity of the mouth of the R. Tapajós.

Further downstream the R. Tapajós is joined on the left bank by a short tributary, the R. Cupari. The cacao population could be at least 200 years old as its existence had been reported in the first half of the 19th century. A number of specimens were obtained from this population but no opportunity has been provided to characterize the families that were obtained. The fruits collected from the parent trees were usually elongate in shape with intermediate surface roughness. The important characteristic noted concerned the frequent occurrence of seed numbers above 60, combined with fairly large seeds.

Several occurrences of cacao in the stretch of the river above its mouth are mentioned in documents dating from the earliest period of European occupation. It is not clear whether these stands were native or established by man and, since they have not been studied, no information can be given about them.

One interesting case concerns a small area at the top of the escarpment overlooking the right bank of the R. Tapajós near its mouth. The site was previously occupied by a tribe of Indians; the evidence for this is shown by the soil being of the type known as 'terra preta do Indio' and by the quantity of artefacts that occur. These facts suggest that the Indians were responsible for the establishment of the cacao plantation. Among the trees on the original site three were selected on the basis of their reactions to diseases. Each of these genotypes is unique in the spectrum of diversity. The most outstanding genotype is vigorous, with small leaves produced on short branches. Its fruit is very short but broad, with a peculiar apex. Another genotype is interesting on account of the delicate flowers of a pinkish coloration. All three genotypes have been used as parents in breeding programmes. The first type mentioned is self-compatible and its progeny indicate that it is homozygous for its distinctive characteristics.

Progenies derived from natural pollination of some of the trees at the site were established there. Several selections were made in this population on the basis of absence of symptoms of Witches' broom disease. Most of these selections belong to a similar type, the fruit shape being short and very broad, with a very short acute apex. Another selection is outstanding on account of its greater vigour and large broad leaves of thick texture. The fruit of this genotype is broad and elongate, with a similar appearance to the fruit from the R. Caquetá (illustrated in Plate 21d). Figure 26 depicts trees of the genotype, as the clone 'CAB 40'.

Before leaving the R. Tapajós system it is pertinent to refer to one variety probably selected on a plantation situated either near the mouth of the river or on the R. Amazon just below the confluence of the rivers, the exact location being unrecorded. The variety was designated 'Mocorongo' (a name applied to the inhabitants of the area) and is represented in germplasm collections by two cloned genotypes that are similar in all respects. Although the genotypes were found to be self-incompatible they possess a superior productive capacity and adaptability to unfavourable conditions. The fruits are dark green in colour and

Fig. 26. Brazil, near R. Tapajós: trees of the clone 'CAB 40'.

broad elongate in shape with an obtuse apex. The fruit shape tends to be dominant in crosses with narrow elongate fruits, the progenies of these manifesting a small degree of surface pigment, not evident in the parents. The husk of the fruit is very woody, so much so that there is a tendency for germination to take place when the fruits are apparently just ripe. Both genotypes carry the anthocyanin inhibitor gene in the heterozygous condition. In spite of the apparent self-incompatibility the occurrence of seeds with white cotyledons from self-fertilization is common. From these it has been possible to establish lines that are homozygous for the recessive allele and that are, apparently, homozygous for other morphological traits.

The River Amazon drainage system

In dividing the Amazon Basin into segments the original name given to the stretch of the river from the confluence of the Solimões with the Negro and Madeira rivers to its mouth applies in this case. In some circumstances the terms 'Middle Amazon' and 'Lower Amazon' are applied to the respective portions of the river as geographical entities. However, the terms are applied indiscriminately by persons not familiar with the hydrology of the river, in a variety of meanings for indicating regions of cacao distribution. Since these applications do not exactly define the areas involved they result in considerable confusion and inaccuracy concerning the specific origins of the genotypes to which the term is used.

The section of the river to which the name Amazon is given differs from the previously described sections of the river in several respects, both with regard to the geographical circumstances as well as in relation to the distribution of the cacao populations. The course of the R. Amazon and the various districts where the species is cultivated are shown in the map of the central and eastern sectors of the basin in Fig. 23, the districts being indicated by the shaded areas.

One aspect that must be considered when describing the cacao populations of the region is the fact that it was in this section that the cultivation of cacao started in the years immediately following the 'discovery' of the tree in the Amazon Basin in the 17th century. For this reason it is nearly always impossible to determine the spontaneous status of an individual cacao tree found at a particular location.

The best manner to explain the situation that occurs is to reproduce the words of Jacques Huber when he described the cacao of the region at the beginning of the 20th century. During this period the cacao enterprises would have been at their maximum development and not affected by the rubber boom. Huber (1906a) wrote:

> It is generally admitted that *Theobroma cacao* L. grows spontaneously in the Amazon region, but up to the present day, one scarcely knows the approximate area of its distribution nor the special conditions under which it is found. With regard to the first question it is noticeable that it has never been found in an indubitably native state neither to the north nor south of the Amazon downstream from the mouth of the Tapajós. It is true that the 18th century historical documents mention the cacao tree as growing spontaneously in the islands at the mouth of the Amazon, notably on the island of Gurupá, and that Martius indicated that cacao was present in the riverain forests of the Amazon near Santarém. However, as important areas of cacao have been cultivated for a long period along the length of the main river from Santarém to the mouth of the Rio Negro, at the present time it is rather difficult to decide whether the occurrences are spontaneous or sub-spontaneous.
>
> However, it is probably not impossible that the primitive 'cultivations' of this zone may have their origin in the natural 'cacauais' and that, at least a part of the cacao trees presently cultivated are the descendants of spontaneous trees found in these parts by the first agricultural colonists.
>
> Whereas, along the lower course of the Amazon the native condition of the cacao tree is rather doubtful, it is another matter with regard to the middle and upper reaches of the river, as well as to the large tributaries: Madeira, Purús, Juruá, Ucayali to the south and the Japurá and others to the north. In these rivers, at least in their upper reaches, the cacao tree has never been cultivated by the whites and, since the Indians of the Amazon do not attach any importance to cacao, it could be considered that no attempt was made to cultivate it on their part.

Where suitable conditions exist for the development of the plants, cacao can be found along the length of the river in populations of a wide range of sizes. This also applies to several of the tributaries (of which the River Tapajós has been treated separately) and sometimes cacao trees may be encountered in the forest in locations at some distance from watercourses. While there are a few areas where cultivation definitely has been practised, and still is, based on planting material from the region it is difficult to determine the status of other occurrences. Some of these could be ascribed to a sub-spontaneous state as survivors from previous attempts at cultivation. It is thought that some of the individuals or groups of individuals may derive from chance sowing of seeds from fruits harvested from cultivated trees that were carried by rubber and balata tappers and Brazil nut and Tonka bean collectors. (Tonka bean is the name given to the seeds of *Dipteryx odorata*, which have been collected for a long time for the extraction of an essential oil used in perfumery.) In these cases it should be possible to relate the genotypes to those forming the plantations in the same area, provided that the genetic constitution of these is adequately known. Because of the possibility that some of the germplasm occurring in them may be derived from previous spontaneous occurrences in the same districts the respective cultivated populations along the banks of the R. Amazon will be discussed in the following sections as components of the ancient diversity.

At the present time a few of the areas in which cacao has been traditionally cultivated have been subjected to examination and collecting for the purposes of germplasm conservation. It must be emphasized that this account deals exclusively with the populations that have been constituted from varieties native to the Amazon Region.

The island of Careiro

The island is situated in the R. Amazon just below its confluence with the R. Negro. Because of its situation, as is the case of many other similar islands, the island of Careiro has suffered during its history from constant erosion and rebuilding and, consequently, has undergone significant changes in its shape. Therefore, this situation, combined with the effects of seasonal flooding, has resulted in the area cultivated at any given time being unstable. It is probable that, as would apply to the surrounding areas on river banks, cacao is not native to the island. The planting of cacao probably would have commenced at the end of the 18th century. The island was an important producer during the middle of the 19th century according to travellers' descriptions. Some of the cacao areas referred to in the reports have since disappeared or are considerably depleted compared to their situation during the 19th century. At the present time the remnants of these plantations exist usually as scattered trees in a variety of conditions and are not exploited commercially.

The planting material that formed the original plantings probably was introduced from nearby regions such as the Solimões, the delta of the R. Japurá or the mouths of the R. Madeira. Some of the trees may derive from a programme of distribution of progenies from 'selected' individuals carried out during the 1930s, the details of which in terms of varieties and destinations have not been obtained.

Collecting activities for germplasm conservation have been conducted on several occasions. These resulted in a total of 17 individual genotypes established in gene banks. The interesting aspect of the specimens obtained is that they represent the same number of completely different genotypes with few phenotypic similarities among them. Most of the clones resulting from the collections have been subjected to use in improvement programmes and, accordingly, more information regarding their genetic composition and inheritance is available than is the case with much of the germplasm from the Amazon Basin. These specimens from Careiro suggest that the island contains what is perhaps one of the greatest concentrations of diversity to be found in a small area. In the probable event that these genotypes are descended from other Amazonian populations they would provide a basis for typifying the populations concerned.

The fruits of nine of these genotypes are shown in Plate 23. Several of the genotypes possess high ovule numbers the most important of which is 'MA 15' (Plate 23b) with 65 ovules per ovary. Although the fruits of 'MA 12' and 'MA 15' appear to be similar the plants are distinct from each other. Fruits similar to those of 'MA 15' occur among the specimens from the R. Japurá. 'CA 2' is distinguished by having large seeds, in which respect it is similar to some of the selected genotypes from the Circum-Caribbean Region. The clone 'CA 2' also differs on account of its relatively inferior vegetative development and sparse canopy. 'CA 5' and 'CA 6' develop into plants of low height. While this trait is constant in their progenies a small degree of heterozygosity is displayed with regard to the fruit characters.

'CA 3' is one of the more interesting genotypes. Although it is effectively self-compatible, fruits resulting from self-fertilization consistently contain few seeds indicating, as has been found in other genotypes from various origins, that some mechanism operates that inhibits the normal development of the seeds. In addition, the plants that develop are dwarfed. Plate 31b compares the dwarf progenies (on the left) with the normal plants that develop from cross-fertilization. In time, the dwarf plants develop normally and when adults show normal fruiting behaviour. The inbred plants inherit the narrow elongate leaf of the parent, which is one of its distinguishing characteristics.

Another outstanding genotype is 'CA 4' with its strange, very narrow elongate fruit. The elongate shape is transmitted to the progenies obtained from self-fertilization as shown in Plate 23d, but segregation for other fruit characteristics evidently occurs. The dominance of the elongate fruit character also appears in hybrid progeny of the clone and these are remarkable on account of their enhanced vigour. The selection 'CAB 45' differs from all the others in its almost spherical fruit, the surface of which is essentially smooth, the colour being a lighter green with noticeable pigmentation.

'MA 16' was collected because of an unusual characteristic that has not been encountered elsewhere. This trait concerns the transformation of the stamens into leaf-like organs that are markedly pigmented. As a result the genotype is male-sterile. The ovary is functional and fruits develop normally but as they usually contain a lower number of seeds (perhaps half of the ovules develop into normal seeds) the resulting elongate fruit shape is similar to what is considered typical of fruits of the *Criollo* group.

The Itacoatiara region

It appears that at some time in the early 18th century attempts at cultivating cacao were made in the area of the town of Silves, established by the early colonists of the Amazon Basin. This area is situated on the left bank of the R. Amazon some distance below the confluence with the R. Negro. Although the extent of the existence of cacao in the vicinity of Silves is unknown, the area near the neighbouring (upriver) town of Itacoatiara had developed a certain importance as a producer of cacao. The origins of the cacao varieties existing in the area have not been determined but, in view of its position near the mouths of the R. Madeira, it may be expected that some seed may have been obtained from this river. The author has no personal knowledge of the area. However, the progenies derived from a large sample of seeds from fruits harvested at random from unselected trees provide some indications as to the genetic constitution of the population.

A few selections were made in this sample of progenies. Their fruits ranged in shape from almost spherical with a dark green colour to elongate and broad of a large size and with a somewhat rough surface. An example of the fruits with the narrower elongate shape is that of the clone 'CAB 56', shown in Fig. 27. Although this fruit is similar to that shown in Plate 21d it has smaller seeds.

One of the last collecting activities carried out in Brazil obtained some specimens from the original fields. To date no observations have been available from these samples since they were established at a location that has not been visited since they were planted. The data from the parent trees sampled are summarized in Almeida *et al.* (1995). These observations reinforce the impression regarding the extensive diversity contained in the population in terms of quantitative fruit and seed characters that was apparent in the first sample. However, no information is provided about the qualitative traits of the trees and their products.

The planting material used for establishing the cacao farms in the beginning of the colonization of the R. Negro perhaps derived from these plantations.

Fig. 27. Brazil, R. Amazon: fruits of one of the genotypes occurring in the Itacoatiara region ($\times 0.27$).

Further east from Itacoatiara is the R. Uatamã. The cacao population appeared to be rather sparse and a few specimens were obtained on the upper valley of this river. However, the complete characterization of these specimens has been impeded on account of the low vigour and adaptability of the plants. Most of the fruit characters conform to those of the types that are considered to be typical of the Amazonian Region cacaos but they differ by possessing high ovule numbers. The flowers seem to be distinguished by possessing very red buds due to intense pigmentation on the outer surface of the sepals while the remaining parts of the flowers are not pigmented. The extent of the distribution of cacao on the R. Uatamã, especially near its junction with the R. Amazon, has not been determined.

The Parintins region (sub-Madeira)

As described in the account of the R. Madeira drainage system, this river joins the R. Amazon through several channels, forming in effect a wide delta shown in Fig. 23. Below the easternmost channels on the right bank of the R. Amazon there is a complex network of waterways and islands formed by the numerous small tributaries of the latter. Cacao occurs throughout this region. Possibly, much of the diversity encountered resulted from the flow of genes from the R. Madeira varieties. Therefore, it could be considered that the populations occurring below its mouth are the same as those in the river valley itself and were at one time established naturally. However, in this area several patches of cultivation occur in association with ancient and more recent villages.

One of the towns that has had a long association with the cultivation of cacao is that of Parintins, situated at the mouth of the easternmost Madeira channel. Cacao farming would have been established in the area from at least the beginning of the 19th century. A large sample of the variability existing in these farms was collected during the Brazilian conservation programme. The number of parent trees from which specimens were obtained exceeded 120 and comprised both vegetative and seminal propagating material from nearly all of the parents.

Observations on the specimens during their juvenile and early adult phases indicated that there was a range of reaction to the growing conditions in terms of vegetative development. This situation prejudiced the gathering of data from the less-developed genotypes regarding the character expressions of the flowers and fruits. Consequently, the lack of conditions to determine the relationships between vegetative development and other traits preclude any conclusions to be described concerning the appearance and composition of the varieties sampled.

Among the more precocious specimens the majority appeared to possess elongate fruits with the surface roughness ranging from slight to intermediate. Many of the specimens possess the short acute, pointed apex that is a feature of much of the cacao along the middle course of the R. Amazon. Figure 28 shows the fruits from two trees which will give an idea of the principal characteristics encountered in the area. The data of the fruits sampled indicate a large range of traits similar to those obtained in the Itacoatiara sample.

Fig. 28. Brazil, R. Amazon: examples of the fruit types of the Parintins region.

Among the traits observed in this sample are trees with high ovule numbers as well as large seeds.

One of the specimens of progenies contained an interesting segregation in the developing seedlings for a combination of small very narrow leaves and a white colour (Plate 31a). The fact that a few of the narrow-leafed plants had leaves that were green in colour indicates that more than one gene is involved in the expression of the mutation. The genotype of the parent tree was not determined but it is assumed that it is heterozygous for the factors that produce the mutation.

It is unfortunate that circumstances prevented the continuation to their conclusion of the observations that were being made on this large sample, which appears to possess several important features. Since the Parintins area is situated in a zone whose geographical characteristics were explained at the beginning of the section, the cacao population sampled should be considered in the context of its place in a wider area of distribution.

The Óbidos-Alenquer region

The area along the left bank of the R. Amazon between the mouth of the R. Trombetas and to the east of the town of Alenquer (Fig. 23) is, perhaps, the most important region in which the cultivation of cacao developed. The area lies across the R. Amazon from the mouth of the R. Tapajós and it would be plausible to expect that the cacao varieties found near the mouth of this river and along the right bank of the Amazon below it are related to those cultivated in the district.

There is a consensus of opinion (such as that expressed by Huber, 1906a) among persons with knowledge of the cacao situation that the extent of the natural distribution of the species along the R. Amazon is delineated by a line drawn between the mouths of the R. Tapajós and the R. Trombetas. Although the town of Óbidos is situated at the mouth of the R. Trombetas no reports of cacao on this river have been found, except those in the vicinity of the town where cacao has been cultivated.

The only occurrence of cacao in the region to which any claim of spontaneity has been made is that along a tributary of the R. Amazon located below the town of Óbidos, called the Rio Branco (no connection with the other rivers of this name cited previously). In 1965 the R. Branco was visited during a collecting activity and specimens were obtained from five trees found in the valley. The description of the population made *in situ* (Vello, F. and Medeiros, A.G., undated, unpublished mimeographed document, Expedição Botânica à Amazônia Brasileira, 16 pp.) indicates that a single variety is involved, the trees encountered varying in age or size. The fruits of this variety were described as having a light green colour when unripe and an 'Amelonado' shape (that is, short and broad to spherical) with size varying from small to large. The variation in fruit size may be attributed to the variability in seed numbers from natural fertilization and other factors commonly encountered in similar situations. The principal interest in this variety was that the trees had light green flush leaves and the seeds were found to have white cotyledons, it being claimed that all of the specimens had these characteristics. It could be concluded that the characteristics observed were the result of the action of the anthocyanin inhibitor gene but this hypothesis cannot be confirmed because of the lack of data on the flowers and the fact that only one specimen survived.

As far as can be determined this specimen (identified as 'OB 52') does not express the traits associated with absence of pigment and no seeds with white cotyledons have been found. If the surviving genotype is correctly ascribed to the population then it may be concluded that the basic cotyledon colour is purple and the anthocyanin inhibitor gene was carried by some of the other trees. As will be shown in the following paragraphs this situation is common throughout the region under examination.

The existing knowledge of the diversity within the Óbidos-Alenquer region comes from specimens collected in the vicinity of the town of Alenquer.

Most of the parent trees involved were sampled in the area bordered by the R. Curuá to the west and the R. Maicuru on the east and inland for several kilometres. Records relating to the area provide precise locations of the original plantations that were established. Many of these were located on river banks and islands as well as on the shores of a large lake. Such locations were subject to flooding during the seasonal periods of high water levels. However, since patches of red soils of high fertility occur at distances up to about 50 km from the backwater of the R. Amazon, cacao was also planted in these isolated areas. Plantations also occur on less fertile, sometimes sandy, soils where the cacao may be subject to near-drought conditions. This large variability in ecological conditions may be unique in the Amazon Basin and has a certain impact on the nature of the diversity in view of the degrees of selection pressure for adaptability that would have occurred over the centuries during which the species was cultivated.

Many of the original plantings have disappeared as interest in cacao declined and the crop was substituted by pastures and other activities. The extent of the original distribution may be judged from the isolated survivors that are found in the pastures and river-bank sites (similar situations being encountered throughout the basin). On the other hand, some attempts at establishing new fields have been made during the past 30 years using seed from the older plantings (in contrast to the fields created with introduced planting material).

After a preliminary survey of the area made in 1976 a more extensive sampling activity took place in 1977. The results of this activity were very encouraging and further specimens were obtained in 1985 and 1986. These specimens were characterized and evaluated and a preliminary synopsis of the results were presented by Bartley et al. (1988), which provides an idea of the scale of diversity created during the evolution of cultivation in the area.

The fields or remnants of populations that formed the older cultivations, or their descendants, vary fairly considerably with regard to their constitution. In some cases in what are very old plantings, as judged from the size of the trees, the variability is significant. In other fields, of newer or surviving plantings, the variability tends to be narrower with genotypic variation being specific to individual locations or farms.

Taking all of the available specimens (comprising parent trees and their progenies) together as representative of the region a large degree of variability is expressed for all plant characters, notably with regard to growth factors, leaves and fruits. The genotypes show marked differences in tree size and growth rates. Leaves vary considerably with regard to shape, from lanceolate to broad elliptic. The leaves of one genotype are sickle-shaped and curled. Fruit shapes vary from almost spherical to very narrow elongate, the elongate shape perhaps being more frequent in some localities. Figure 29 is a composite image of several fruits from the specimens acquired and demonstrates the ample range of variability for their characteristics. Fruit colours range from very light green to dark green, the latter apparently associated with some of the short and broad fruits. One tree was encountered with very noticeable anthocyanin pigment on the surface and progenies of other parental genotypes segregate for fruits with pigmented surfaces, such as the fruit second from right in the lower row of Fig. 29.

As mentioned above, the anthocyanin inhibitor gene is common in the region, being encountered in genotypes that differ with regard to other characters. An example of one of these trees is shown in Plate 32b, the fruit having the typical short acute apex but a fairly large size.

The variability encountered within the fairly small sample from the region indicates the probability that the planting material derived from various sources. These are, of course, unrecorded. Considering the area once occupied by cacao it may be assumed that introductions of seed into the region were made at various periods but also some plantings were developed or enlarged by using seed from the existing plantings. One field was visited in 1977, which

Fig. 29. Brazil, R. Amazon: a collection of fruit types from the Alenquer district to show the range of variability.

had been established a few years earlier and where the trees were commencing production. Several specimens were obtained from this field on account of the high level of production and because the variety concerned differed from those that had been seen in other fields. This variety has fruits with a broad elongate shape and which are distinguished by being broader near the base (peduncle end), as well as the surface being rougher than the other types.

This variety, which possesses traits unique for the region, is similar to some of the genotypes occurring in the Solimões and R. Japurá delta region. Several of the genotypes with elongate fruits exhibiting surface pigment resemble those of progenies of crosses between the 'Parinari' types from the R. Marañon. One of the progenies from the first collections has combined leaf and fruit characters that are virtually indistinguishable from those of the progenies of one of the 'PA' clones.

Production records from the Óbidos-Alenquer region indicate that during the first decades of the 20th century, when the effects of the slump caused by the competition with rubber were not yet pronounced, average net yields were among some of the highest known at that time. The performance of many of the specimens obtained attests to the potential productive capacity of the diversity that the region's germplasm possesses.

The 'Lower Amazon'

For the purposes of this account this title is being applied to the stretch of the R. Amazon starting below the mouth of the R. Maicuru and ending at the island of

Marajó. This term embraces what is a recognized geographical entity. It must be emphasized that the title is not being employed in the sense in which it has been customarily, and erroneously, applied in attempts to classify the species' diversity. It is also necessary to point out that the region is not especially important in terms of the quantity of cacao that may occur. Little, if anything, is known about the composition, nature and origins of the occurrences of the species that have been reported.

These reports indicate that certain areas of cacao are associated with the towns on the left bank of the R. Amazon below Alenquer but these reports have not been substantiated through personal examination.

The R. Xingú (Fig. 23) is the remaining major tributary of the R. Amazon, also draining lands to the south of the basin, which has not been discussed so far. There are virtually no reports of the existence of cacao in this drainage system and it could be assumed that it does not belong to the natural distribution area of the species. The only occurrence of cacao in the valley that has been verified is one in an area of colonization situated about 15 km to the north of the town of Altamira. It is probable that the trees seen were progenies of seeds from a single fruit introduced from an unspecified source. This group of trees was unique in that some had fruits with an entirely smooth surface without indications of ridges. The specimen obtained possesses some superficial ridges on a dark green surface. The characteristics of the parent genotype (the clone 'CAB 47') may be appreciated from the fruit at the top of Fig. 30. The fruits of some of its progeny are shown in the bottom row. From their resemblance of the fruit of the parent its homozygous nature may be deduced as well as the probability that it is self-compatible. This genotype appears to be distinct from any variety encountered. It is one example of a genotype that would be descended from a population that evolved in isolation.

Fig. 30. Brazil, R. Xingú: (top) fruit of the accession 'CAB 47' and (bottom) fruits of some of its inbred progeny to demonstrate their uniformity.

The Xingú drainage system includes one major tributary, the R. Iriri, which flows from the southwest. Again, the existence of cacao in this valley has not been confirmed. However, the opportunity was taken to acquire some specimens from a small group of trees that were reported as being grown from seeds of a fruit brought from an unknown Indian village along the river. The existence of cultivated cacao in Indian villages along the R. Iriri is possible on account of the proximity to the R. Tapajós. The trees in the group exhibited a degree of variability and had fruits of an intermediate green colour, were of intermediate size and had a broader elongate shape.

The left-bank tributaries of the River Amazon

The remaining tributaries of the R. Amazon of importance with regard to the distribution of cacao are the R. Parú and the R. Jarí, both on the left bank of the river (Fig. 23). The 17th-century description of the town of Almeirim situated at the mouth of the R. Parú, the westernmost of these rivers, reported that the inhabitants made annual excursions up the river to gather the cacao produced. The exact location of the cacao populations is as yet unknown but, in view of the nature of the terrain through which the river runs, it is unlikely that there are large concentrations of trees along its banks. It is probable that the major populations would be found near the headwaters of the river and, in terms of their variety composition, fall within the distribution area that includes the Guiana Shield, which will be discussed in the succeeding section.

Cacao was also known to exist in the valley of the R. Jarí. This river also has its sources in the range of the Guiana Shield, which separates the waters that flow into the Atlantic to the north from those that flow into the R. Amazon. It also forms the border between the Brazilian States of Pará and Amapá. A few specimens were collected in 1968 on lands on the left bank of the river in the latter State. Four locations were sampled, providing 12 specimens of the parent trees of which ten exist. It was reported (F. Vello and L.F. da Silva, unpublished manuscript, CEPLAC, Brazil, Relatório de Viagem à Região Amazônica, Comunicado Técnico No. 22, 19 pp.) that cacao was abundant in the area and occurred in locations distant from watercourses, the area surveyed surrounding a small range of hills called the 'Serra de Tigre'. The inadequate attention paid to the individuals after their establishment in the field resulted in uncertainties regarding their identities and a lack of observations on their development. The specimens were resuscitated and identified as best as was possible from the original descriptions.

The clones derived from these specimens may be separated into four groups on the basis of their growth characters, the distinctions being related to the respective locations of the parent trees. Fruits of two of the types are depicted in Plate 26a. A third type has narrower elongate fruits, differing essentially from the two types illustrated. It may be assumed that the cacao of the region contains two basic fruit types, one having a short and broad shape and the other with a broad elongate shape with ridges that are arranged in closer pairs. Within these fruit types there are distinct differences with regard to leaf and flower characters. Leaf shapes include some with elongate blades. One of the more unusual flower characters is the combination of purple staminodes with light red (pink) guide-lines. Variation is observed with regard to the intensity of anthocyanin pigment of the cotyledons, some genotypes being homozygous for intermediate shades of purple pigment.

The northern perimeter of the Amazon Basin – Upper Orinoco and Guianas

The intention in this section is to describe the cacao populations that occur in the areas on the periphery of the Amazon Basin, which are considered to form part of the Amazonian Region as defined in Chapter 1. As shown in Fig. 23, some of the areas included in this Guiana/Orinoco region have direct fluvial connections with the rivers of the Amazon Basin. One of these is the Casiquiare Canal, which connects the Rio Negro drainage system with the Upper Orinoco, as described in the discussion of the former drainage system, in which the constitution of the cacao population of the area is discussed.

The other connection is located to the east of the Orinoco valley through the R. Takutú (or Tacutú), one of the rivers that form the R. Branco tributary of the R. Negro. As stated in the appropriate section, the R. Takutú forms part of the boundary between Brazil and the country of Guyana. The remaining areas concerned in this section are not geographically part of the R. Amazon system but in most cases the cacao populations are distributed on both sides of the watersheds.

One of the problems regarding the description of the cacao diversity in this extensive area is that, although information of the occurrence of the species throughout the region is available, there is a paucity of information regarding the characteristics of the genotypes that exist at each of the locations. The available details are limited to a few of the locations where specimens have been collected and characterized. Consequently, the state of the knowledge about individual populations in the countries that belong to this section of the Amazonian Region is uneven and prevents us from obtaining a comprehensive picture of the genetic variability that exists in it.

Several rivers flow into the R. Orinoco on its left bank from their sources in the Andes (Fig. 21). Reports of the existence of cacao on some of these rivers have been published but, in view of the absence of the mention of cacao in the reports during the earliest period of European occupation, it could be conjectured that such occurrences are cultivated areas established during more recent times. Most of these rivers drain the open lands known as the 'Llanos', where conditions would be unsuitable for the support of cacao except for small patches. In any case no information is available regarding the varieties that occur on these rivers since the possibility exists that the planting material was introduced from the cultivated areas in other parts of Colombia and, therefore, are not necessarily related to the varieties of the Amazonian Region.

The one tributary of the R. Orinoco that is of interest for our purposes is the R. Guaviare, the most southerly of its left-bank tributaries. The watershed between the R. Guaviare and the R. Vaupés marks the separation of the

R. Orinoco drainage system from the Amazon Basin. The R. Guaviare has one long tributary that joins it on its right bank just above its confluence with the R. Orinoco. This is the R. Inirida, part of which was explored by the Anglo-Colombian Cacao Collecting Expedition, but no signs of cacao were seen. It would be expected that any cacao encountered on the R. Guaviare would not be related to the varieties in the R. Caquetá/Japurá system, unless through the region where the two river systems have their headwaters. Extensive areas of cacao have been known to exist on the R. Guaviare for a long time and, on this account, the river has enjoyed a certain reputation. The cacao of the Guaviare system is also known as 'Amanaven', the name of a branch of the river where the principal stands of cacao occur near its confluence with the R. Orinoco. From published descriptions of the cacao of the region it may be concluded that this population comprised a single variety on account of the apparent lack of variability expressed.

A specimen fruit from this population was acquired by the Anglo-Colombian Cacao Collecting Expedition. The progenies established have been subjected to detailed study and use in breeding programmes on account of the interesting and valuable characters of the genotype. The plants have broad leaves that are of a smaller size with a wavy appearance and a hard texture. The flowers are unique and easily recognizable on account of the strongly pigmented guide-lines that are visible from the exterior of the petal pouch, thus giving it a red appearance rarely found in other varieties. The fruit is short and broad in shape with a fairly smooth surface. The most outstanding character is the hard husk of the fruit, the mesocarp of which is lignified to such a degree that rodents find it difficult to gnaw into it. The seeds are of relatively small size and have dark purple cotyledons. The genotype is self-compatible, hence the ability to reproduce itself. No variants from this genotype are known.

The origins and botanical status of the genotype are unknown. Although the trees abound in the forests and appear to be spontaneous it is probable that the population derives from cultivation established as long as 300 years ago. The plantings would have been abandoned, leaving them in a sub-spontaneous state, during which they expanded through natural multiplication. The logical assumption would be that the original planting material was brought from the cacao occurrences in the upper Orinoco valley but the absence of sufficient knowledge about these and their history precludes the confirmation of this association.

Further to the east lie the headwaters of the R. Orinoco. It will be assumed that, for the purposes of this treatment of the diversity of the area, the natural distribution of the species encompasses all the land above the falls at San Fernando de Atabapo. Prior to the arrival of the Europeans cacao had been growing in the valleys of the right-bank tributaries of the R. Orinoco. Patiño (1963) reproduced the information contained in reports by several members of the commission sent, in the latter half of the 18th century, to establish the boundary between the possessions of Spain and Portugal. These reports described the existence of sizeable populations of cacao on the upper R. Orinoco and some of its tributaries, such as the R. Padamo. Claims were made that the fruits and seeds were of larger size (than an unspecified variety) but it was also stated that the seeds were 'slightly smaller than those of the cacao of the Caracas variety'. This same variety was that encountered in the region by von Humboldt (1821), who, as stated in Chapter 1, was the first to observe the differences between the upper Orinoco cacao and that of Caracas. However, naturalists who visited the region subsequently do not mention the presence of these populations of cacao. Also the populations are not mentioned by the geographical surveys carried out in the 20th century.

The River Orinoco drainage system

Only in the past 20 years has any attention been paid to the cacao of the upper R. Orinoco region and collections made of living material that provide the opportunity to know the characteristics of the populations. The available information is based on collecting activities that were undertaken by Lanaud (1986) and Sanchez and Jaffé (1989). The reports provide us with an interesting picture of the situation in the region. Overall, specimens were obtained from several locations in the wide area occupied by the various tributaries that flow into the R. Orinoco as well as some along the Casiquiare Canal. A few of the populations were sampled by both collecting activities and confirmed the existence of the large populations of cacao, mentioned in the 18th-century reports. Also encountered were plantations established with seed obtained from the, possibly, natural populations.

The collecting activity reported by Lanaud (1986) took place principally around the Cerro La Esmeralda and it was found that the natural stands of cacao were located in swampy terrain. It would appear that the presence of cacao stands in environments that are waterlogged for all or part of a year is a normal feature of the dispersal of cacao in the Guiana/Orinoco region. Obviously the process of establishing secondary stands using seed from the local trees has been practised over a considerable period so that it becomes impossible to determine whether a particular occurrence is spontaneous or sub-spontaneous.

The above reports give very sketchy information regarding the characteristics of the individual parent trees that were sampled and, therefore, lack the specific details that would provide a basis for determining the constitution of the populations. Although Sanchez and Jaffé (1989) refer to the apparent uniformity of the trees observed with regard to their vegetative, floral and fruit characters from which it could be inferred that all of the individuals sampled belonged to the same variety, they do not describe the characteristics observed. Accordingly, the reports do not provide a basis for determining the position of the variety in the spectrum of the species' diversity.

Although living specimens from these locations have been established no information about them has been made available that would provide a more acceptable characterization of the variety than is furnished by the collection data. The author had the opportunity to observe the flowers of the only progeny genotype available from the specimens of Lanaud (1986). These are of small size with short sepals. The stamens are not pigmented. The most distinguishing character was the intense pigmentation of the guide-lines, which make them easily visible from the exterior of the petal hood. It would be recalled that this character also distinguishes the variety of the R. Guaviare. The petal strap is also intensely pigmented.

Most of the information that is provided relates to the characteristics of the fruits and the seeds. The combination of the data contained in the two reports indicates that the fruits observed in the R. Orinoco system were green in colour and had a broad elongate shape, the length:diameter ratios of the fruits with higher seed numbers being in the range of 1.6–1.8. The fruit surface was almost smooth with low ridges. All fruits appear to be of small size. The colour of the cotyledons was uniformly light purple. Lanaud (1986) reported that one fruit contained some seeds with apparently white cotyledons. The possible explanation for this is that the parent tree possessed the anthocyanin inhibitor gene, but this has not been confirmed from the progenies of the specimens that were established. Another observation made was that the husk, although apparently thin, was lignified. This characteristic would also link the variety with that on the R. Guaviare.

One of the locations where cacao was encountered was a native village situated almost at the headwaters of the R. Orinoco. This location is close to the frontier with Brazil. Sanchez and Jaffé (1989) suggested that the Yanomami, who currently are dispersed over a large area embracing the upper R. Orinoco valley and adjacent parts of Brazilian Amazonia in the States of Roraima and Amazonas, was the only tribe that knew cacao prior to the arrival of the Europeans. In the State of Roraima the R. Catrimani, a right-bank tributary of the R. Branco system, which was described above, has its source at a short distance from that of the R. Orinoco. Similarly, the source of the R. Demini is just south of that of the R. Orinoco. This river is a left-bank tributary of the R. Negro, situated to the northwest (upriver) of the mouth of the R. Branco. These circumstances lead to the possibility that cacao from the R. Orinoco may also be encountered in these rivers, having been transported by the Indians towards the south and east. Conversely, the R. Orinoco variety may have originated on the R. Negro tributaries, as Sanchez and Jaffé (1989) suggested that the variety found in the upper R. Orinoco would have been introduced. However, the proof of such a hypothesis would depend on finding the same unique genotype in the R. Negro system.

The rivers that flow into the Atlantic Ocean

As mentioned above, the R. Takutú (see position in map in Fig. 23), as a formative stream of the R. Branco, forms part of the border between Brazil and the country now known as Guyana. In fact it can be considered to be the baseline for all of the cacao that occurs to the east in the Guianas and north of the hills that separate them from Brazil.

In Guyana itself cacao occurs sporadically in certain locations, some of which appear to have fairly ancient origins. The first reference to cacao in the south of the country was made in 1763, at the time of the Dutch Essequibo and Demerara colonies (Harris and de Villiers, 1911). It is to be noted that this report predates the Spanish accounts of cacao in the upper R. Orinoco valley. The cacao referred to was described as follows:

Up in Rupununi [the R. Rupununi is a major tributary of the R. Essequibo draining a large area of the south of Guyana] there are found whole woods of cacao, some of which has been brought down on various occasions, and found to be as good as any other, and of which the monkeys and other animals now get the benefit.

The account did not state whether this cacao differed from that which was being cultivated near the coast at the time. The fact that the presence of cacao was reported in the plural indicates that there were several occurrences known at the time and it could be assumed that a similar variety was concerned. However, it is not possible to relate such occurrences to those encountered during explorations made since the beginning of the 19th century or mentioned in other documents.

The problem with attempting to identify natural and perhaps ancient populations of cacao in the south of Guyana and determine the extent of their distribution is that, since the establishment of the Dutch colonies, introduced varieties have been cultivated in other parts of the country. Interest in the crop has undergone cycles of activity and decline with many plantations being abandoned at various times. This resulted in the farms reverting to forest and creating a situation in which it may be difficult to distinguish between the spontaneous and the formerly cultivated populations. The identification could be made worse in cases where planting material from the pre-European populations was used in the cultivated areas. The consequence of this situation is that very confusing signals are given concerning the spontaneous or sub-spontaneous status of the various occurrences of cacao that have been encountered. Accordingly, all reports of such occurrences must be analysed in this context.

The first report of a specific occurrence of cacao in the south of Guyana was made by Robert Schomburgk (1836). During his journey up the R. Essequibo he arrived at an offshoot of the river called Primoss Inlet (3° 50' N, 57° 52' W). The inlet was the beginning of a path used by the Indians connecting the R. Essequibo to the R. Courantyne (and, therefore, the upper reaches of the R. Berbice). His account contains two references to cacao. In one case, Schomburgk wrote, 'We found a lime tree in bearing, and a number of cocoa-trees, hog plums, etc. This proved that its former possessor had, unlike the Indians in general, planted and raised useful trees.'

Later Schomburgk added:

> While penetrating through the woods in search of the path, we saw numerous cocoa-trees (*Theobroma cacao*) loaded with fruit in all stages. They extended more than a mile from the river's bank, and though they were overshaded by larger trees, they had reached, nevertheless, a height of from 30 to 40 feet; and from the luxurious growth, and numerous fruit, proved that the plant was satisfied with the soil. It is not to be doubted that the trees were originally planted by the Indians; but from their number, and distance from the river, I judged that they were propagated by animals. When the fruit has reached its maturity, it falls to the ground, and is eagerly sought by the peccary or bush hog.

In the following year Schomburgk (1837) ascended the R. Berbice and found the Indian path that led him back to the cacao population at Primoss Inlet and the R. Essequibo. In this report he makes it clear that the land was marshy during the rainy season. Unfortunately, no description was given of the fruits in this account. Reference was also made to an abandoned settlement called Cumaka, 3 miles below the Primoss Inlet on the R. Essequibo. According to Schomburgk the Indians considered the path to have existed for a long time when the area was more densely inhabited. From this information it could be assumed that the establishment of cacao in the locality would have been associated with the movement of the inhabitants of the upper reaches of the rivers, probably before the arrival of the Europeans.

In recent years the locality of Primoss Inlet has been visited and some studies undertaken (Watkins *et al.*, undated; Jennings *et al.*, 1999). The population was estimated to contain about 350 trees. The conditions at the time of the survey were unfavourable for making reliable observations so that any attempt at describing the population would be provisional. The initial impression is that the population consists of trees with very similar traits but that some variants may be present. It would be expected that the population derived from seed of a single fruit brought to the site at some time in the past. As yet it is not possible to state whether it is related to any of the other populations scattered throughout the distribution area described in this section.

The leaves appear to be of a fairly small size, with a broad shape and a crinkled surface. The fruits in general are of small size and their shape falls in the range of short and broad to broad elongate. All fruits are green in colour having a smooth surface texture with low ridges that are in nearly equidistant pairs. It was reported that the husk is hard. The seeds apparently are of small size. One interesting feature of the seeds is the apparent occurrence of cotyledons in which pigmentation is unevenly distributed; in overall terms the expression of anthocyanin would be in the intermediate class.

As described by Myers (1930), Schomburgk also encountered cacao in the valley of the R. Quitaro (or R. Guidari) to the west of Primoss Inlet. This river is a tributary of the R. Rupununi and, therefore, may be within the area of the occurrence of cacao in the R. Rupununi region described in the Dutch reports. Schomburgk stated that the cacao abounded in the numerous swamps: 'It is remarkable that cacao in its wild state is only found in swampy, or, at best, moist situations.' Of the characteristics of the trees he added:

> The trees which I observed, although of a peculiar growth, almost shrubby, and the trunk less developed than in large forest trees, often attained a height of 50 feet. The capsules were large (previously described as 'melon-shaped' of a green colour), and contained from 60 to 70 seeds, which were larger than in the cultivated kind, but not so thick.

The significant aspect of this description is that of the high seed numbers, which, if correct, differentiate this variety from the others found in the region. The other significant detail of this account is that Schomburgk ascribed the binomial *Theobroma bicolor* to this population and others in the region. Although this binomial is applied to another species he probably used this name to distinguish the populations he encountered in the south of Guyana from the *Criollo* derived types that were cultivated.

Since the surveys of Schomburgk, many years passed without further information becoming available about the spontaneous or sub-spontaneous cacao populations in the southern Guianas distribution region. The discovery of apparently spontaneously occurring cacao in the south of Surinam, which will be discussed in due course, was followed by a visit by Myers to the area and interest in the general existence of similar populations in Guyana. As a result, Myers (1930) summarized the information regarding the existence of cacao in southern Guyana and reports of cacao populations in other parts of the country. He pointed out that no specimens from these populations existed.

In 1932 Myers himself had the opportunity to rectify this lack of up-to-date knowledge and representative specimens of the cacao populations, as described in Myers (1934). In this paper he referred to additional information regarding the occurrence of cacao near the head of the R. Essequibo valley. Acting on the report of the finding of a previously unknown cacao population in the Kanuku Mountains, Myers made a journey to the site. When this population was reached it was found that it extended along the valley of a stream but it was possible to examine only a few trees. The trees were large and it was concluded that they were very old. They had long and spreading branches, one branch was 45 feet (14.6 m) long – probably an effect of the shade. Although flowers and young fruits were seen no description was given of them. The lack of sound ripe fruits precluded the collection of living material for propagation. The exact location and name of the stream was not given, but it may be assumed that it may have been a tributary of the R. Rupununi.

Myers appears to have returned to the region a few years later as there is a herbarium specimen of *T. cacao* attributed to him from December 1935, but no written information accompanies it. The location in which the specimen was collected is given as the Mataruki river, Upper Essequibo, British Guiana. The specimen is described as being from a small riparian tree, a fact that may indicate that the location is not the same as that visited in 1932. The leaves of the specimen are large and fairly broad. There appears to be no accompanying fruit specimen.

A few years later another occurrence of cacao was reported to the east of the above area. This was in the Marudi mountains in the south and southeast of the Pakaraima range. The report stated:

> There was also what had evidently once been a cocoa plantation, now half smothered with secondary bush but there were a few pods on the trees. The cocoa was of an unusual type, unlike the forastero, and may have been introduced from the Amazon Basin. The pods were of a pale green colour ripening to pale yellow with a remarkable glazed surface.

It was also observed that the symptoms of witches' broom disease on these trees differed considerably from those usually found elsewhere.

Two cacao populations situated in the Marudi mountains were studied by Jennings *et al.* (1999). One of these populations consisted of a single tree, presumably a remnant of a cultivated field in which the other trees had been destroyed, but its characteristics were not described. The other population was somewhat larger and 76 individuals were studied. The fruits of these trees were described as being light green in colour and, therefore, the population may be the same as that discovered previously. They were of a small size, with an almost spherical shape with low ridges and possessing what was described as a slightly mamillate apex. The lack of information about the other characteristics of the trees and their fruits as well as the inability to propagate from them precludes any attempt to classify the variety concerned. It is assumed that the population is composed of a homogeneous variety that is distinct from the others described from the region.

In view of the fact that several other reports of cacao populations in the south of Guyana have been made, but, as yet, unsubstantiated, indicates that there is considerable scope for discovering additions to the known germplasm from the country. It must be emphasized that, in spite of the existence of details about the locations at which the populations described above occur, none of these genetic resources has been conserved and adequately studied. From the scanty descriptions that are available concerning the characteristics of the populations that have been encountered, it could be concluded that several varieties may be involved, all of which are homogeneous in constitution. In the light of these circumstances it may be appropriate to consider that different varieties have been introduced from various sources over a long period of time.

The same sources of information, including herbarium specimens, indicate that spontaneous or sub-spontaneous populations of cacao exist in the valley of the R. Courantyne, which forms the boundary between Guyana and Surinam to the east. The sources of information would originate in both countries and, especially, would be connected with the cultivation of cacao in the latter country and the exchange of planting material between the two countries that is referred to in several documents. However, no specific information is available regarding the characteristics of the varieties that occur in the valley.

As far as Surinam is concerned there is only one substantiated report of the occurrence of cacao and this, in fact, proved to have been given considerable attention as the first genuine case of a spontaneous population in the Amazonian Region. The population was discovered in 1923 by G. Stahel on the Mamaboen Creek, a secondary or tertiary tributary of the R. Coppename (Fig. 23) in the south of Surinam. Myers later visited the same population and his report (Myers, 1930) contains all the details of its discovery. This cacao population was found to contain a large number of trees and appeared to belong to a single homogeneous variety. Figure 31, from a photograph taken at the time Stahel visited the site, illustrates the characteristics of the fruits of the Mamaboen Creek population, which also occurred on marshy land.

Progenies from seed of the genotype were established in the Agricultural Experimental Station at Paramaribo. Later, seed from these progenies was supplied to the Imperial College of Tropical Agriculture in Trinidad, where the plants were incorporated into the 'Museum Collection' and identified as 'M 8'. These plants would represent the second generation from the original trees. The clone developed as 'M 8' received a degree of attention; first, on account of its origin and, second, because of its homozygous status, which set it off from the mostly heterozygous genotypes that formed the bulk of the germplasm available for research at the beginning of the improvement programme. The genotype is self-compatible and this factor, combined with the homozygosity,

Fig. 31. Surinam, R. Coppename valley: the fruit of the variety found on the Mamaboen Creek of the R. Coppename.

offered opportunities for research on the genetics of cacao that the other available germplasm did not provide so readily.

From the point of view of the characteristics of the genotype, most features are described as being intermediate in status. The cotyledon colour is classed as a less intense purple than had been encountered in many of the genotypes from the Amazonian Region. The plants are also in the lower vigour category. One of the more consistent findings was the frequent occurrence of twin embryos in single ovules in several of the fruits. The normal maximum ovule number is 50.

Although it would be expected that cacao would occur within the upper valleys of some of the other rivers to the east of the R. Coppename, no reports of the presence of the species have been found.

The River Oiapoque and adjacent drainage systems

However, further towards the east the R. Maroni forms the border between Surinam and French Guiana. The upper tributaries of the river and those belonging to the upper R. Oiapoque (Oyapok in French) system have been known since the beginning of the 18th century (principally in French Guiana territory) to possess significant areas of cacao. The history of the discovery of the cacao populations of southern French Guiana was reviewed by Froidevaux (1894). There is little doubt that the original occurrences of cacao were largely of spontaneous origin and were distributed over a considerable area. The R. Oiapoque drainage system is shown in Fig. 23.

With reference to the size of the cacao population in the valley of the R. Camopi, one of the tributaries of the R. Oiapoque, Froidevaux cites the following statement made by one of the explorers of the region:

> Le nombre des cacaoyers est indefini; l'entendûe du terrain qu'ils occupent est immense, l'on sait que la culture ne change la qualité du fruit et qu'il n'y a point de differences pour la boite entre la cacao naturel avec celui qu'ont a cultivé; celui des Espagnols et des Portugais n'est produit generalment que par les arbres qui croissent naturellement dans differentes parties de l'Amerique meridionale. [The number of cacao trees is unlimited (unknown), the extent of the terrain which they occupy is immense, it is known that the cultivation does not alter the quality of the fruit and that no differences are observed between the natural cacao and that cultivated, that of the Spaniards and the Portuguese is usually produced on trees which grow naturally in different parts of South America.]

Evidently the author knew something about the exploitation of the species in other parts of South America but did not distinguish the genotypes of the R. Camopi from those of the other regions. He proposed that the Camopi population could be exploited based on an estimated 50,000 trees and a yield of 2 lbs of cocoa per tree, which would compensate for the difficulties in transportation. The proposal was acted upon, the cacao produced being exploited during some time but this was eventually discontinued for various reasons.

To place the region of presumably spontaneous occurrences of cacao in its proper geographical perspective, it is necessary to recognize that the area where the boundaries of Brazil, French Guiana and Surinam meet is another focal point in which are located the sources of rivers that flow towards the north and east into the Atlantic Ocean and towards the south into the R. Amazon. This situation may have an important bearing on the formation of the populations and the dispersal of germplasm within the area drained by these rivers.

The major effort in the conservation of the germplasm of the region as considered within this context has taken place in the populations that occur in the south of French Guiana. This effort commenced only in 1986 although surveys of the area made some years previously provided information about some of the characteristics of the cacao genotypes of which the population was composed. Subsequent collecting, principally in the upper R. Oiapoque, and on its tributaries such as the R. Camopi, resulted in accumulating a sizeable collection of germplasm that has been established in field collections. The details of the first collecting activities are given in Lachenaud and Sallée (1993) and later activities described by Lachenaud et al. (1997).

For purposes of consistency and agreement with the usage in geographical literature the name Oiapoque will be used instead of the French spelling of Oyapok (or Oyapock).

Some of the fruit types encountered during these activities, particularly those on the upper R. Camopi and upper R. Oiapoque, are illustrated in Lachenaud and Sallée (1993). These provide a means of assessing the scale of variability that is exhibited in these populations. These samples indicate the existence of two basic fruit types. One type appears to have a short and broad shape with a rounded apex and a smooth surface with the length:diameter ratio of the order of 1.5 or 1.6. The other fruit shape is more elongate with a degree of surface roughness and a short, acute apex. In Lachenaud *et al.* (1997) the second type is named 'Forme Guyanaise' and stated to have a length:diameter ratio of 2.0–2.2. In either case the fruits collected from the parent trees appeared to have been of a relatively small size. The seed numbers of the contents were low but the seed sizes apparently fairly large.

Lachenaud *et al.* (1999) studied the flowers from the specimens resulting from the various locations at which the established genotypes derive. The principal characteristic that appears to have distinguished the descendants of the specimens from the various localities was the shape of the petal ligule. The paper contains sketches of the ligule shapes encountered. A ligule of intermediate width with an acute apex appears to be the most common feature of the R. Camopi germplasm, although some of the specimens from a few of the other localities showed a wide range of shapes. Very broad ligules were found in only one site, an occurrence that perhaps distinguishes it as being genetically different from the other sites sampled; why this should be the case in this area is not immediately apparent. The specimens from the Kérindioutou area of the R. Oiapoque were distinguished from the others by their very narrow ligules.

Ovule numbers in all of the specimens from these activities appear not to exceed 50, which probably accounts for the low seed numbers obtained. Although the fresh seed weights seem to be fairly large the dry weights are relatively smaller.

Most of the parent trees sampled in the R. Camopi area are represented as families of progenies. The data published so far are confined to family averages and do not provide any information as to the variability within families or the associations among the attributes measured on the parent trees and those of the families.

The specimens that have been observed by the author are, in the case of the R. Camopi and, perhaps, the location at the French Guiana–Brazil border, individual plants from some of the families. Several of these progenies have produced fruits that differ from the descriptions of the fruits of their parents. This situation indicates a possible greater variability than can be conjectured from the collection data and a certain degree of hybridization. The fruits of the progenies that are available for examination are predominately of the narrower elongate high-ridged and rough-surfaced type regardless of the shapes of the parents' fruits. These results may be explained by the dominance of the characters of the elongate fruit type. This would be consistent with the populations concerned being constituted by two varieties, of which the elongate fruit type was a more recent introduction.

The observations on the floral characteristics as reported by Lachenaud *et al.* (1999) do not include some of the more important features of these populations. Although some variation in this respect appears to occur, the flowers are generally of a small size, a conclusion confirmed by the data in

Lachenaud *et al.* (1999). The staminodes are mostly short so that they are only slightly longer than the gynoeceums. One interesting feature of the flowers is the tendency for the pigmentation of the basal part of the guide-lines to be significantly darker than the remainder in which the distribution of pigment is interrupted. The heavy pigmentation at the base of the guide-lines produces an intense coloration of the corolla ring, producing a contrast with the remainder of the flower. The petal ligules of most of these specimens are light yellow in colour and the presence of pigmented veins is common.

In the artificial conditions under which the plants that are observed are growing the plagiotropic branches tend to be long with the spaces between the nodes at marked distances. Mostly, the leaves appear to be narrower elongate in shape with a finer texture. The characteristics of these leaves are shown in Fig. 32. In contrast, the plants from the R. Kérindioutou produce leaves that are rough on the under-surfaces of the blades.

One additional item related to the populations on the R. Oiapoque that is of considerable interest concerns the report in Lachenaud *et al.* (1997) of their exploration of the R. Euleupousing. This is a short tributary on the left bank of the R. Oiapoque below the R. Camopi. The cacao population was considered by the authors to be isolated from other occurrences of cacao. It was found that the fruits had pigmented surfaces and seeds with pale purple cotyledons. These traits make the variety distinct from the others in the region.

The reports of the collecting activities in French Guiana tend to produce claims regarding the uniqueness of the cacao populations that have been

Fig. 32. French Guiana, R. Oiapoque valley: the narrow elongate leaves evidenced in genotypes from near the source of the river.

sampled. The truth is that they belong to a much larger area of distribution, especially that of the Brazilian State of Amapá, which borders on the south of the area sampled. The R. Oiapoque also forms part of the frontier between the two countries. Populations of cacao are known to exist in significant sizes immediately south of the frontier. There are other aggregations of trees in other parts of Amapá. These include a sizeable plantation established by the Jesuits on the R. Araguary. However, no details about these populations are available and no samples from them exist in living collections.

Similarly, the French Guiana populations are contiguous to those of the upper R. Parú and upper R. Jarí, which have been referred to previously in the discussion of the distribution of cacao along the R. Amazon. Since the eastern part of the R. Jarí belongs to the State of Amapá, the whole valley also falls within what is one contiguous distribution area. Until sufficient information is at our disposal regarding the Amapá cacao populations it would be premature to accept the claims made that the varieties collected in French Guiana belong to a separate and unique assemblage of cacao genotypes. The observations made on the French Guiana genotypes may provide an indication of the scale of variability and characteristics that may be applicable to the entire population of the distribution area defined above, if it is genetically homogeneous, a hypothesis that remains to be proved.

In common with the other occurrences along the southern extension of the Guianas and the headwaters of the drainage systems the upper R. Oiapoque populations are usually found in marshy terrain or along river banks that are subjected to flooding. It may be admitted that at least some of the populations in the Guianas may have been established spontaneously. However, it is a fact that a part of the cacao in the area was subjected to human exploitation during a long period. Therefore, the possibility must be considered that human activities of an accidental, if not deliberate, nature, may have been responsible for the establishment of part of the existing populations.

The position of the R. Oiapoque germplasm in the spectrum of the species' diversity would be determined through the acquisition of knowledge about its relationship to the other populations inhabiting the area between the R. Oiapoque and the R. Amazon, such as those of the R. Jarí. This area would naturally include the, as yet, unknown cacao population of the island of Gurupá and the varieties existing on the adjacent shores of the river. This is the point that we can consider as being the easternmost limit of the distribution of the species in the Amazon Basin when the Europeans arrived on the American continent.

PART 2: THE CIRCUM-CARIBBEAN REGION

In accordance with the discussion of the subject in previous chapters the spontaneous origin of the populations in the Region, as defined in Chapter 1, is debatable. However, even though the diversity that existed in pre-Columbian times did not emerge from spontaneous origins in the region, it is appropriate to consider it separately from the Amazonian Region on account of the special characteristics of the varieties that it comprises. The fact that cacao was cultivated, or occurred in other forms, in the Region prior to the arrival of the Europeans in the New World is undisputed. It is also necessary to take into account the importance of the Region as a producer of cacao. In the first instance this involved, primarily, consumption by the original inhabitants and later as an important, and sometimes the only, supplier of cocoa when the product became an object of consumption outside the Region.

The basic distribution area of cacao in the Circum-Caribbean Region as we can conceive as existing in the pre-colonization period could be divided into three geographical zones. One of these comprises the area in the north of the South American mainland, which appears to have been located to the west and southwest of the Lake of Maracaibo in what originally came under the government of Nueva Granada. It is essential to understand that prior to the independence of the countries in northwestern South America and the formation of the present day states the entire region was under a single government. Therefore, the pre-Columbian distribution of cacao and the development of cultivation before independence apply to the whole of the Reino de Nueva Granada and this fact needs to be appreciated in order to dispel any misconceptions regarding the role of the individual countries.

The other distribution areas that are components of the Region are those that occur in Mesoamerica (Mexico and Central America), for which we should consider a possible southern limit of the pre-Columbian distribution as the southern shores of Lake Nicaragua and the rivers that empty into it. These distribution areas are divided by the central cordilleras into a Pacific coastal zone and an eastern or Caribbean seaboard zone. It is intended to give a separate description of the situation in each zone on the assumption that their formation may have involved genetic elements from different sources or different evolutionary paths.

In view of the fact that virtually no reliable descriptions exist regarding the characteristics of the cacao varieties that existed in these zones at the time of the arrival of the Europeans, any attempt to describe the nature of the populations that inhabited each area would be speculative. It would be possible to arrive at a general picture of the situation as it may have existed by means of the genetic analysis of the present-day varieties that are presumed to be descendants of the pre-Columbian populations.

The difficulties that are presented in the determination of the nature of the parental varieties are based on two situations. One is that, through the expansion of the areas of cultivation in order to meet the increasing demand for cocoa, the definition of the primary distribution areas has become blurred. Although the varieties established in the areas into which cultivation expanded would be descendants of the primary diversity they need not necessarily reflect the variability that existed in the parental populations. It would be expected that, on the one hand, variability may have been increased through mutation and recombination and, on the other hand, alterations in the genetic base would have resulted from natural and directed selection in the long period during which the expansion took place.

The other aspect that affects the diversity existing in the present day populations in the Region is related to the importation of other varieties, particularly those associated with the Amazonian Region, either in their original form or as the products of hybridization. It will be demonstrated in subsequent discussions of the composition of the present day cultivated populations and inheritance that it is sometimes impossible to distinguish, on the basis of phenotypic characteristics, between the prototype varieties and descendants of their hybrids that possess similar features.

While the areas in which the prototype *Criollo* populations are distributed have undergone substantial changes in their structure, principally on account of the superior qualities of imported germplasm, it is important to consider the fact that some of the basic diversity has been conserved in other countries. This conservation has come about owing to the importation of varieties from the *Criollo* populations, primarily to satisfy interests relating to product quality. The scant attention paid to the descendants of these introductions by the persons involved in research on the *Criollo* diversity is regrettable.

The basic situation regarding the genetic diversity of cacao in the three distribution zones is that the archetypal varieties possess several common features. It is because of this situation that it is appropriate to consider the populations of the Circum-Caribbean Region as a genetic entity. In accordance with the definition given in Chapter 2 the title of *Criollo* will be used throughout the book to identify the populations and individual genotypes that may be assumed to belong to this variety concept.

In common with most of the concepts concerning cacao diversity, that of *Criollo* is subject to different interpretations based on the information available to, or on the whims of, the persons who propose to identify the components of this section of the species' diversity. The subject of an acceptable definition of a 'variety' within the concept of *Criollo* has preoccupied many persons since studies of cacao varieties began in the second half of the 19th century. J.H. Hart set out to define the concept in an article discussing the characteristics of *Criollo* (Hart, 1908). Perhaps this article and similar publications defining the common varieties that existed in Trinidad during that period, as will be described in the discussion of the evolution of the cacao populations of that island, form the basis of the concept that has been used since then. Various other authors, such as Cheesman (1944), have dealt with the subject and evolved additional interpretations from their own experience or other sources of information.

With regard to the characteristics that are applied to define the varieties that belong to the *Criollo* group, it is commonly assumed its members have elongate fruits with long acute or attenuate apices. Other characteristics included in the definition are the rough and ridged surface, a thin or soft husk and large seeds with white cotyledons (Bailey, 1947).

The principal characteristic of the varieties that may be attributed to the concept of *Criollo*, which is common to all distribution zones and which distinguishes them from the previously held notion of an Amazonian variety, is the presence of anthocyanin pigment in the fruits. The factors that produce this character have pleiotropic effects on the pigmentation of other organs of the plants. The main effect associated with red fruits is the presence of pigment in the leaf axils and pulvini, which is normally expressed in the leaves of developing flushes. It is to be expected that these factors also produce the development of anthocyanin pigment in the flush leaves, but the expression of flush leaf pigment is not exclusive to genotypes with red fruits. Some variations in pigment of flower parts are associated with the presence of the fruit pigment factors. However, the penetrance of the pigment factors may depend on other genes carried by particular genotypes that result in variations in the intensity of pigment and the homozygosity status of the loci concerned. Other factors include the reaction of the genotypes to environmental conditions, such as the intensity of light. The presence of the fruit pigment factor also has an influence on the level of pigment in the cotyledons (an expression that is not usually appreciated). This relationship will be explained in the section of Chapter 7 that deals with the inheritance of these expressions. Additionally, the presence of the allele responsible for the pigmentation in the fruits is exhibited as pigmentation in the hypocotyls of the germinating seeds.

However, it is necessary to emphasize that, as demonstrated in the discussion of the Amazonian Region populations, fruit pigmentation also occurs in those populations, sometimes to a significant degree. Accordingly, the red fruit phenotype is not exclusive to the *Criollo* populations but in the absence of studies of the inheritance of the Amazonian phenotypes it would be premature to suggest any relationship between the phenotypic appearances of the varieties from the two Regions.

An understanding of the nature of the pigment expressions in the Circum-Caribbean aboriginal varieties is important in recognizing their contribution to the ancestry of many hybrid genotypes. On this basis it can be concluded that, apart from other evidence, the ancestor of any genotype possessing red fruits would be a red-fruited *Criollo* genotype even though it exhibits other traits that may be completely different from any of the *Criollo* traits (for an example, see Fig. 36).

Research on the system of reproductive compatibility that operates in cacao conducted during the second half of the 20th century revealed that the *Criollo* genotypes used in breeding programmes possibly possessed a genetic mechanism that differed from the Amazonian Region genotypes with which they had been combined in hybridization. The research on this subject has lacked the continuity of effort required in order to elucidate the genetic differences between the two groups of diversity involved. Further details on the subject will be provided in the discussion of the inheritance of the compatibility systems.

The other trait that appears to occur exclusively in varieties belonging to the *Criollo* populations is that in which the plants do not form the normal whorl of five plagiotropic branches (*jorquette*) from the terminal buds of the orthotropic stems but, instead, a single plagiotropic branch. So far, this trait has not been recorded in the Amazonian populations.

Flowers of many *Criollo* genotypes are often of large size. In these flowers the appearance is that of laxity with the presence of long straps connecting the petal hoods to the ligules. However, such characters are not universally expressed and it is possible to encounter genotypes that undoubtedly correspond to the *Criollo* group but that produce very small flowers.

One feature that previously was considered to be unique to some of the *Criollo* populations is the formation of fruits with the arrangement of the ridges in the fruit form known as 'Pentagona'. However, as has been demonstrated, this character also occurs in the Amazonian populations and can no longer be considered a unique characteristic of the *Criollo* group. In fact, the character does not occur uniformly in all of the *Criollo* populations, in which case it is not specific to the group.

Other features that have in the past been considered as specific to the *Criollo* populations include the soft husk of the fruits, large seed size and unpigmented cotyledons. The large seed size is not universally encountered. It is usually associated with genotypes having large fruits and low ovule or seed numbers. In these cases the negative correlation between seed numbers and seed weights applies. The high seed weights are not transmitted to the hybrid progenies of such genotypes in the same way that the factors carried by Amazonian varieties with seed weights of similar magnitude are transmitted to their progenies. It has been shown in the discussion of the Amazonian Region diversity that in certain populations genotypes can be encountered whose phenotypes express all of these traits, either singly or in combination.

The subject of seed weights and seed numbers in the populations belonging to the *Criollo* group involves the general occurrence of low ovule numbers. These probably never exceed 50 and it seems likely, although no reliable data have been obtained that will show the range of ovule numbers, that some varieties have numbers inferior to 40. Low ovule numbers would set serious limitations on the reproductive capacity of the genotypes concerned, decreasing their aptitude for use for commercial purposes.

It is evident that a considerable degree of genetic variability occurs in the archetypal varieties of the Circum-Caribbean Region when taken as a whole. The variability expressed by genotypes that could be ascribed to these varieties, or to their relatives with similar phenotypes, renders it inappropriate to consider all of these varieties as being strictly related. Cheesman (1944) suggested that it would be inappropriate to apply the term 'variety' to these populations and suggested that the term '*Criollo* group' would serve the purpose. He proceeded to describe the characteristics of the types of the regional populations known to him and made it abundantly clear that the diversity was so marked that no definition of the group name would be satisfactory.

However, in spite of this evidence, Cuatrecasas (1964), in his monograph of the genus *Theobroma*, presented a definition of the cacao types attributable to the concept of *Criollo*. This definition, which has no botanical basis, is unfortunate in the sense that some persons assume that only genotypes that exhibit the traits so defined belong to the group. Unfortunately, as a result of the unnecessary adherence to the narrow definition, given by Cuatrecasas, which is based on fruit characters only, a tendency exists to disregard other genotypes that are also members of the *Criollo* group but which do not exhibit the characters covered by his definition.

In the following sections an attempt will be made to describe the history and variety situation in each of the distribution zones. It is necessary to take into consideration the paucity of reliable details concerning the prototype genetic materials that formerly constituted the cultivated and sub-spontaneous populations of the Region. This situation prevents the development of a comprehensive and accurate description of the situation in each zone. In addition, it should be appreciated that much of the available information is derived from sources that involve personal interpretations based on observations made in a limited range of the populations belonging to a given distribution zone.

Further, much of the personal knowledge about the diversity of the Region is based on material introduced into other countries where the genotypes have been subjected to investigation of one kind or another. It can never be assumed that these genotypes represent the true range of the diversity concerned. From the author's point of view the personal knowledge of the populations in their native habitats is fragmentary or lacking, as is the case of the Mesoamerican populations. Consequently, it not claimed that the following account is correct or complete. It is intended to provide the best interpretation of the information available, including the author's own experience.

Distribution Zone A: Northern South America

The origins of the cacao populations that developed in the zone are unknown and this lack of knowledge also applies to the location and area of dispersal of the primary source of the germplasm that formed the basis of the plantations that evolved. The species is not referred to in the reports of the earliest European explorations of the shores of South America. If it did occur in the zone at that time it would not have been identified as 'cacao' by the Spaniards and others who arrived in the region without them having been acquainted with the plant in Mesoamerica. Again, it is not known exactly when interest in cultivation and the beginnings of the cacao trade with Europe began.

It may be supposed that the existence of cacao dates from well before the arrival of the Europeans and, also, that the plant was used by the indigenous inhabitants. Information culled from various documents concerning the early history of cacao in the region is presented by Patiño (1963), who also summarized the history of the expansion of cultivation in Venezuela. Ciferri and Ciferri (1950) also described some of the history of cacao in the sub-Andean region of Venezuela including its subsequent distribution. The position of the sub-Andean area and the regions to which cultivation spread in the colonial era are shown in the map in Fig. 33.

From these reports it can be concluded that the area of occurrence of the prototype *Criollo* population during the pre-Columbian period would have been located to the southwest of the Lake of Maracaibo. It is likely that its original distribution may have been confined to the northern foothills of the Venezuelan Andean range as far as the Sierra de los Motilones. It would appear that it was in this area that the species was first exploited and that the planting material for establishing the subsequent cultivated fields was derived from this source.

However, Hernández (1968) stated that the historical records suggested that the Indians would have cultivated cacao in the river valleys on the southern flanks of the Venezuelan Andes. If ancient indigenous cultivation had existed in the area, it is likely that these trees provided the planting material

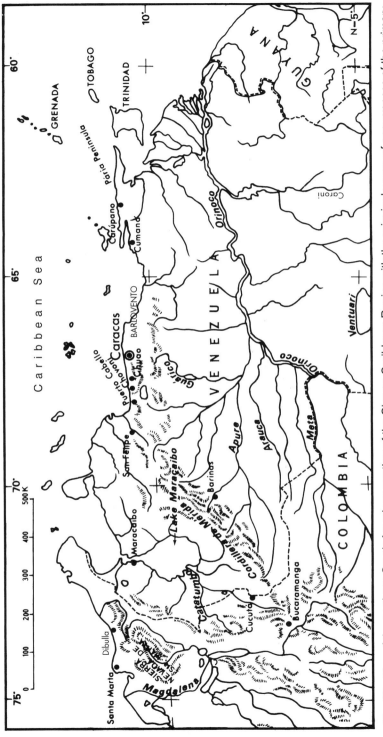

Fig. 33. Map of the northern South American section of the Circum-Caribbean Region with the principal zones of occurrences of the primary and cultivated populations.

used for the plantations subsequently established during the colonial era. Fairly large areas of cacao are known to exist on the foothills of the Cordillera Oriental in the upper valley of the R. Sarare. Several historical references relating to the occurrence of cacao along this river were cited by Patiño (1963). The above area would be situated on the logical route by which cacao may have migrated by land from the Amazonian Region along the eastern foothills of the Andes. Hernández (1968) mentioned the possibility that the cacao occurring in the Lake Maracaibo area could have been introduced from Mesoamerica. Pittier (1934) had previously expressed the same opinion but without citing the evidence to back his suggestion.

In the current work it is not the intention to debate the subject of the origins of the cacao populations in the zone. The objective is to present the details that will enable us to construct an idea of the genetic structure of the primary population from an analysis of the characteristics of such remnants of the original cultivated populations that can be identified as being descended from it. The only aspect of the knowledge that can be gained from the cultivated genotypes that would provide a concept of the primary population is the fact that both red and green fruits occur. This indicates that the population would have been genetically heterogeneous.

The first attempts to cultivate the plant would have taken place at short distances from the localities in which the primary population was encountered. On the basis of this conjecture the oldest plantations would be found around the shores of the Lake of Maracaibo and in the valleys in the foothills of the Venezuelan Andes. Several visitors to the region reported on the extent of the plantations and the trade that was carried on. In their survey of the cacao plantations in the zone Ciferri and Ciferri (1950) described the varieties observed in some localities as well as indicating the principal areas of production. They suggested that the missionaries would have been the most important promoters of cultivation, as is repeated elsewhere in South America. Among other observations they refer to localities in which the majority of the trees had green fruits in contrast to the predominance of red fruits in other areas. From the other details given in the report of the survey it would be concluded that, from the genetic point of view, a certain amount of variability is to be found in the zone. Some confusion is apparent in the report, caused by speculations as to the origins of the types that differed from those that were usually encountered.

The cultivated fields that were first established would have been located in isolated areas, such as narrow river valleys, in situations favourable for the evolution of distinct varieties according to the origin and genetic composition of seeds used in each case. These distinctions would have been the consequence of the hypothetical heterogeneity in the primary population and the probability that small samples of seed were involved, perhaps from a single tree or from adjacent related trees. For example, a plantation established with seed from a tree with green fruits would be likely to produce a population in which green fruits might have predominated, especially if the population sampled mainly consisted of trees with green fruits. On the other hand, if the sample had been taken from a tree that was homozygous for red fruits the resulting population would have consisted entirely of trees with red fruits. These examples are basic to the understanding of the genetic composition of the *Criollo* varieties that evolved throughout Venezuela and Colombia during the existence of the Viceroyalty of Nueva Granada.

After the Ciferris' survey of the sub-Andean region no further investigations appear to have been carried out until the last quarter of the 20th century, when a programme was undertaken with the objective of conserving samples of the gene base of the surviving *Criollo* varieties. This programme is important in the context of the areas in which the diversity was sampled, probably being in proximity to the region where the primary diversity existed.

The results of the programme, based on the initial observations, were summarized in Reyes *et al.* (1993). In this summary the varieties of the area are designated as *Criollo Andino*. The sub-Andean region is probably the one cacao-producing region in Venezuela that is least affected by contamination by introduced genetic material (that is, until the arrival of recent introductions), implying the continued existence of fields containing the original varieties. With regard to the characteristics of these varieties the information contained in the summary indicates that considerable differences occur involving almost all the observable characters of the fruits of the trees that were sampled. It would appear that the main criterion for considering a tree to belong to the *Criollo* group was the occurrence of fruits entirely containing seeds with white or slightly pigmented cotyledons. There is the rather frequent occurrence of fruits with very narrow elongate shapes, the diameters of which were often less than 70 mm. Fruit surfaces varied from smooth with low ridges to rough with high ridges. Significant variation was also recorded for the thickness of the husk. It was alleged that trees with fruits with a 'pentagonal' shape had been discovered but the corresponding illustrations given do not support this claim; the fruits concerned appear to be distorted and not comparable with similar types from Mesoamerica. However, since some of these genotypes are reported as being called locally 'four ridges' it is possible that variations in fruit shape and other characters do exist. Fruit colours ranged from light green to very deep red. Several specimens were recorded as having a slight blush of anthocyanin pigment on a light green base colour.

It could be assumed that the specimens collected belonged to the *Criollo* group of varieties, as judged on the basis of fruit and seed attributes, but this needs to be confirmed by observations on the progenies and by studies of their inheritance. No details have been provided regarding the vegetative and floral characteristics of the specimens that were collected, the number of these being insignificant in terms of the size and importance of the cacao populations of the zone.

It should be noted that, in addition to these specimens, some genotypes from the area have been preserved in other countries. This has come about through the introduction, from various growing areas of Venezuela, of seeds in order to reproduce the varieties for which the country was famed under the trade name of 'Caracas' on the basis of its quality. One such introduction was established in the Dominican Republic from seed brought from the State of Mérida at the beginning of the 20th century. The progenies have been maintained intact and form a homogeneous variety entirely of trees with green fruits of the type in Plate 29c. These fruits have a relatively smooth surface and appear to be self-compatible. The progenies descended from these trees are also uniform in their characteristics. Similar varieties, whose origin is attributed to seed obtained from the State of Mérida, occur in isolated plantings in Puerto Rico. The fruits are green in colour with slightly rough surfaces and are also, probably, self-compatible. Hence, these populations would represent the pure form of *Criollo* present in the respective source locations.

To the north of the sub-Andean region and west of the Lake of Maracaibo there is an area in which cacao is cultivated that is of special interest for two reasons. First is that the cacao that is traded with the name of 'Maracaibo' has been specially favoured for its quality. The second reason concerns the occurrence of a 'variety' that is distinctive in its characteristics.

This variety may have been distributed throughout the plains between the Andean foothills and the lake but it appears that the cacao of this area has declined to such an extent that it has virtually disappeared. At the present time the variety, as far as is known, is confined to the valleys of the Rivers Catatumbo and Encontrados in a rather unique environmental situation. The lower reaches of these rivers would have been marshy and subject to flooding so that special measures were taken to ensure the continued existence of the plantations.

In Chapter 6 reference will be made, in the descriptions of the cultivated populations of South America, to the area of cultivation that existed in the upper reaches of the Catatumbo valley in which the particular variety may have existed. For this reason and for simplicity it is proposed to apply the term 'Catatumbo region' for the distribution of the renowned variety on the western shores of the Lake of Maracaibo. The relationship of the R. Catatumbo to the Lake is shown in Fig. 33.

The variety of interest is distinguished by the fact that the fruits have a smooth surface that is glossy when of a red colour. This appearance has led to the variety or, perhaps more correctly, phenotype being given the name of 'Porcelana' (= 'Porcelain'. It has been the practice to apply the term to any fruit that has a smooth and glossy surface so that it is probably more accurate to use a designation such as 'Catatumbo Porcelana' to distinguish the particular variety). On account of the fact that the fruit shape differs from that of the usual conception of *Criollo* the status of the variety caused some confusion regarding its classification.

However, the author (Bartley, 1964a) recognized that the 'Porcelana' fruit type was a phenotypic expression within the *Criollo* range. The variety possesses a somewhat poor degree of adaptability and the plants develop adequately only in optimal environmental conditions. It has become particularly adapted to the special conditions that occur in the Catatumbo region's flood plain. Although a few specimens representing the type variety exist in some germplasm collections no studies on the inheritance of their characteristics have been conducted to provide a notion of its place within the *Criollo* group. On account of the segregation that appears to occur in families of progenies of the 'Porcelana' type it could be suggested that the phenotype results from a particular combination of factors in the recessive state.

When the lower Catatumbo region was surveyed it was supposed that the term 'Porcelana' had been associated with trees having green-coloured fruits.

However, it was found that, in certain properties, the populations mainly comprised trees with red fruits. The estimated ratios of segregation implied that the parent trees would have been heterozygous for the red colour. The green fruits have a lighter green base colour on which a slight trace of anthocyanin pigment is superimposed. It was found, as occurs with almost all varieties in the *Criollo* group, that the green fruits contained seeds with cotyledons that were uniformly unpigmented. Where red-coloured fruits were concerned a light pink coloration was observed in some cotyledons, the proportions of these to white cotyledons depending on the homozygous status of the individual trees. The fruits of trees in the Catatumbo region are illustrated in Plate 28c and d. It may be concluded that the term 'Porcelana' may be applied more appropriately to the red fruits than to the green fruits.

Another character in which segregation was observed was in the branching habit. Some trees had the normal whorl of five plagiotropic branches while others exhibited the characteristic single plagiotropic branch system that occurs in most varieties of the *Criollo* group.

To complete the account of the Catatumbo flood plain cacao population it is necessary to emphasize that the 'Porcelana' type is not the only variety that can be encountered. Trees were observed that had the rough elongate fruits that perhaps belong to the sub-Andean types. It is also important to note that trees of apparently greater age exist in nearby areas with large elongate green fruits. These trees appear to have survived from a population that was more generally distributed throughout the zone that encompasses the Lake of Maracaibo drainage system and the Venezuelan Andes.

The Anglo-Colombian Cacao Collecting Expedition surveyed the area in Colombia west of the border with Venezuela, which belongs to the upper valleys of rivers comprising the Catatumbo region that is being considered, for the purposes of this account, as a specific distribution region. In the area concerned the only type found in the river valleys consisted of trees in what was probably a sub-spontaneous state, with large elongate green-coloured fruits similar to those seen in the area west of the Lake of Maracaibo. The expedition also encountered a type in the forests on the range of hills that forms the border between the countries, the fruit of which (illustrated in Baker *et al.*, 1954) was dark red in colour with a smooth surface. Such a tree could be related to the 'Porcelana' type through migration downstream on one of the rivers of the region.

To the north of the Catatumbo region cacao cultivation was extended along the foothills of the Sierra de Perijá. Although no details are available regarding the present situation in the area in terms of the cacao population and diversity to be found, the area has been demarcated as an area of high frequency of *Criollo* genotypes. These plantations would have been established with seed from the sub-Andean population and later abandoned. Specimens have been obtained from trees in the forest but no descriptions of them are available.

To the southeast of the Venezuelan Andean range cacao was cultivated in what is now the State of Barinas. The area around the towns of Barinas and Pedraza was an important producer of cacao in the 17th century but declined in more recent times. Plantations were established on the slopes of the range and sub-spontaneous remnants of these can be encountered as attested by herbarium specimens collected in the area. All of these specimens appear to have been collected from trees with green fruits but possessing other *Criollo* characteristics. (For example, see L.J. Dorr, No. 7864, 1992, from the Estado Barinas, Districto Pedraza, Trail from Mesa de Canagua to Pozo Negro, with green fruits.) The seed for the plantations was most likely to have originated from the population on the northern Andean slopes.

The region to the west of the Lake of Maracaibo in Colombia, especially that along the Caribbean coast, is of considerable interest on account of the varieties of cacao that are found there. The northern foothills of the Sierra Nevada de Santa Marta, in an area approximately defined from Dibulla in the east to the suburbs of the city of Santa Marta, are noted for the significant presence of cacao that is a feature of the forest (Fig. 33). The first published description of the cacao of the area resulted from a survey of the area carried out by Robert Thomson (1894) and is noteworthy because of the details contained in the report. The population is of special interest on account of its size, the area of distribution covering several hundred square kilometres, perhaps larger than any contiguous area of cacao that existed in the colonial period.

The most appropriate manner to describe that variety is to reproduce the description given by Thomson (1894):

> The matured cacao trees attain a height of from 35 to 45 feet, with slender trunks devoid of branches to within a few feet of the top; and these trunks are as straight as those of a palm tree. All the mazorcas (pods) with few exceptions are borne among the sparsely foliaged branches at the summit…
>
> Among the cacao trees there is no variation whatever in the general form and size of the fruit. The predominating colour is yellow, though mazorcas of a reddish hue are not uncommon. It is interesting to note that the seeds in section are perfectly white. All are undoubtedly one true specific type.

The above report stimulated some interest in various quarters regarding what was known as 'Cacao Peñon' (in the sense of 'cacao of the hills') but no practical use of the variety seems to have been made as a result of Thomson's information, except for a few plantings in the area. Thomson (1894) reported that small plantings existed at the time at about 1000 m altitude exposed to full sunlight with the capacity to produce satisfactory yields. Such plants had a different appearance to those encountered under dense forest shade. However, notice must have been taken of Thomson's report on this unique variety because de Wildeman (1908) refers to cacao of a type called 'Peñon', which was being imported into France at the beginning of the 20th century.

The same area was visited by the Anglo-Colombian Cacao Collecting Expedition and the observations recorded confirm most of the details described in the above report. A sample of the fruits seen is illustrated in Baker *et al.* (1954) attesting to the divergence of the fruit type from the usual definition of *Criollo*. Unfortunately, no living specimens of this unique variety were obtained to enable its characteristics to be adequately known. The origin of a distinctive variety that is essentially uniform in genetic terms and distributed in a specific area on a large scale is a matter for considerable speculation. The questions regarding its origin devolve upon the possibility of introduction by the former inhabitants of the Sierra Nevada when they migrated from the south or during the colonization period from an unknown source. The Lake Maracaibo region does not appear to be the source of the seeds involved in the establishment of this area of cacao.

In the valleys of the western and southwestern slopes of the Sierra Nevada cacao also occurs cultivated in isolated pockets. These occurrences are derived from plantings made during the last two decades of the 19th century and during a later phase. They seem to belong to a variety whose genetic constitution similarly appears to be homogeneous but distinct from that along the northern foothills. The variety is composed of trees with red- and green-coloured fruits that are small in size with a very narrow elongate shape. The fruit surface is very rough. It was stated that the seeds used in the more recent plantings were introduced from the area of Don Diego on the west of the Sierra Nevada coast. If this information is correct it would appear that the cacao of the northern foothills is composed of two distinct varieties belonging to the *Criollo* group. Some plantations of cacao are known to have existed at the western end of the northern foothills during the beginning of the 19th century. These would have derived either from introduced types or the more ancient population along the coast.

From its beginnings in the sub-Andean region of western Venezuela cacao was transported east and northwards to form the basis for cultivated areas that extended as far east as the area around Cumaná and Carúpano (the Pária Peninsula) including the area of Nueva Barcelona as described by Humboldt (1821). The town of San Felipe was the centre of an important production area, mostly carried out by the missions according to historical accounts. However, nothing is known about the varieties cultivated during the period prior to the introduction of new germplasm.

The most important areas that developed during this expansion period were the valleys along the Caribbean coast that lie, mainly, in the State of Aragua and the Districto Federal. The most important valleys mentioned are shown in Fig. 33. Considerable attention was paid to these valleys, perhaps on account of the proximity to Caracas, which made them popular among visitors from abroad. The impression has been given that the production of these valleys constituted the product commercialized under the name of 'Caracas' but, in fact, cacao from all regions including the Sierra Nevada de Santa Marta massif was exported under this title.

On account of the familiarity with the valleys near Caracas some details about the constitution of the cacao populations are available through accounts of visitors and samples obtained from them. In addition, the fame of these plantations resulted in the exportation of genetic material to several producing countries in the Americas and in the Old World. As a result the cacao varieties of the valleys are better known than those cultivated elsewhere in Venezuela. Since the valleys along the coast would have been isolated from each other populations may have developed in the individual valleys with their own attributes as a result of natural and artificial selection.

The valley of Ocumare de la Costa, which is the most westerly of the important valleys, was, probably on account of the ease of access and its port, the area best known to visitors. The first of these visits was made in 1853 (Lisboa, 1866). This author visited the Hacienda Santa Rosa and stated that two varieties were cultivated, one of them named 'Colorado' was said to be native to the country and would have been a member of the *Criollo* group with red fruits. It is interesting to note that the author called the other variety *Preta* (= black), which would have been his translation of *Criollo* since this word, in the respective forms, is generically applied in the Caribbean and Brazil to persons of African origin. The characteristics of this variety are not described but it was stated that it had been introduced from Trinidad. This account is intriguing in that it suggests, if the author reported the information correctly, that the title *Criollo* was not applied to the Venezuelan varieties at that time.

One of the more reliable descriptions of *Criollo* cacao in the coastal valleys, prior to large-scale contamination (this was already taking place as the above report indicated) by foreign germplasm, was given by Preuss (1901). This account requires some editing and interpretation owing to repetitive statements and the fact that his observations would have been influenced by the information acquired in the countries he had visited previously. The descriptions given of the varieties in the coastal valleys of the State of Aragua refer to the places Preuss visited, and may not be entirely accurate regarding all the situations in the region at the time. Preuss would have stayed in the Ocumare area as his book contains a photograph of the hacienda in this valley that belonged to a Señor Fonseca.

In his description Preuss stated that three varieties were recognized on the basis of the fruit surface colour. These were: 'Cacao Legítimo' with dark red fruits; 'Cacao Amarillo' with yellow (green when immature) fruits; and 'Cacao Mestizo'. The last type would have had fruits of a green base colour with traces of red pigment. Their hybrid nature is indicated by the name and such trees were probably second-generation progenies where the anthocyanin pigment was diluted and expressed by factors of low penetrance. The red-fruited trees constituted 99% of the populations seen. The extremely low presence of green-based fruits may be ascribed to later introductions to the area. These trees would have been authentic members of the *Criollo* group as they had purely white cotyledons, indicating that such trees would have been self-compatible.

From the statements made with regard to the seed characters it could be inferred that the cotyledons of the genotypes with red fruits had a noticeable degree of purple pigment. The shape of the seeds was described as having an elliptic and almost round cross-section. The *Criollo* trees were distinguished by their apparent vigour or height but having sparse and open canopies and small leaves. The fruits were described as being of intermediate size, with deep furrows, an asymmetrical shape and a thin husk. Preuss' impression was that the base of the fruit was broad without constriction. The acute apex of the fruit was slightly curved so that it pointed towards the trunk.

Later Stahel visited Ocumare in 1923 and wrote an account of his visit to a hacienda named 'Guayabita' (Stahel, 1924). Here, about 50% of the population consisted of 'pure (*Legítimo*) *Criollo*' trees, derived from seed brought from the Chuao valley.

Although Preuss' description gave the impression that the 'Cacao Legítimo' was very uniform as regards the phenotypic characteristics, there is little doubt that significant variability existed within the cacao population in the area considered as a whole. The author visited the Chuao valley in 1961. The valley was considered to have been the least affected by introduced germplasm and the fact that the population was composed mainly of genotypes belonging to the *Criollo* group was confirmed. Also, it was evident that red-fruited types predominated, as was the case previously in Ocumare, the estimated proportion being 75%, most of the trees with green fruits being descended from foreign varieties. Plate 29a shows a field of progenies of genuine *Criollo* trees from the Chuao valley. Nearly all of the trees in this sample of the germplasm in the valley had red fruits but their shapes varied considerably, some having essentially smooth surfaces. There was a significant proportion of trees with the single plagiotropic branch character but even in these genotypes the trees attained heights taller than those of other varieties of comparable age.

One of the local varieties that had been picked out as possessing 'repute' was one from the Ocumare valley, described as belonging to 'the celebrated Ocumare estate', although which one of those named above is a matter for conjecture. However, it may be assumed from the account of Stahel's (1924) visit to the valley that this estate may have been the Santa Rosa, which he also visited. The variety concerned was first established in Trinidad at the beginning of the 20th century. From these plants seeds that were reputed to represent this type were sent to other countries, including the former West African colonies, where plants were established under the name of 'Ocumare'.

If the variety introduced into West Africa came from either of the farms mentioned above, from what is known about the fruit type, the progeny would have been fairly uniform having a red fruit with a broad and short shape. The fruit shape and general characteristics of the drawing of Venezuela-*Criollo* in Preuss (1901) resembled those of the 'Ocumare variety'. While the reports quoted above indicated that fairly large-sized seeds would have been the norm, those of the progeny seem to have been rather smaller in size. Since the seeds would have been produced in naturally pollinated fruits, in what perhaps had been a mixed population, it is possible that they were hybrids. However, the resulting progenies turned out to be poorly adapted to the new environments.

The only extant illustration of cacao from the Carúpano area, dating from the last half of the 18th century, shows a fruit of a light red colour, ovoid in shape, with a pronounced apex and a rough surface, but no accompanying information concerning its location is known.

Distribution Zone B: the Pacific Coast of Mesoamerica

The treatment of the complex of populations that are found in this zone cannot be accurate and will be at best sketchy in view of the author's lack of personal knowledge of the situation in the field, which prevents him from analysing the populations competently. In addition, a considerable problem has to be faced in this respect by the absence of reliable descriptions of the varieties cultivated during the historical period and of authentic specimens and illustrations of the archetypal varieties that existed prior to the introduction of genotypes from other origins.

The principal areas of pre-Columbian cultivation, which belong in this distribution zone, extend from a northern limit on the R. Tonalá in the State of Chiapas, Mexico, to a southern limit located to the south of Lake Nicaragua. These areas are actually confined to a narrow strip along the western foothills of the Sierra Madre and the Central Cordillera. Formerly, as was specially the case in Guatemala, cultivation was practised in a few areas further inland, such as around Lake Atitlán, but it is likely that these have declined to insignificant importance. These areas of cultivation do not by any means constitute a homogeneous region since cultivation was practised in isolated valleys and sites where conditions were suitable for the crop and the needs of the human inhabitants. The locations of the cultivated areas in the zone can be seen in Fig. 34.

Other areas of minor importance occur along the Pacific coast to the northwest of the limit of the main area of cultivation. One of these is at the Isthmus of Tehuantepec, where it seems that cacao is grown on a communal basis by the indigenous inhabitants. There was also a sizeable area of cultivation further west in the vicinity of Omotepec (Tecuanapa) on the border of the States of Guerrero and Oaxaca.

The most northerly section of the main Pacific coast distribution region is that of the Province of Soconusco. This section was the most reputed producer of cacao in the pre-colonization period and was the main supplier to the Aztec kingdom, a situation that continued for some time after the arrival of the Europeans. For various reasons the area declined in importance although the name 'Soconusco' remained as the standard for a superior quality product. During the centuries of Spanish occupation Soconusco was a Province of the Kingdom of Guatemala, being occupied by Mexico after independence. For this reason it is inappropriate to consider the region, from the point of view of the constitution of the cacao population, as being Mexican but as an integral part of the cacao population of northwest Guatemala.

The paucity of information regarding the characteristics of the cacao varieties that were present in the region during pre-conquest times or about developments during the colonial period, does not provide any basis for analysing the diversity existing in pre-Columbian times. The first known description of cacao in Soconusco derives from observations made during Friar Alonso Ponce's journey through Mexico and Central America in about 1586 (Anon, 1966). The tree and its product were described in comparison with trees known to the Spaniards. The account mentioned two varieties based on the existence of trees with fruits that were longish and others being round with pointed ends. This suggests that some variation in fruit shape existed in a range from elongate to short and broad. The seed number in average-sized fruits was 28–30.

In his treatment of the *Criollo* group varieties Cheesman (1944) gave the impression that the predominant fruit colour in the population he saw in Soconusco (as part of the State of Chiapas, Mexico) was green, this for all purposes being a very light shade of green. The only specimens from the area comprise several selections made on a farm at the southeastern end of Soconusco, the cacao population of which appears to have consisted of trees from a single origin. The genotypes concerned exhibit little variability although segregating for fruit colour, and some of them are so alike that it is difficult to

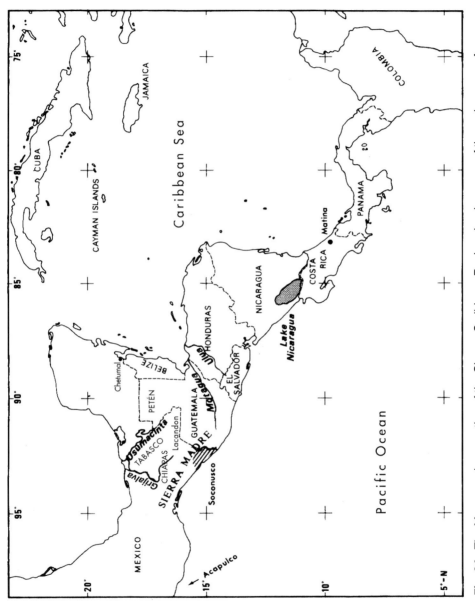

Fig. 34. The Mesoamerican section of the Circum-Caribbean Region showing some of the major areas of pre-Columbian and present-day cultivation.

distinguish between them on the basis of their phenotypes. It is suggested, perhaps because of the occurrence of seeds with pigmented cotyledons, that this population originated from hybridization. However, progenies obtained by inbreeding from some of these genotypes do not segregate for any known non-*Criollo* genotypes. In fact the progenies of seeds with white cotyledons possess all of the characteristics conforming to the current definition of the *Criollo* group.

In addition, trees having fruits of the 'Pentagonum' type, which similarly exhibit all of the characteristics of *Criollo* genotypes, were encountered in the same area. Fruits of one of these genotypes are shown in Plate 28b. The genotype is homozygous for anthocyanin pigmentation (the fruits illustrated appear to be green – resulting from exposure to low light intensities) as well as for slightly pink cotyledons. It also has the single plagiotropic branch gene.

Production of the aboriginal varieties in the principal zones of Guatemala (such as Suchitepequez, Mazatenango and Escuintla) has virtually disappeared owing to several factors, including the introduction of hardier varieties of cacao. As stated above these zones could be treated as extensions of the Soconusco Province and the cacao diversity would be the same. The existence of variability in fruit forms within these populations (at the stage when the introduced varieties would not have become generally established) was noted by Bernoulli (1869), who described three fruit forms to which he gave species status. The fruit form considered to be typical of the *Criollo* was identified as *Theobroma cacao*. A type with fruits having smooth surfaces was denominated *Theobroma leiocarpa* and the third form with 'pentagonal' shaped fruits was called *Theobroma pentagona*. As noted in Chapter 2 the term *leiocarpa* (*-um*) has caused some confusion among names attributed to cacao varieties but there is no doubt from the drawing in Bernoulli's paper that the type also belongs to the *Criollo* group.

Preuss (1901) also described the cacao situation in Guatemala as it would have been at the time of his visit and found a significant degree of variability in the composition of the populations that he saw. The descriptions do not show clearly the extent of the influence of the germplasm that had been introduced from Ecuador some 40 years previously. One of the varieties seen was said to have possessed fruits with smooth surfaces without ridges – similar to the 'Calabacillo' of Trinidad. Since Preuss wrote that all of the types seen had white cotyledons, this variety probably was the same as Bernoulli's '*leiocarpa*'. At the other end of the range of shapes were fruits that were very elongate with acute apices and very rough surfaces – corresponding to Bernoulli's illustration of *T. cacao*. On the basis of Preuss' statement that all of the types he saw had seeds with white cotyledons it could be concluded that they belonged to the *Criollo* group. Among the variety names used the term 'Lagarto' occurs. The term was picked up during his stay in Nicaragua, prior to visiting El Salvador and Guatemala. The term will be explained when the Nicaraguan varieties are described.

Even by the 1930s it was reported that the occurrence of trees identifiable as belonging the *Criollo* group on the western slopes of Guatemala was rare. A programme of selection in Guatemalan cacao farms was carried out in the 1940s. Of these selections known to the author only one genotype represents the aboriginal varieties. This genotype has a green fruit with a broad shape and is self-compatible. Progenies of this genotype, obtained by self-fertilization, segregate for all the characteristics that are found in the group, including the single plagiotropic branch habit and pentagonal-shaped fruits. Most of these progenies produced fruits of green colour that have shapes identical to the fruit of *T. leiocarpa*, illustrated in Bernoulli (1869), as well as the smooth surface.

The next country in the sequence of cacao producers is El Salvador. This country was also part of the Kingdom of Guatemala prior to independence and it could be assumed that there is a common thread among them in terms of the germplasm that was traditionally cultivated. This supposition cannot be confirmed, however, since no information exists regarding the varieties in that country. There were two main areas involved, Izalco and Sonsonate in the north and San Miguel in the south. Cultivation extended to other areas in the interior of the country. Several of the older areas were destroyed by volcanic eruptions and vestiges of cacao material on these sites have been revealed by archaeological investigations. The report of the journey by Friar Alonso Ponce in 1586 (Anon, 1966) described the production in the area of fruits of *Crescentia cujete* that were exported to Mexico to make the cups used by the Mexicans to drink the cacao beverage. In this report it was stated that they resembled the fruits of cacao. Since the *C. cujete* fruits tend to be round in shape with very smooth surfaces the description may be interpreted as suggesting that cacao varieties were present with the same characteristics.

On the other hand, Preuss (1901) had the opportunity of seeing cacao in El Salvador, including the Sonsonate region. On one property visited in the highlands the population comprised trees with red and green fruits that had very rough surfaces. In his description of another property located west of Sonsonate, he referred to the presence of a variety having fruits of the 'Lagarto' type and another similar to *Criollo*. The reason for the distinction was unexplained but it is probable that the fruits of the former type had very rough surfaces. Both red- and green-coloured fruits were present. The Salvadorean cacao had the characteristic of possessing seeds that were of small size but plump and rounded in shape. Preuss also noted the association between fruit surface colour and the presence or absence of pigment in the cotyledons. The preceding descriptions indicated that the composition of the populations seen would have been similar to those encountered in other zones of the distribution region.

Just further south of the San Miguel cacao area of El Salvador is the coastal region of Nicaragua situated to the west of Lake Nicaragua. Oviedo y Valdez spent some time in this area and he was the first person during the post-conquest period to give a comprehensive description of cacao, its cultivation and uses (Oviedo y Valdez, 1855). In this account he states that the Chorotegas, the original inhabitants of the region, did not cultivate cacao. The *Nahuatl*-speaking Nicarao tribe were the only cultivators of cacao. Therefore, it could be concluded that this tribe had migrated to the region from an origin related to the Mexicans and would have brought the planting material with them, probably in the form of fruits from which the seeds were extracted for planting. The system of cultivation and the uses would likewise have been introduced with them. The variety described by Oviedo y Valdez would have derived from the diversity that existed in the sub-region, unless it was introduced from the Caribbean coastal region.

Oviedo y Valdez (1855) states that the names given to the fruits were 'coco' or 'cacao' or 'cacaguat', information of possible use in identifying the origin of the tribe. Presuming that the planting material employed derived from a few fruits (or even a single fruit) from a given plantation it would be expected that the progenies, from which fairly large areas of cultivation were developed, would have comprised a single type with a narrow genetic base. In Oviedo y Valdez's account (author's interpretation of the Spanish used) the colour of the fruits was described as green, partially tinted with red. The fruit size was given as a hand's breadth in length and the diameter that of a man's wrist – in the largest fruits. There are four rows of seeds numbering about 20–30. A sketch of a cacao plantation is appended to the volume concerned, this being referred to in the text. It can be concluded that, if the sketch was made at the time the work was written, it would be the oldest historical illustration known of the plant.

From the above description it may be inferred that the fruit characteristics of the variety were uniform, the shape being narrow elongate. From the details regarding the fruit colour it appears that the original fruit(s) was/were heterozygous for the colour combination.

Although several persons who had spent time in Nicaragua refer to cacao in their writings, the first observer who could be considered to have had the ability to provide a competent assessment of the cacao varieties was J.H. Hart, who worked in Trinidad at the time. Hart (1893) reported on the visit he paid to Nicaragua in 1893. He noted that, at the time of the visit, the cacao population was very heterogeneous with fruit types occurring that were similar to those of the complex that occurred in Trinidad. However, it appeared that, regardless of the external fruit characters, the seeds were larger and 'fuller' than those in the Trinidad population. The cotyledons were all white or slightly tinged with a purple colour. This range of expression was associated with the colour of the fruits.

The fruit type that caught Hart's fancy was the one that was not present in Trinidad, namely, the one called 'Alligator'. The specific characteristics that differentiate this fruit type from the others were described in detail, as follows:

> The Alligator cacao has, however, only five ribs, and instead of these being the prominences between the indented lines they are actually that portion which carries the indented line itself raised and formed into a prominent rib, at the same time showing the indented line down the full length of each rib.

Hart's (1893) report, therefore, indicated that although the individual trees that comprised the populations he observed were heterogeneous in terms of external fruit characters, they were all related to the general concept of *Criollo* through the possession of the genes responsible for the presence of unpigmented cotyledons.

Hart took samples with him to Trinidad, as seedling plants (presumably from seeds with white or slightly pigmented cotyledons) of the 'native' Nicaraguan varieties that he itemized as:

> Best Nicaraguan variety 'Nicaraguan Creole' – 150 plants.
> 'Alligator variety' – 75 plants.

The plants from these samples that were established in Trinidad, and their descendants, constitute much of the present knowledge about the Nicaraguan population as it existed at the end of the 19th century. Before discussing these samples further it is necessary to emphasize that other varieties had been introduced into Nicaragua from Trinidad prior to 1893, and that the introductions' performance was superior to that of the 'native' varieties. The importance of the presence of the introduced varieties concerns the possibility that some of the plants, of which the samples were made, since they resulted from natural fertilization, had been contaminated by foreign pollen.

The observations of Hart were augmented by those of Preuss (1901), whose account of his visit to Nicaragua helps us to place the cacao of the country in relation to the regional diversity. It is obvious from Preuss' report that conditions in the area were marginally favourable for satisfactory cacao cultivation and the total population at the time was relatively insignificant. He added that several varieties were cultivated, of which two could be considered to have been indigenous. The main variety was called 'Cacao del Pais', which would be the equivalent of Hart's 'Nicaraguan Creole' belonging to the *Criollo* group and including the type named 'Lagarto'. The 'Lagarto' is synonymous with the 'Alligator' described above and both names derive from the knobby appearance of the fruit surface. In some cases this phenotype is associated with the 'Pentagonum' fruit type but Preuss mentions that 'Lagarto' was also applied to fruits with similar surface protuberances other than those with the 'Pentagonum' fruit shape. There is no doubt that this type is merely a variant of the traditional *Criollo* variety.

The trees of the 'Cacao del Pais' had small leaves and canopies with sparse foliage. They produced fruits having both red and green colours, with the cotyledons of the red fruits being pigmented and those of the green fruits unpigmented. The husks of the fruits were very thin. Preuss found that, in many respects, the Nicaraguan variety was similar to those seen in Venezuela except for a more attenuate apex to the fruits and seeds of considerably larger size, as well as being larger than the seeds of other varieties known to him. The *Criollo* variety was less productive and less adapted to the conditions than the introduced varieties.

Preuss went on to describe the population cultivated on a small property near Matagalpa, where he observed a significant degree of variability in the fruit characters. These phenotypes ranged from a small fruit with a smooth surface of a green colour with an acute apex to various kinds of elongate fruits of red and green colour and rough surfaces. However, all types had large seeds with white or light purple cotyledons and thin fruit husks. If this population derived from the 'Cacao del Pais' the variability would indicate that the trees of the variety of this name would have been, at least in part, heterozygous. The fruit with the smooth surface may have corresponded to Bernoulli's *leiocarpa* fruit as illustrated in his paper. A Guatemalan population similar to the one in which Bernoulli found the variants he described is a possible source of the diversity observed at this location. Or, to extend the argument further, the same type having round fruits and smooth surfaces with pointed apices described as occurring in Soconusco was distributed throughout the region.

Once Hart's sample of plants had established the novelty of the 'Alligator' type and the expectation that, being an authentic member of the *Criollo* group, its cultivation would improve the quality of the Trinidad cacao population, considerable efforts were made to disseminate the variety in the island. These efforts were not restricted to Trinidad and seed from the descendants was exported to several countries in the Caribbean and the emerging Old World producers. The fact that the progenies of the plants established from the introduced sample exhibited the same characteristics as the parents demonstrated that the type was essentially homozygous for the characters that produced the 'Alligator' phenotype. Since it is obvious that the plants that comprised the introduced sample had fruits of green colour all known progenies possess this trait.

The plantings formed from the first descendants of the 'Alligator' sample performed erratically under Trinidad conditions and few appear to have succeeded. Certainly no pure stands are recorded as existing 30 years later. Since all of these plantings had green fruits the inference was that the Nicaraguan population was composed entirely of green-fruited trees. However, Hart had mentioned in later publications that the 'Nicaraguan Creole' contained plants with red fruits. Presumably seed from such of the trees of this sample that had been established were also distributed. However, in the absence of details concerning the sites of these plantings it is difficult to distinguish such individuals from those that would be related to the varieties that originated in Venezuela and their descendants.

When the first programme of selection of individual trees in the populations of Trinidad and Tobago was carried out one of the criteria was the seed size. The progenies of the Nicaraguan introductions involving two or more generations from the parents, which comprised the original sample, possessed seeds that were significantly larger than those of the other components of the base germplasm. As a consequence of the preference for the large seeds the Nicaraguan genotypes make up a considerable proportion of the first series of selections. Subsequently, other genotypes were chosen within these lines and added to the resulting assemblage of selections.

Most of the selections that are descendants of the Nicaraguan varieties were chosen on the basis of vigour as well as their seed size, and they would have been the genotypes that were better adapted to the conditions prevailing in Trinidad. Several of the selections were outstanding in terms of production and, consequently, received a great deal of attention both for purposes of cultivation and use in the breeding programme and they continue to be important components of many germplasm collections. These selected clones, on the whole, although all have fruits of a green colour, segregate for white and purple cotyledons of a range of intensity. For this reason it has been suspected that these clones are of hybrid descent. However, all of the clones in this category are self-incompatible and possess the genetic mechanism for the trait associated with the *Criollo* group. The descendants of inbreeding or hybridization with self-compatible genotypes derived from the Nicaraguan introductions, segregate for the 'Pentagonum' fruit type and the single-branch habit. They also exhibit the deterioration, as they get older, of the vegetative parts of the trees that is a feature of the genotypes belonging to the *Criollo* group.

The selections made in Trinidad include one clone ('ICS 38') with the typical 'Pentagonum' fruit type but, in contrast to the narrower fruits of other genotypes, those of this clone are broad and shorter in shape and the 'knobby' surface is more pronounced. Another clone of the same origin that has been subjected to study and evaluation is 'ICS 45'. This genotype is more typical of the Nicaraguan population as described and is self-compatible. The inbred progeny segregate for a variety of fruit shapes and characters ranging from elongate fruits with smooth surfaces to the 'Pentagonum' type. All of the progenies have seeds with white cotyledons. Most genotypes of Nicaraguan *Criollo* descent have fruits with a darker green surface colour than is the case with the populations in the northern zones of the distribution region.

The situation in Costa Rica differs from that in the other countries in this distribution zone on account of its small size and essentially narrow shape. Here, except for the plains in the north of the country, conditions for growing cacao are limited to narrow strips inland from the shores of the Caribbean Sea and the Pacific Ocean. For this reason it is convenient to consider the two growing zones together.

Oviedo y Valdez (1855) does not mention the occurrence of cacao in western Costa Rica and references to the species in documents from the colonial era are very ambiguous about the subject. It may be assumed that cacao was not present in pre-Columbian times except along the rivers that drain into the south of Lake Nicaragua and perhaps in the district of Talamanca in the southeast of the country where *Criollo* varieties are still found. The question whether cacao was native to Costa Rica was discussed by Pittier (1902b) but he did not arrive at any definite conclusion.

Pittier (1902a) described the *Criollo* varieties grown at that time and indicated that trees with red fruits were most common. He mentioned an unusual type that was called 'La Palma', the name presumably being given on account of the possession of the single plagiotropic branch habit. The description of the fruits agrees in every detail with that of the Costa Rican 'Pentagonum' type in Plate 28a.

Distribution Zone C: the Eastern Sector of Mesoamerica

Similar to the situation that applies to the Pacific coastal areas the absence of reliable information about the varieties that are encountered in the zone hinders the ability to arrive at an analysis of the aboriginal cacao populations in this zone. Whatever information is available in published records does not provide the detailed characterization or observations that are required to make a satisfactory and complete analysis of the populations concerned.

For the purposes of this account the cacao populations that existed in the Eastern Sector in the pre-Columbian era in Mesoamerica are considered to have been distributed in the plains between the eastern foothills of the Sierra

Madre and the Caribbean Sea. The northern limit of cultivation could, perhaps, be situated near the Tascalan capital of Cempoala (certainly in the State of Veracruz) and the southern limit the region around Lake Izabal. Cempoala is included since it is the place where the use of cacao is first mentioned in Díaz del Castillo's (1908) history of the conquest of Mexico by Cortés, but it is likely that the seeds used by the Tascalans were imported.

Cacao cultivation in the Lake Izabal region was described in some detail by Cortés in his fifth letter to the King of Spain, the text of which is appended to Díaz del Castillo (1908). This region is located in Guatemala near the border with Honduras and the entire region also embraces the neighbouring areas of the latter country, such as the valley of the R. Chamelecon. Although the Lake Izabal region appeared to have been important during the 16th century cacao disappeared in time and was substituted by sugarcane following the invasions by pirates.

In addition to these main areas, cacao was cultivated to some degree in other areas. The coastal region of the Caribbean, in what is now Belize, and which, in some way, is a continuation of the Lake Izabal region also has remnants of the cacao plantations established by the Mayas. In the same letter (in Díaz del Castillo, 1908) Cortés also recorded that during his march to Guatemala he visited some cacao farms in the area of what is now southern Belize.

The first notice of cacao in the region, as reported by Oviedo y Valdez (1855), is contained in the account of the voyage made in 1510 by Alonso Pinzón along the Caribbean coast of Yucatán to Guatemala. Alonso Pinzón described the production of cacao in the Chetumal region of southeastern Yucatán. The region concerned is most likely to have included the valleys of the rivers New and Hondo, the latter forming the border between Mexico and Belize. Although the Chetumal region was noted as a major producer of cacao in the Maya era it has not been possible to obtain information concerning the present day situation.

Alonso Pinzón was unable to land on the coast south of Chetumal. Otherwise he would probably have encountered the continuation of cultivation in the valleys to the south. Among these, in the sequence, is that of the R. Sibún, where specimens have been collected dating from the 19th century. In the same sequence would follow the valley of the R. Toledo, which is one of the rivers draining the Maya Mountains in the hinterland of Belize. Through historical records and studies by archaeologists a few of the sites where cacao occurred are known. Some of the reports from more recent times refer to cacao trees lost in the forests. These would be remnants of the original cultivated varieties around which the forest would have regenerated. During the past centuries the cacao trees may have been exploited by the inhabitants who occupied the territory during intermittent periods.

From the scanty information available from previous reports on the Belize cacao and more recent collecting activities it can be concluded that the populations at such sites were composed of genotypes that could be attributed to the *Criollo* group in the broad sense of the term. However, it is evident that a certain degree of variability occurs in the sub-spontaneous populations of the region taken together, but it is not possible at this stage to describe distinctions among local populations. Fruit colours include both red and green. Most fruits seen in illustrations are elongate in shape but variability occurs with regard to their diameters and to the shape of the basal half.

A brief description of a sample from 'indigenous' cacao in Belize (British Honduras) given by Hart (1910) sums up the characteristics of this cacao population as it applies to the material collected in recent years. In his description Hart wrote that the trees grew in the forest and were very prolific. The fruits were light green in colour but some had a pinkish surface. They had a length of 5 inches and a diameter of 3 inches. The seeds had light purple cotyledons.

One of the first persons to give some idea of the cacao populations in the south of Belize was Gillett (1949), who made some useful comparisons with other cacao types. He described trees with fruits that, although elongate, had a more obtuse apex than that considered to be generally characteristic of the *Criollo* group varieties. Some fruits were of very large size and narrow elongate in shape. Husk thickness seemed to vary considerably between the varieties that were seen at various sites, some of the fruits having very thick husks. He also came across trees with unusual vegetative features. Gillett (1949) stated that the varieties seen appeared to be similar to some of the types he observed in Grenada. In Chapter 6 the *Criollo* element of the cacao population of that island will be described as probably originating from Venezuela.

Additional information about the constitution of the cacao population is derived from a few herbarium specimens, some of which were collected in the 19th century and presumably prior to any introductions. One of these sources refers to cacao trees growing in apparently virgin forest which had red fruits with a smooth surface (cf. G.R. Proctor, No. 36134, 1976, Belize, Toledo District, Columbia Forest Reserve). However, the specimen examined possesses other characteristics on the basis of which it appears to belong to the *Criollo* group. The specimens resulting from recent conservation activities have not been characterized to a sufficient degree to provide a satisfactory basis for analysis of the diversity.

With the decline in importance of the cacao populations that were cultivated at the time of the conquest along the Caribbean coast and its hinterland, the major area of the occurrence of cacao lies further north, principally in the Mexican State of Tabasco and the zones contiguous to it. For the purposes of dealing with the cacao populations it is convenient to consider their area of distribution as a single ecological zone with the title of the 'Tabasco region' or Gulf of Mexico hinterland. The area covered by these names is shown in Fig. 34. In pre-Columbian times, when cultivation was practised by the Mayas, the crop would have been planted throughout the region without the political divisions that exist today.

For the purposes of defining the distribution of cacao cultivation the ecological zone would cover the lower stretches of the R. Grijalva and inland from the coast and the entire drainage system of the R. Usumacinta. Therefore, the distribution area goes beyond the borders of Mexico and includes parts of

Guatemala, especially the Departments of Alto Verapaz and the Petén. In view of the possibility that some cultivation of cacao may have been carried out in the Yucatán Peninsula in ancient times, especially prior to the great drought that is considered to have affected the area around AD 700, the vestiges of the original populations should be included.

Coe and Coe (1996) proposed that the Olmecs were the first people to prepare the drink made with cacao beans, known as chocolate. If this conclusion is correct, there are two possibilities to explain the presence of cacao in the region during the Olmec period: either the species was present when the Olmecs began to occupy the hinterland of Tabasco and Veracruz; or they introduced it to the region. Cultivation of the species by the succeeding Maya tribes would have continued through the centuries until the area was colonized and later by their successors. Such cultivations must have been established throughout the Usumacinta drainage system to supply the numerous ceremonial centres that are to be found in it.

In accordance with what is known about the agricultural systems traditionally practised cacao plantings would have been confined to the ponds, or 'aguadas', scattered throughout the lower reaches of the valleys and also in the 'cenotes' or subterranean water holes found to the north in Yucatán. The report of the visit to Soconusco made in 1586 (Anon, 1966) also refers to the small, scattered cacao farms in the lower Yucatán and hints that in comparison with the Soconusco area production was low. Unfortunately, at this time it has not been possible to consult the part of the report that deals exclusively with the Yucatán and Tabasco region.

In terms of the distribution of variability in the population the possible consequences of this system of cultivation may have been the creation of isolated pockets of variability derived from small quantities of planting material, and maintained over several generations. This is in keeping with Martínez' (1912) statement that the cacao cultivation in the region, up to the stage when expansion of cultivation was promoted, was carried out in properties with small numbers of trees, the maximum perhaps being 1000 trees. The seeds utilized in the continuous process of replanting and expansion of cultivation produced by the small and isolated producers must have descended from the original varieties through many generations of inbreeding that would have occurred during the centuries since the establishment of the crop in the region. The consequence of this situation would be the existence at the present time of populations or varieties that are distinct but individually composed of homozygous individuals.

The fact that, at the time of colonization, cacao cultivation was based on a range of varieties is emphasized by the description of cacao in 'New Spain' given by Hernández (1959). This account of the natural history of Mexico is said to have been written in the period between 1571 and 1576 (less than 50 years after Cortés expedition and, therefore, referring to an unaltered state of the cacao populations). It would be pertinent to assume that the description related to the cacao of the Tabasco region. The general description of the species refers to the fruit having an oblong shape (in appearance compared to a 'large melon'), ridged and of a red colour.

Four varieties were described, these being:

1. 'Quauhcacáhoatl', which had the largest trees as well as the largest fruits.
2. 'Mecacacáhoatl', the trees of which were of average height and size and having smaller fruits than the previous class.
3. 'Xochicacáhoatl', the trees being smaller than the preceding classes and with smaller fruits, the seeds being red on the outside but the interior (cotyledons) similar to the others.
4. 'Tlalacacáhoatl', that is to say 'small' (trees) and also giving the smallest fruits, but of the same colour as the preceding types.

He added that all of these varieties had the same characteristics and were used for the same purposes except that the last class would be more appropriately restricted to preparing drinks while the others would also serve as money.

Although the classification of cacao variability described by Hernández has been cited innumerable times no attempt appears to have been made to compare the varieties that exist at present to the descriptions in this book. Hence, it would be an interesting exercise to relate the existing varieties that can be identified as descendants of the aboriginal *Criollo* to the four classes described.

Martínez (1912) gave a summarized classification of the cacao varieties that were recognized by specific names in the State of Tabasco. These varieties obviously included some that were introduced, without specifying which varieties belonged to each origin. From this classification it may be concluded that, at the beginning of the 20th century, the scale of variability existent in the region was significant. The main classification of the types that could be attributed to the *Criollo* group indicates that the population was made up of trees with both red- and green-coloured fruits as well as differences in surface rugosity. There were also differences in the intensity of anthocyanin pigment in the red fruits. From the weights given in the descriptions of the various fruit types, some of the 'variety' classes produced fruits of a very large size. However, the weight of fresh seed contents of the fruits appeared to have been the same regardless of the colour/surface roughness class. It is interesting to observe that the weight of seed contents of the fruits with a dark red colour was significantly lower than that of the other types that comprised the same population.

Martínez (1912) also included a type named as 'Cacao Lagarto'. This was described as having fruits that were broad with very rough surfaces, of a small size and with colours ranging from light green to dark red. Elsewhere he relates this type to the 'Pentagonum', stating that it had been introduced from Central America and that it was grown only as a curiosity. In other words, the 'Pentagonum' type is not a natural element of the diversity of the Caribbean zone. A coloured drawing of the type in the article shows a very dark red fruit.

The author goes on to deduce from the observations made that the largest fruits were those of the green and rough class while the trees with red fruits having rough surfaces were the most productive. He also stated that the farmers tended to select the parents from which they obtained seed for planting on the basis of fruit colour and not on productivity. Selection based on this criterion would lead to the formation of fields in which one colour or the other would predominate, resulting in populations that would have appeared to be homogeneous. Although the preferred fruit colour was not indicated it could be presumed that preference was given to the red fruits. There would have been unconscious selection for higher productivity if the relationship stated above had been true.

Preuss (1901) did not visit Tabasco but was able to analyse a sample of fruits sent to him. These were very large in size and green in colour. They were distinguished by their very rough surfaces with high ridges and exceptionally thick husks. The cotyledons were white but it appeared that, in spite of the large fruits, the seeds were smaller than those of the other Mesoamerican varieties seen by him. The fruits in this sample may not have represented the principal type grown in the Tabasco region as they could have been specially selected but there was no doubt that the variety differed substantially from those cultivated in the Pacific zone.

The extensive variability of the cacao populations in the Tabasco region at the time of his visit with regard to the elements that could be attributed to the *Criollo*, or prototype, diversity, was pointed out by Soria (1961). In an abandoned plantation on the R. Teapa, he observed all forms of fruits and seeds in a population composed entirely of genotypes with white cotyledons. The most common feature was the acute apex, some of the fruits exhibiting the curved apex associated with many of the types belonging to the *Criollo* group. Also frequent was the soft, or fragile, husk regardless of the thickness. In addition, trees having rounded fruits occurred but whose seeds had white cotyledons. In the absence of an exact description of these fruits it is not possible to relate them to any other variety, but, as will be explained in the discussion of the cultivated populations of the Mesoamerican zones, there are other genotypes that possess similar combinations of characters. Similar descriptions of trees in the same area were reported by López-Baez *et al.* (1989).

Soria (1961) gave a fuller description of the characteristics of the *Criollo* genotypes, stating that it was easy to distinguish them from the introduced varieties by their appearance. The *Criollo* type trees were of small size with rounded canopies and inferior growth. The leaves were small and ovate in shape, with a thick texture, the colour being light green. The flowers were distinguished by their very wide ligules that had a dark yellow colour and by the lighter shade of purple colour in the staminodes.

During the last decades of the 20th century interest has been focused on the cacao populations that inhabit the forested area of the R. Usumacinta drainage system. Historical accounts of the area to the west of the river make it clear that until fairly recent times it was untouched in terms of development because of the savage opposition to outsiders by the inhabitants, the Lacandon tribe, after which the area in Mexico is named. The location of the Lacandon area may be determined from Fig. 34. In practice, the cacao distribution area extends into the Petén on the east of the river in Guatemala where the species was exploited by the Carib Indians brought to the east coast in the late 18th century. As explained in the introduction to this section the Lacandon/Petén region abounds with sites of Maya occupation and for this reason it must be accepted that the cacao was once cultivated in it and that its occurrence in a seemingly 'wild' state is sub-spontaneous.

The first useful description of the diversity encountered in these sub-spontaneous cacao populations was provided by López et al. (1989). The fruits of the trees seen in occurrences of low density were described as ovoid in shape and their size as being small to average. The husks were thin with not very rough surfaces that were green in colour. All seeds had white cotyledons. On the basis of their ability to produce fruits in their low density and isolated state they were considered to be self-compatible. These observations of the fruits were confirmed by the information provided by H.C. Evans (personal communication), who visited the region in 1999. Some of the fruit types seen are depicted in Plate 30a. At the location at which these fruits were harvested segregation occurred for fruit colour from dark green to dark red even though it appears that there was little variability regarding their shape. The most striking features of the particular sample are the shiny surface with the absence of pronounced rugosity as well as the shape that is broader near the base.

Trees at another location seemed to produce only green-coloured fruits that ripen to an unusual shade of yellow. Again, the fruit size was small and the husks slightly rough and thin. The trees producing this type of fruit, growing under heavy forest shade, developed into thin tall trunks with sparse branching. The description of the trees as well as their fruit characters is strikingly similar to the description of the 'Cacao Peñon' of the Sierra Nevada de Santa Marta in Colombia, described in the section on the South American range of the Circum-Caribbean Region.

Although the above observations made in the plains of the R. Usumacinta system indicate that a fairly uniform variety is distributed in the area it remains to determine whether it occurs in the region to the south below the headwaters of the rivers that form the R. Usumacinta drainage system. These sources lie in hilly terrain located principally in Guatemala, in the Departments of Alta Verapaz and Petén. Although various reports refer to cacao having been planted in this area as far back as the 17th century there is no information about its existence prior to the advent of the missionaries or the sources of the seeds used in planting. In the light of the topographical conditions it is probable that whatever occurrences of cacao may be encountered would consist of small, isolated populations. Indications concerning the cacao varieties that compose these populations are lacking. The only information that provides any clues on the subject is obtained from a report of a rather superficial collecting activity that produced few specimens carried out in the Alta Verapaz Department described by Rivera de León (1986). The scarcity of cacao in the areas surveyed is evident as well as the isolation of the fields.

All of the six specimens described possess characteristics that place them in the range of the *Criollo* group, the seeds having white cotyledons. The description of the fruit of one specimen suggests that the tree may be related to the 'Lacandon' variety. The other trees sampled appear to have had rather similar fruits but of another type. Some were of a small size but their shape was narrow and elongate. All were described as possessing acute apices that were curved in the manner considered as being one of the features of the varieties composing the traditional Mesoamerican populations.

With the disappearance of the areas of cacao cultivation in the R. Polochic drainage system, including Lake Izabal, no means are available of knowing the variety situation in the border region of Guatemala and Honduras. Although the species had been cultivated during certain periods in the various zones of Honduras no details about the origins and characteristics of the cacao populations have been found. It may be pertinent to consider the genotypes found in the upper valleys of the R. Usumacinta system as relics of the original R. Polochic populations.

Only recently has it been possible to become acquainted with a genotype belonging to the *Criollo* group that occurs in southeastern Honduras in the valley of the R. Patuca. That such a variety is indigenous to the valley is questionable since the historical records refer to the valley being occupied by foreign pirates during various periods. In addition, the indigenous tribe that inhabits the valley at the present time belongs to a Chibcha-speaking group. This situation combined with the apparently homogeneous status of the variety raises the possibility of its introduction from the South American mainland. The colour of the genotype's fruits is a darker shade of green and their shape is broader elongate. The appearance of the fruits is similar to those of the variety from Mérida found in the Dominican Republic or the original variety of western Nicaragua. The seeds have white cotyledons and this trait is associated with light green flush leaves. Although there are indications of segregation for the single plagiotropic branch character this has not been proven so far. The genotype differs from most of the other varieties ascribable to the *Criollo* group in producing small flowers.

From the above account it can be deduced that a small number of samples have been collected among the populations in the Caribbean Coastal Zone. Consequently, they constitute a poor representation of the variability that existed in the zone when its history began. These collected genotypes are absent in genetic resources centres and other institutions where they can be analysed in breeding programmes and their genetic constitutions determined. As a result we are unable to determine the composition of the populations to which they belong and attribute them to their place in the spectrum of the diversity. In spite of the variability expressed in the aggregation of the varieties ascribed to the *Criollo* group it may be pertinent to point out that these differ from those in the Pacific Coastal Zone by the absence of the natural occurrence of the 'Pentagonum' fruit type. A degree of association with the northern South American genotypes appears to exist. This association may not have occurred entirely as a result of migration of the species in prehistoric times. The possibility cannot be discarded of cacao from the Lake Maracaibo Sub-Andean area being introduced into Mesoamerica during the period when the 'Caracas' type made up the bulk of the cacao that was commercialized.

6 The Cultivated Populations as Secondary Depositories of the Diversity

Introduction

The origins of the domestication of cacao are unknown but evidently of considerable antiquity. From the references to the subject in Chapter 5 the logical conclusion is that domestication would have begun in the natural distribution range as the inhabitants began to discover uses for the plant. The initial spread of domestication would have taken place within this range. It has been made clear that the present distribution of the diversity involves spontaneous and domesticated populations whose status we are unable to determine. In the course of time, as the inhabitants migrated to other areas, the species was extended to regions of the Americas beyond the range of spontaneous occurrence. Further expansion occurred some time after the conquest both within the Americas and the tropical regions of the Eastern Hemisphere. The map in Fig. 35 demonstrates how this expansion took place with indications of the movements of planting material that are known or presumed to have occurred.

Because of the difficulty in distinguishing the spontaneous occurrences in the natural distribution range from those that derive from human intervention, it was necessary to deal with all of the variability encountered in the locations within the range of natural distribution as being directly derived from the primitive diversity. This applies to the populations in which cultivation was known (or presumed) to have been practised within the areas concerned, some of which continue to be exploited in modern times.

Because of the ample variability in the conditions and circumstances in which cultivation has been practised it is impossible to generalize on the processes involved in the development of these populations. The basic factors that are concerned in the creation and development of cultivated populations and their genetic constitutions were discussed in Chapter 4. In this context it is worthwhile reiterating that the genetic composition of any individual secondary

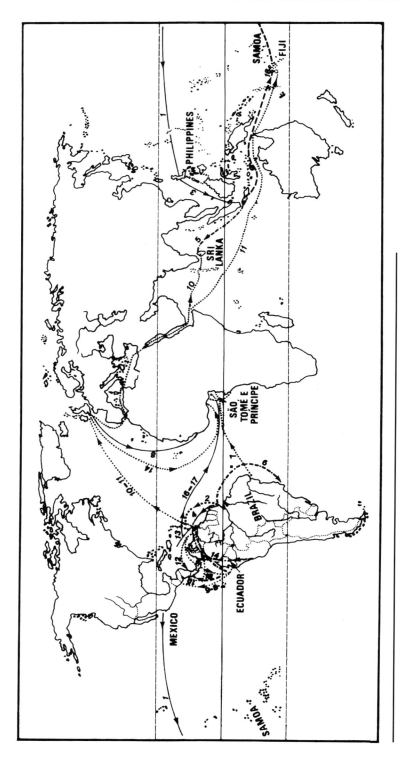

Fig. 35. Map of the world showing the principal routes of the movement of cacao germplasm within the American continent and islands and from the American continent to the Eastern Hemisphere. **1.** 1660–1670, Mexico to Philippines; **2.** 1664, Amazon to Martinique; **3.** Philippines to Indonesian Archipelago; **4.** 1757, Amazon to Trinidad; **5.** Early 19th century, Indonesian Archipelago to Ceylon; **6.** 18th and 19th centuries, Amazon to southeastern Brazil; **7.** 1822, Brazil to Príncipe; **8.** 1840s, Dublin to Sierra Leone; **9.** 1861, Ecuador to Guatemala; **10.** 1880/1881 Trinidad (via England) to Ceylon; **11.** 1883, Trinidad (via England) to Fiji; **12.** 1892/1893, Trinidad to Nicaragua, Nicaragua to Trinidad; **13.** 1898, Trinidad to Costa Rica and Colombia; **14.** 1890, Venezuela to Ecuador; **15.** 1930s Ecuador to Costa Rica and Panama; **16.** 1880s Trinidad, Venezuela and Ecuador to São Tomé; **17.** 1899, Trinidad, Venezuela, Ecuador and Central America to Cameroon; **18.** End of 19th century, Indonesian Archipelago to Samoa.

population would depend on that of the source population and that the secondary populations would consist of descendants of small samples from the source populations.

The cultivated populations have been developed in various ways with regard to the genetic materials from which they have been constituted. The simplest situations involve a single variety, which, if composed of homozygous individuals, would result in homogeneous populations. If the individuals introduced to a given area are heterozygous the resulting population would be heterogeneous, comprising individuals that differ genetically from each other. In the cases where another variety has been introduced into an area in which a variety already existed hybridization would occur between them. Material for planting taken from the hybrid population would consist of the types descended from inbreeding in either variety as well as the hybrid progenies.

In some cases the genetic composition of a cultivated population may have been the result of material introduced from more than one source population, subsequent introductions being added to the variety planted when the original cultivation was established. Some examples will be encountered where several varieties were introduced on a single occasion. Where these varieties were planted promiscuously they would have provided planting material of a highly heterogeneous nature, resulting from the various combinations among the individual varieties. Many of the complex populations produced from multiple introductions were formed during the early part of the 20th century as a consequence in the rapid expansion of the demand for cocoa. Such populations have genetic structures with more ample diversity than those of the original cultivated varieties and the source populations. In addition to the greater diversity of the genetic components and the products of recombination from interbreeding among them, the cultivated populations formed in this manner usually contain larger numbers of individual genotypes than the source populations.

It is necessary to take into consideration the fact that all of the cultivated populations established prior to the use of seed from controlled pollination would have been created with seeds from fruits produced by natural, random fertilization. The genetic composition of the planting material would depend on the structure of the field in which the fruits were produced and the outcome of the reproductive processes involved. When it is present, self-incompatibility would be one of the most important factors determining the mating systems that occur. The seeds obtained from a given parent tree may or may not be derived from self-fertilization; but they could also result from hybridization with any of the neighbouring trees. Consequently, the various possible mating systems that function in the specific circumstances would produce a mixture of genotypes from the available sources.

The introduction of planting material in the form of seed, regardless of the historical period in which a specified transfer would have taken place, had several consequences. One was the possibility that, as mentioned above, the phenotypic expressions of the progenies may have differed substantially from those of the parental trees in the source populations. This would have happened especially in the situations in which the source population had been stable and heterozygous in constitution. Additional introductions into areas where the species was already being cultivated would, in time, result in hybrids being produced between the original and introduced varieties. The recombination occurring in the descendants of the hybrids during the subsequent generations would produce genotypes substantially different genetically from the parental varieties but possibly having phenotypic ranges encompassing those of the parents.

In the outline of the situation in the basic populations, attention was given to the role of mutation in producing the diversity of the species. The same process applies to the cultivated populations. In the first instance recessive mutations that are not expressed in the base populations will appear in the progenies comprising the subsequent generations. Mutations would also be produced in individual genotypes of these generations. Since many of the cultivated populations involve large numbers of individuals, larger than many base populations, as well as successive generations of reproduction they provide fertile grounds for mutation. Mutations are more easily distinguished in the populations that are composed of nearly homozygous genotypes. More often than not such mutations consist of types that have unusual, frequently deleterious or useless for practical purposes, attributes. However, they do indicate that mutation may be a potent factor in creating diversity and have an application in the determination of the inheritance mechanisms of the species. Mutations affecting attributes of agronomic or commercial interest, especially those involving quantitative characters where small genetic effects operate, are more difficult to detect on account of the large effects of environmental factors in determining the expressions of individual alleles.

One consequence of introductions of new varieties has been the partial or total elimination of the native varieties in their original state during the course of time as a result of the aggressiveness and better adaptability of the introduced varieties. In these circumstances the decline of the native varieties is attributable to both natural and artificial selection. The native varieties could be maintained in the face of the adaptive superiority of the introduced germplasm by means of directed selection for the purpose of maintaining any specific attributes that they may possess. There may also be examples of germplasm having been introduced on the basis of the possession of specific characteristics but, owing to poor adaptability to the new environment or through neglect and lack of interest in them, they failed to be established.

The utilization of the diversity that evolved in the individual cultivated populations as they became established depends to a great degree on the manner in which they were constituted. While it is possible that many of the populations of heterogeneous genetic constitution, which have provided most of the genotypes used in basic improvement, derive from random recombination, it should be considered that this is not always the case. The practice of establishing plantations from seed obtained from a single individual, as described by Preuss (1901), was explained in Chapter 4. The result of this practice would be the formation of fields containing trees that are related by descent or, at least, groups of trees within which there would be a degree of consanguinity.

However, in cases where the parent trees of the seeds used belonged to heterogeneous populations genetic variability would have occurred among the progenies, the scale of which would have depended on the genotypes of the parents and factors such as cross-fertilization. During the course of time additional variability may have been added through the planting of other types for the purposes of replacing stands that had not survived owing to mortality and other causes. Since this was a normal practice some alteration of the constitution of the original groupings would have occurred.

The application of intentional or random selection can be observed in many examples, either involving rows of trees with outstandingly similar characters or in successive plantings descended from a single ancestor in several generations or sites. In the latter the sequence of the occurrence of a given type may be easy to trace. Thus, the procedure described by Preuss (1901) is not exclusive to Trinidad but occurs in several other situations and applies to most of the cultivated cacao populations that evolved up to and including the first decades of the 20th century.

The first attempts at propagation by vegetative means were made at the beginning of the 20th century. The basis of the concept was that the improvement of the existing heterogeneous populations should be achieved through identification, selection and reproduction of the best genotypes. Once the methods of propagation were established the concept became a practical reality when selection programmes were undertaken. These were carried out in most producing countries during the following decades and resulted in the availability of many selected genotypes as clones that were initially destined for application in the improvement of the local populations. A stage was reached when proposals were advanced to encourage the exchange of the selections between the countries involved and their distribution to other countries in which such selection programmes had not been conducted. These activities led to the creation of clone collections composed of a wider range of germplasm than was available in the individual countries.

The history of the selection programmes was described by Bartley (1995), who discussed the consequences of the application of the results. From the results of the utilization and study of the selections it could be concluded that they were exceedingly superficial in their concepts and effectuation. The lack of application of genetic principles was very evident. Consequently, apart from a small proportion of effectively superior genotypes, few of the selections that were made have been utilized as cultivars or in improvement programmes and a very significant number of selections are maintained in germplasm or clone collections as museum pieces.

Because the criteria used in these selection programmes were limited to a few chosen traits the resulting selections cannot in any way be considered to represent the diversity of the populations from which they derive. Many of the selection operations were conducted in a limited number of chosen farms where the populations would probably have been composed of related genotypes.

The practical application of the concept of single plant selection as a method for improving the level of performance of the cultivated populations took place around 1930. Accordingly, that year will be taken as the benchmark for describing the genetic diversity on which the individual selection programmes were based. The adoption of 1930 as the date is convenient but may not be generally applicable in all circumstances since there may be individual cases in which the base populations were evolving after 1930. The selection programmes that were adopted were variable with regard to their scope and the countries and regions involved. In Peru no attempts at selection in local populations were made until 1986 and this to some extent was carried out in fields that had been established with recently introduced germplasm. The only local varieties adopted as cultivars prior to this period were those from the collections made in 1938 and 1943 as described in Chapter 5. Bolivia is an unknown factor in the utilization of local germplasm. Selection in Brazil, the most important source of diversity, was conducted later than in the other major cacao producers of Latin America.

The story of systematic selection of 'superior' genotypes in the producing countries outside the Americas is very different since in many of them the cultivated populations were being established after 1940. In any case all of these populations are derived from germplasm introduced from the base populations of the Americas. Consequently, the populations that evolved or were established in several countries and regions must be considered in relation to the specific sources of germplasm.

PART 1: SOUTH AMERICA – POPULATIONS DERIVED FROM AN AMAZONIAN REGION GERMPLASM BASE

Ecuador – Pacific littoral zone

The classic population

The cacao populations that occur in the humid lands of Ecuador bordering the Pacific Ocean that occupy the coastal valleys to the west of the Andes are of special interest for several reasons. They, perhaps, comprise the oldest populations cultivated for commercial purposes outside the Circum-Caribbean Region, exportation of the produce having been introduced as early as the 17th century. This product developed a certain position in commerce on account of its special flavour characteristics. Primarily, because of these qualities the variety that was cultivated gained a certain interest for introduction to other countries for the purposes of improving their germplasm. The interest in the Guayaquil cacao applied especially to the Mesoamerican producers where the native populations were in decline and seed from Ecuador was introduced officially from 1861 to counteract this decline.

The populations that were cultivated up to the end of the 19th century are considered in general terms as belonging to a single variety with a small genetic range. It would seem that the 17th-century cultivation was centred in the valley of the R. Baba. From there it would have spread along the coast and towards the north. The northern section of the distribution range involves mainly the Guayas Basin (Fig. 5) and to some extent smaller areas in the Province of Manabí. At least until the 19th century the island of Puná, at the mouth of the R. Guayas estuary, was a significant producer until unfavourable environmental conditions led to the disappearance of the plantations.

Following the introduction of foreign germplasm near the end of the 19th century, as will be described later, the name 'Nacional' was introduced to identify the variety that had existed previously. The use of this term implies that a single variety with a uniform genetic composition was concerned. (Some farmers apply the term 'Criollo' to the varieties they recognize to be the traditional ones; this is perhaps a more appropriate manner of identifying them.) The fruit type of the variety to which the name 'Nacional' is mostly applied is illustrated in Plate 27a. The occurrence of large areas of such a uniform variety, which had been brought into cultivation in relatively recent times, is consistent with their origin from trees at a single location and reproduction of the same narrow range of genotypes during successive generations.

The emphasis given to this type masks the possibility that other varieties may have been present in the ancestral population. Even in the type to which the name 'Nacional' is applied, variability occurs with regard to certain concealed attributes, such as compatibility.

Plate 27b shows the fruits from a plantation of trees that were claimed to the author as representing the genuine 'Nacional' variety. These fruits, in addition to being rather broader than the type in Plate 27a, had very thick husks. Preuss (1901), who studied the Ecuador cacao situation prior to any influence of introduced germplasm, indicated that the feature that distinguished the variety seen from other fruits with the same general characteristics was the thick husks. Although this feature is more commonly encountered, trees are found whose fruits have the same general appearance of the 'Nacional' variety but are distinguished by their thin husks. In addition, the impression was obtained from visual observations that the types grown in the south of the region differed from those encountered in the Guayas Basin and other areas. Plate 27c shows a fruit of a type grown in the Province of Manabí, stated to have been brought from the area of Balao, in the south of the Province of Guayas, and named the 'Balao' type.

The occurrence of other varieties, considered to be within the concept of 'Nacional', was observed by Pound (1938), who was in a position to obtain a wider picture of the situation in the region than is possible at the present time. In this respect he refers to the presence in the Balao zone of 'an unduly large proportion of full white-based calabacillo trees (*blanco calabacillo*) together with other types higher in the quality scale also with this full white pod colour.'

The 'full white pod colour' translated into the scale of the basic fruit surface colour refers to a very light green. Later, Pound wrote that the 'Nacional' cacao of the Balao zone was 'distinctly more "*blanco*" than in the Arriba (Guayas Basin) zone, and many of the "*blanco*" trees have distinctly smoother pods than types occurring further north,' attributing the presence of such types to, possibly, a more recent introduction. Also, some of the earliest illustrations of cacao in the Ecuadorian littoral depict fruits that are more elongate than the type in Plate 27a, but possessing similarities with regard to other fruit characteristics.

The complete lack of knowledge about the true range of variability within the classic Ecuadorian cacao population, a situation resulting from the low level of attention given to research on the subject, is an impediment to the ability to define the variability expressed in the concept of 'Nacional'. The possibility that variability did, and does, exist in the population precludes the discussion of the characteristics of its component elements. The descriptions given in the literature confuse the real distinguishing features of the varieties concerned with their attribution of terms used to classify the type such as *Amelonado* and *Forastero*.

The type to which the variety name 'Nacional' is usually attributed has a fruit of a green colour with a fairly large size. The base of the fruit is slightly constricted and the surface somewhat rough. As mentioned above, there is a tendency for the husks to be thick. The features that differentiate the 'Nacional' type from the other varieties with fruits of similar characteristics are the larger size of seeds and the intermediate purple pigment of the cotyledons. The flowers of genotypes with affinities to the 'Nacional' type frequently are distinguished by having reflexed sepals and staminodes that bend outwards near their extremities.

Pastorelly *et al.* (1994) reported the results of investigations on some traits of a group of trees identified as belonging to the 'Nacional' concept. It was found that only 25% of the individuals could be considered to be self-incompatible, a result consistent with random segregation in the progenies of a self-incompatible parent. There was a strong correlation between the self-compatibility status of an individual and its level of fruit production. This would suggest that trees selected for reproduction on the basis of yield would result in the creation of plantings composed of self-compatible individuals.

In this book the discussion of the population in the Ecuadorian littoral that appears to contain a narrow range of variability is based on the proposition that the variety is not spontaneous to the region. This implies the need to determine its probable origin. Admittedly, the population must have existed for several centuries prior to the arrival of the Europeans, but the date of its introduction would never be known. There is no evidence that the species and its product played any part in the lives of the inhabitants of the coast during the Chorrera phase of the Valdivia culture (2000 BC).

The most probable source of the material from which the cacao population of the Ecuadorian littoral zone derives would be the foothills of the Andes drained by the tributaries of the upper River Marañon. Pound (1938) noted the similarity of the fruits encountered in some parts of the upper River Marañon valley to those of the 'Nacional' type, as referred to in the discussion of the River Marañon populations. The recent collections made on some of these tributaries reinforce these observations and reveal the existence of genotypes with fruits having a close similarity to those of the 'Nacional' type. The phenotypic similarities may be recognized by the comparison of the examples of fruits from the two areas in Plates 10 and 27. It has to be noted that those shown in Plate 10 do not include all of the fruits observed on genotypes from the upper River Marañon system, some of which bear a closer resemblance to the varieties of the littoral zone.

It could be argued that the diversity within the classic populations of the Ecuadorian littoral zone might have evolved from various introductions from the upper River Marañon tributaries made on several occasions. The type possessing fruits with a smooth surface, which Pound (1938) referred to as *blanco calabacillo*, may have been one of these introductions that contributed the large seed size found in genotypes with similar fruits that belong to the cacao populations of the valleys of Ecuadorian Amazonia.

The study of Lerceteau *et al.* (1997) of selected genotypes attributed to the 'Nacional' population revealed that, in terms of the phenotypes characterized, some of the genotypes from the south of the zone diverged in what would be a secondary population from the Province of Manabí. However, the paper does not indicate the traits that are involved in producing the differences.

Foreign germplasm

The cacao population that existed in the Ecuadorian littoral zone in 1930 contained a strong element of germplasm that was descended from introductions made during the previous 50 years. The history of these introductions, in common with that of the classic varieties, is based on extremely scanty and nebulous information. It is taken that the first of these introductions was made 'about 1890' but this does not exclude the possibility of planting material having been introduced by unrecorded private initiatives independently of the material said to have been introduced about 1890.

A degree of confusion concerning the origin and characteristics of the source population ensues from the lack of reliable information regarding the germplasm involved in the introduction of about 1890. Some authors (e.g. Pound, 1938) give the origin of the seeds as 'Trinidad'. Although it is recorded that one of the Ecuadorian proprietors visited the island in 1894 it is not stated whether he took any fruits or seed with him. Certainly, such material does not seem to have been provided from the official variety collection (this will be described in the discussion of the cacao population of the island) but it may have been obtained from a private farm.

However, it is evident that no planting material had been introduced from Trinidad prior to the end of 1894, which leaves the alternative source of the introduction as Venezuela. Van Isschot (1900), in his survey of the situation in Ecuador, stated that 'several' farmers had introduced *Criollo* from Venezuela, the source of his information being one of the persons responsible for introducing this material.

This statement confirms the likelihood of an origin other than Trinidad, which is to be found in the use of the term 'Venezolano' to designate all genotypes that did not belong to the classic population. It could be assumed that the original plantings of this introduction were established on the farms belonging to the proprietor mentioned above. The connection of these plantings to varieties that existed in Venezuela at the time was made by Preuss (1901), who stated that he came across 'only the inferior varieties, Carúpano and Trinitario', the descriptions of the respective varieties being found in the reference to his account of the situation in Venezuela. Preuss, who was the earliest writer to observe the planted introductions, gave the origins as Venezuela and Trinidad and that (as would be supposed) small quantities were involved.

Later observers of the Ecuador populations who were familiar with the Trinidad scene, such as Stell (1933), concluded that the genotypes that were claimed to be descendants of the 'Venezolano' introduction bore little resemblance to the types known in Trinidad. Stell was concerned with identifying the trees that were not infected with *Crinipellis perniciosa* (Stahel) Singer at the time of his survey. Among four 'varieties' found to be in this category, two have some association with the Venezuelan varieties described by Preuss (1901), being named 'Cojón de Toro' and 'Zambo' (perhaps the same as 'Sambito' mentioned by Preuss).

Stell (1933) also referred to the presence in Ecuador of a variety named 'Soconusco'. This suggests that an introduction had been made from Central America, but details of this and the characteristics of the plants established are not described. If this introduction consisted of fruits from trees belonging to the *Criollo* group the progenies may not have been distinguishable from the descendants of members of the group (or hybrids) among the Venezuelan introductions.

At the present time it is unlikely that any of the survivors or direct descendants of the introductions can be identified as such.

It is known that, in addition to the above introductions described in the reports cited, certain properties introduced planting material from Colombia. The origins of these introductions were unspecified and no details are available regarding their variety status.

In summary, the lack of the requisite details prevents a satisfactory analysis of the introduced germplasm from being made and only some deductions regarding the possible genotypes involved can be made on the basis of the segregation occurring in progenies of hybrids with the 'Nacional' variety. The one certain factor related to the introductions is the fact that they added genes for anthocyanin pigmentation to a population that had been entirely composed of individuals with green fruits. The analysis of the introductions and their hybrids is also complicated by the possibility of unrecorded introductions having been made from the eastern foothills of the Andes. The relative contributions of germplasm from Venezuela, Trinidad and Central America and that from the Oriental Amazonian region will be discussed in the following section.

The compound population

By the beginning of the 20th century the basis had been set to convert one of the more homogeneous cultivated cacao populations then in existence to one with a very diverse genetic base. The intermixing of sources of germplasm was given an impetus from about 1920, when two of the major diseases of cacao in the Americas made their appearance and attention was focused on the introduced varieties as sources of resistance. At the time all reproduction involved the use of seed from uncontrolled hybridization and inbreeding. After the discovery of the occurrence of *Crinipellis perniciosa*, a search was made for trees that escaped infection, these being called '*refractario* trees', in an effort to reduce the level of susceptibility. Nurseries were planted with seeds from these trees and the seedlings that, likewise, escaped infection were distributed in their thousands to various farms. A consequence of the production of seeds from uncontrolled fertilization was the creation of an extensive array of recombinant genotypes that were disseminated on a large scale. In this way, the creation of the complex population was accelerated. In some areas the original plantings of the 'Nacional' types continued to exist and are still encountered, especially on account of the interest that was maintained in the quality product for which the variety was renowned. The present populations would consist of a large segment of hybrids between the 'Nacional' types and their descendants. At the other extreme there are genotypes that appear to possess nearly all of the attributes that are associated with the introduced varieties.

The links between the original plantations, on which the introductions were established, and the modern populations have been largely broken because the farms in which the former varieties and their first generation hybrids were established have been converted to the cultivation of bananas and other crops. There was also a period of replanting with varieties introduced in the expectation of resistance to *Crinipellis perniciosa*, which was mostly unfulfilled. Some of these plantings may still be found (contributing additional sources of diversity) but most were, presumably, abandoned in favour of more lucrative activities, including the substitution of cacao by other crops.

Although the widespread cultivation of the new germplasm by the larger properties was instrumental in accelerating the amplification of the genetic base their subsequent abandonment of cacao cultivation contributed to the loss of part of the diversity that had been created. Another factor that contributed to this loss was the substitution of elements of the base population by selected foreign genotypes, which began to be introduced, starting from the late 1940s, and by the promotion of cultivation of selections made in the local populations in the same decade. In spite of these trends, interest in the pre-1940 gene pool was kept alive, especially by small farmers, who continued to use planting material descended from it. This practice still continues.

As a result it is possible to encounter throughout the littoral zone areas cultivated with genotypes that owe their origin to the 'Nacional' and 'Venezolano' variety groups but arising from casual reproduction. The local populations that owe their derivation from this gene pool exist in a variety of circumstances, being formed from diverse origins and exhibiting wide variation regarding their genetic composition. In the region-wide context it is possible to encounter local populations made up of more closely related individuals, evidently obtained from a single source genotype. At the other end of the scale extremely wide phenotypic variability is found. In some areas, such as the cacao developments near Malimpia, it seems that each tree has its own characteristics, certainly exhibiting combinations of fruit shapes, surfaces and colours not found in other countries.

Contrasting examples of the various groupings of 'varieties' that are found on individual farms are shown in Fig. 36. The fruits illustrated were harvested from two properties in a small area in the south of the littoral zone. The fruits in the top row are of unmistakable *Criollo* ancestry. Those in the bottom row, besides the almost spherical shape, were distinguished by the very glossy and smooth surfaces. However, in the latter, the presence of red fruits indicates the descent of the variety from a *Criollo* ancestor. Neither group shows any traces of descent from a 'Nacional' ancestor.

As it possesses a more distinctive phenotype the presence of individuals descended from the 'Nacional' group is often marked, with the genotypes having fruits showing most of the salient features of this variety but with evident traces of traits from other varieties. Trees belonging to this category would owe their origin to segregants in the F_2 and later generations of the hybrid genotypes or to backcrosses of these with the 'Nacional' genotypes. Accordingly, local populations occur in which all or most of the individuals have fruit phenotypes similar to that depicted in Plate 27a, a condition perhaps more common in the south of the region. Lerceteau *et al.* (1997) indicated how such genotypes classified as being similar to the 'Nacional' type actually differ from it in their morphological and molecular characteristics.

The presence of genotypes directly attributable to any *Criollo* element of the variety complex is difficult to detect. It has to be remembered that neither Preuss (1901) nor Stell (1933) observed genotypes attributable to the *Criollo* group in the 'Venezolano' complex. However, the reference to 'Soconusco' as being present in the 1930s suggests a possible source of *Criollo* genes in addition to those in the hybrid forms in the 'Venezolano' introduction. Some individuals have been observed that have fruits with an elongate, narrow shape, rough surfaces and attenuate apices that tend to be curved. It may be inferred that, when such fruits are red in colour, the genotype may be descended from an individual belonging to the *Criollo* group (such as the fruits in the top row of Fig. 36). As stated earlier, it could be taken for granted that any genotype with red fruits or a member of a family in which segregation for red fruits occurs is descended from the 'Venezolano' introductions. In these families considerable variation occurs with regard to other characters, probably a consequence of breeding for the several generations that comprise the successive stages of the dispersal of germplasm from this source.

In contrast with the genotypes with green fruits that manifest a close relationship with the 'Nacional' types similar relationships are less evident in the case of red-fruited genotypes. Many trees with red fruits are encountered that do not show any indication of descent from *Criollo* genotypes. An

Fig. 36. Fruits from two farms in the south of the Pacific littoral zone of Ecuador. Top row, types exhibiting distinct *Criollo* ancestry; bottom row, fruits with a contrasting shape but also probably descended from *Criollo*.

example of such situations is that of individuals seen in the Province of Esmeraldas said to be descended from fruits of a type grown in the Province of Manabí (see Fig. 21). These progenies are so similar that it was inferred that they were essentially homozygous but they all had dark red fruits with a short, broad shape and smooth surfaces. This variety shows no indications of descent from the *Criollo* group or similarity to genotypes derived from this source found in other countries. Similar types occur elsewhere in Ecuador, belonging to generations not far removed from the original genotypes that were introduced. Therefore, it can be inferred that the introductions from Venezuela, in addition to possible elements belonging to the *Criollo* group, included descendants of the hybrids that had been introduced from Trinidad. (These introductions will be described later in this chapter.) The hybrid genotypes may have lost most of the characteristics of their *Criollo* ancestors and, therefore, were distant from them.

The subject of the degree to which genes from the *Criollo* group may be present in the germplasm introduced into Ecuador would depend to some degree on the appearance of the features of the fruits associated with the group. These features would include characteristics involving fruit shape, surface roughness, husk hardness and cotyledon pigmentation, in addition to fruit colour. However, it is now evident, as can be concluded from the description of the populations that occur in the Ecuadorian Amazonian region, that genotypes are present that exhibit phenotypes which are considered to be typical features of the *Criollo* group. The traits for which these populations are distinguished are fruit surface roughness and unpigmented cotyledons.

Progenies of hybrids with the River Napo genotypes produce fruits that have elongate shapes and very rough surfaces, which can, if taken out of context, be confused with the individuals belonging to the *Criollo* group. However, it cannot be claimed at this stage, since the subject has not been investigated, that the same genes are responsible for the phenotypic expressions common to both groups. The ambiguity concerning the ancestry of individuals expressing these characteristics applies principally to trees with green-coloured fruits, since the trees with red-coloured fruits could be assumed to possess genes from a *Criollo* ancestor.

One type of fruit that is found almost universally in populations of mixed ancestry has a short, broad shape with an apex that is short acute or slightly rounded and a smooth surface of a green colour. It could be assumed that fruits of this type, the trees of which need not necessarily be related, derive either from the *blanco calabacillo* described by Pound (1938) or the 'Cojón de Toro' of Stell (1933) – unless they are different names for the same variety. These types exhibit a range of seed sizes, some being very small.

A large proportion of the knowledge about the diversity resulting from the hybridization of the 'Nacional' with 'Venezolano' and other introduced varieties is based on obscure collections and progenies descended from material exported to other countries.

What could probably be the first accumulation of Ecuador germplasm is to be found in the section called 'La Buseta', which contains a collection of varieties created by the United Fruit Company (UFCo) at its farm, 'Tenguel' (shown in Fig. 21). The origin of this collection could be contemporary with Pound's survey of the littoral zone in 1937. Its existence was unnoticed until fairly recently and no information is available regarding its formation and the varieties represented. It would appear that the collection was established in an area that originally was planted with the 'Nacional' variety. The varieties observed by the author appear to have been planted in progeny rows but, since no complete survey had been carried out, it is not possible to report on the entire collection of varieties other than those seen.

Among the types seen was a row of trees that appeared to be uniform with regard to their fruit characteristics, the most outstanding being the narrow elongate shape, a very light green colour and a rough surface. The fruit size was also relatively large as were the seeds, the cotyledons of which had a light to intermediate purple pigment. The pigment of the flower stamens and the high seed numbers indicate a possible origin from the Ecuadorian Amazon. Another outstanding type had rather large fruits of a dark red colour and an almost smooth surface, the trees being rather vigorous. Apparently, this genotype caught the attention of the managers of the collection early. Seeds were supplied to other countries, where the outstanding features of the progenies were noticed. Thus, selections made within the families have provided genotypes of which some use has been made in breeding programmes. The progeny of the genotype are also outstanding for the dark green leaves, which tend to have a narrow elongate shape, and the intensely pigmented flowers. However, this variety has a practical disadvantage in that the husk is extremely thick, which results in a low ratio of seed weight to fruit weight.

Another of the varieties in the collection at the Tenguel farm is a family whose offspring possess several common features, among which are the fruits of a dark green colour and a broad elongate shape. In some individuals there is a trace of pigment on the surface. The ridges are fairly well marked and of intermediate roughness. The appearance of the fruits suggests a relationship to the 'Nacional' varieties but it is evident that they are the products of hybridization. The members of the family would have been outstanding as they developed and several of them have been selected and cloned and have been used as cultivars and parents in hybridization. The genotypes concerned are outstanding with regard to their large seed size and on this account it has been claimed that they are related to the *Criollo* group. The greater probability, in the light of the recently acquired knowledge of the diversity in the Marañon and Napo systems, is that an ancestor from these populations is involved. The genotypes are self- and inter-compatible and their progeny segregate for trees that have short, broad fruits with smooth surfaces; hence related to Pound's '*blanco calabacillo*'.

In the late 1930s (as far as can be deduced) the collection at the Tenguel farm was used to supply planting material to the cacao enterprises of the UFCo in Costa Rica and Panama. (A few selected types were also sent to Trinidad at the same time, a fact that needs to be recorded in view of references to these varieties in the literature.) Further information about this transfer will be given when the situation in Costa Rica is described.

The selections that derive from the UFCo enterprises, including those clones and families of progenies planted in Ecuador, should be considered to be of hybrid constitution. As far as can be determined the base germplasm does not include any pure 'Nacional' genotypes. In general, this germplasm exhibits a significant degree of variability. The most commonly occurring genotypes are those with green fruits. Some of the trees in this category have fruits similar to the 'Balao' type while other genotypes have fruits of a shape that is short and broad with fairly prominent ridges. One interesting feature that is found in the Ecuador-derived genotypes is the short height of several of the genotypes, a characteristic often associated with leaves of smaller size and a thick, hard texture.

Other privately conducted programmes aimed at the genetic improvement in this hybrid complex were carried out in the 1930s. These include research conducted at the Hacienda Clementina, a large foreign-owned farm. Some of this work was related to the search for resistance to diseases and Pound (1938) gave a few details about the programme. It would appear that the genetic base involved was narrow, being confined to some lines of hybrid progenies. However, some of the more outstanding genotypes selected possess features akin to those of the 'Nacional' genotypes but differing in respect of the low height that the plants develop.

The other main aggregation representing the diversity that was present in the Ecuadorian littoral in the 1930s is that comprising the offspring of some 70 trees established in Trinidad, these being derived from the selection programme carried out in 1937 and described by Pound (1938). The total quantity of genotypes established in Trinidad amounted to over 1400 individuals from most of the 70 trees involved. The number of genotypes in existence at the time of writing is a fraction of the original number and it is possible that several families have disappeared in the course of time.

The descendants of the 1937 introduction are considered to be a sample of the hybrid germplasm that developed in Ecuador. However, although they have existed for about 60 years, relatively little has been done with regard to characterizing the individuals in relation to the families to which they belong. Few individual genotypes have been studied or used in breeding programmes. As a result, as was the case for the UFCo collections, it is inappropriate to attempt to describe the germplasm represented. A large proportion of the genotypes belonging to this collection derives from the work conducted at the Hacienda Clementina. The possibility of the existence of the parents of some of the families that were established in Trinidad, which belong to the UFCo and Clementina programmes, provides an opportunity for studies of parent–progeny relationships.

The total collection embraces a diverse range of types. No pure 'Nacional' individuals are known. However, some types do indicate a close relationship to the variety, especially with regard to fruit shape and colour characters and the thick husks. On the other hand, many individuals have red fruits; their 'Venezolano' ancestry, of course, being undisputed, but there is considerable variation in this category of genotypes. Some of the genotypes are outstanding as possessing very large flowers (perhaps the largest known in the species), of which an example is included in Plate 8a. The possible introgression of genes from the *Criollo* group can be postulated from the presence of plants with leaves of a lighter green colour with a soft, velvety texture.

The foundation of an organized improvement programme in the region, under the control of the Estación Experimental Tropical, Pichilingue, during the 1940s included the selection of individual trees along the lines established in Trinidad. The programme could be considered as a sequel to those already undertaken by the UFCo and the Hacienda Clementina. An extensive and diverse area of the region was sampled, with trees in several farms being included. The variability range of the resulting collection of selections is fairly large and included genotypes attributed to the 'Nacional' variety on the basis of phenotypes as well as hybrids.

One of the interesting features of the programme is the fact that several individuals were selected in each of several properties. This would provide a basis for determining the composition of each population with regard to the scale of the variability it contained. The selections also included some of those made by the UFCo and the Hacienda Clementina, which were referred to above. Some of the selections from other farms show affinities to them and constitute a group of genotypes with similar traits, which, in addition to the green colour and shape of the fruits, include the large seed size.

The Estación Experimental Tropical, Pichilingue germplasm collection that resulted, as it concerns the regional diversity, has not been adequately studied, besides having been subjected to a loss of a significant proportion of the initial collection. Consequently, the diversity contained is imperfectly known. Among the genotypes that have been observed are a few, perhaps related, types with fruits that have a short and apparently near spherical shape except for the fact that the ridges are very prominent, perhaps the highest known in the species for fruits of this shape.

In summary, the diversity contained in the cacao population that evolved in the Ecuador littoral zone during the first half of the 20th century is extensive. After all, it is the melting pot of several of the basic sources of germplasm, combining that from the western sector of the Amazonian Region with elements of the Circum-Caribbean Region, which also, as will be discussed later, contain elements of the eastern Amazon gene pool. The population provides excellent opportunities for studying the contributions of these sources of genes as well as the inheritance of their characteristics.

Brazil – the Amazon Extension Zone

The area to which this title will be applied is located to the southeast of the region of the natural distribution of the species in the Amazon valley, which was dealt with in Chapter 5. Its approximate extent is demarcated in the maps in Fig. 1 and Fig. 23. Although the zone may appear to be part of the Amazonian Region and is included in it because it possesses the same forest cover, in fact it does not belong geographically to the Amazon Basin.

In reality a large part of the Amazon Extension Zone belongs to the River Tocantins drainage system, which is independent of the Amazon Basin. Except for the lands bordering the rivers that receive alluvium from flooding and backwash from the R. Amazon, soils are poor in nutrients and structure. As a result, they are generally unsuitable for the natural establishment of cacao populations and, for this reason, cannot be considered to be areas where the species evolved naturally. The extension zone also includes lands to the east of the River Tocantins system and its mouth, which belong to other systems and where cacao has been introduced into cultivation on a minor scale.

Although millions of cacao trees probably exist in the area it must be considered, in terms of the distribution of the species, to be an artificially created extension to the Amazonian Region. Consequently, all of the germplasm encountered in the area would have been introduced from the populations that occur in the natural area of distribution. Our ability to determine the origins of the varieties encountered in the southeastern segment is hampered by the paucity of the knowledge about the characteristics of the diversity contained in it as well as by the fact that the natural distribution area has not been completely surveyed.

The area covered by the cacao populations of the Amazon Extension Zone is large and ill defined. The structure of the diversity in the zone, which has accumulated since the first plantations were established, can be considered to be relatively stable. This is one reason why efforts at its conservation have not yet been made on a systematic basis. Another reason for the task being delayed is that the populations concerned are accessible to the centre for the Amazonian genetic resources established in the zone. Consequently, the knowledge about the diversity that exists in the extended region is extremely scanty, only an insignificant proportion of the genotypes having been sampled for conservation and characterization.

However, in view of their immediate presence a visual assessment of the component genotypes of the populations contained in it indicates that ample genetic variability appears to be represented. Probably, the range of variability that can be observed has been constructed in stages during the 400 years that have elapsed since its foundations were established and the first intentional introductions of planting materials arrived. As will be seen, the situations in which cacao genotypes are found in the area vary considerably. These situations range from solitary trees planted in back gardens in the towns to the larger conglomerations on farms where the species has been cultivated for commercial purposes. In some instances records of the attempts to establish cultivation are available. Some of these are still exploited; others have been abandoned for various reasons, including the fact that the prevailing conditions are inadequate to support these undertakings.

The first attempts at cultivation in the area would have been made about 1660, associated with the process of colonization by Europeans, and suitable crops to support the incipient agriculture were being sought. At the time the hinterland was undeveloped except for the activities of the missionaries. It is very likely that the first commercial cultivation of cacao would have been established by the Jesuits, who had a cacao farm on the R. Mojú, known to have been in full operation at the beginning of the 18th century. Since the Jesuits were deeply involved in the exploitation of cacao during the extractive phase it is likely that the planting material for the R. Mojú farm would have been brought from one of the other missions. The mission at the mouth of the R. Madeira, where cacao was an important source of revenue, based on the ample populations of the species that existed, is one probable source. The Jesuits also had a cacao farm on the left bank of the R. Amazon below Óbidos within the area of cultivation that developed.

Other missionaries were also involved in cultivation. The Carmelites had a farm on the island of Marajó, the planting material possibly originating from the Solimões sector, where the order had been assigned the responsibilities for the missionary work.

The largest and most important area of commercial cultivation in the extended region is located just above the mouth of the R. Tocantins, the centre of the area being the town of Cametá, which gives its name to the cacao-growing zone. In this region cacao cultivation is confined to the banks of the river and the islands where alluvial soils occur. The Cametá area has perhaps been the largest producer in the Amazon since the start of the history of the commercial production of cacao and from where the first significant quantities exported date from the last quarter of the 18th century.

In spite of the importance of the Cametá area no attempts have yet been made to evaluate the genetic resources that exist in it. Some slight knowledge of the situation has been gained by observations on the varieties said to have been introduced as seed in other areas that have taken up cultivation in more recent times, as will be mentioned in due course.

The only samples of the R. Tocantins cacao populations that are available for evaluation come from a few trees that derive from an attempt made in 1979 to conserve that germplasm that existed in the area south of the Tucuruí hydroelectric dam. Since the cacao plantations would have been located on the banks of the river they would have been lost when the lake was formed.

The nine genotypes selected on three sites were multiplied in the form of seed progenies but three of them were propagated vegetatively. They appear to belong to a single variety and are distinctive on account of their vegetative habit. The height of the seedling trees is kept low and they have relatively small, open canopies formed of horizontal branches with a tendency to geotropism. The fruits are of fairly large size, either short, broad or broad elongate in shape, the latter tapering towards the apex. The flowers that are white in colour are unusual because the guide-lines' colour is less intense than that of the staminodes. Segregation for compatibility occurs and also for cotyledon pigment ranging from pale purple to dark purple.

The dissimilarity of the pigmentation of the guide-lines and staminodes and the segregation for cotyledon colour constitute some of the special features found in the populations of the zone. It is usual to encounter genotypes that have pale-coloured guide-lines or staminodes, or both expressions combined. Pale green (unpigmented) flush leaves are also frequent. Sometimes, plants with this trait are produced from pale purple cotyledons. However, cases are known of dark purple cotyledons producing progenies that uniformly have non-pigmented flush leaves. The combinations of pigment distribution are reminiscent of the situation in the R. Napo and R. Zamora drainage systems but the eastern Amazonian genotypes differ from the populations on these systems in many respects. In particular, the pigment distribution patterns observed in genotypes from the zone seem to be produced by factors acting independently, while the similar expressions of variability in the eastern Andean varieties would be the result of pleiotropic gene action.

One striking example of the associations of pigment variations in the leaves, flowers and cotyledons is that of the clone 'BE 2'. This plant has pale green flush leaves that are associated with a pink colour of the staminodes and guide-lines (Plates 9c and d). The cotyledons of this genotype are uniformly of a pink colour, somewhat reminiscent of the shade found in some members of the R. Zamora drainage system varieties. Another genotype from the same area, but having narrow elongate fruits, has pale flush leaves associated with white cotyledons. The clone 'BE 2' has another distinguishing feature in the broad leaves with an undulating surface.

Among the other varieties that are encountered in the Amazon Extension Zone are those that possess fruits that are almost spherical in shape with smooth surfaces. Some of these genotypes have very hard husks. A similar type with small fruits but having a more elongate shape and some surface roughness is also common. However, although such types are prominent in the population it should not be assumed that all of the trees belong to two varieties with the same fruit characteristics. Segregation occurs for other traits, such as the hardness of the fruit husk (mentioned above), fruit surface colour and the shapes of the leaves. Some of the genotypes in this category that have been studied appear to be homozygous for the characters by which they are distinguished.

It has been the custom to typify the zone as one having a population consisting of a single variety, the fruits of which possess the characteristics described above. This concept has been based on the varieties from the region (or supposedly attributed to it) that have been transported to other regions. The above examples show that this concept is a fallacy based on inaccurate knowledge of the composition of the germplasm of the zone.

In addition to the types already described, other outstanding varieties could be mentioned. Along the lower reaches of the R. Mojú and R. Acará (adjacent to the mouth of the R. Tocantins) the predominant fruit type has a broad elongate shape with a noticeable degree of surface roughness. Although these characters are specific to the variety, variability is also encountered for other traits, such as the taste of the pulp. Genotypes with fruit phenotypes so similar that it is difficult to distinguish between them differ considerably with regard to their breeding behaviour.

One unique fruit type was found among progenies raised from seed at Tomé-Açu among those reputedly brought from the Cametá area. When first seen it was recognized as unusual on account of close paired ridges, suggesting a similarity to the 'Pentagonum' fruit shape (Plate 25a). The progenies, from natural fertilization, of this genotype segregate for several fruit shapes but all of them have the same dark green colour. The fruits of some of the progenies are very narrow elongate in shape, as shown in Plate 25b. The flowers of this family are also distinguishable by the red colour of the sepals.

Other trees on the same farm produce fruits that have broader shapes and perhaps resemble the similar fruits in the progeny of the type of Plate 25a. One of these genotypes has non-pigmented flush leaves and flowers with pink staminodes and guide-lines.

In 1937 F.J. Pound had the opportunity to visit some of the plantations near the city of Belém and also the Cametá area. He probably saw a limited range of the diversity but his comments and descriptions (Pound, 1938) give some indications of the variety composition of these populations, from which the following remarks have been extracted:

The principal type of cacao tree occurring around Belém produced an unpigmented half blanco pod, usually with a bottle neck, a long oval body, slightly or normally warty with a definite but not pronounced point.

There was a good deal of tree to tree variation about this central type. Some trees bore pods which were greenish and others pods which were full blanco. Some bore smooth pods and others warty pods while in shape some pod types were short and quite devoid of a point like the 'Amelonado' of Surinam. However with all this variation no pods were pigmented.

In all types the beans possessed very dark purple cotyledons. This population for convenience will be referred to hereafter as the 'Para' type. It appears to be a hybrid mixture between an Amelonado type like that in Surinam and what will be described later as the warty 'largarta' type.

The importance of these statements is that they establish that the populations of the region are composed of a heterogeneous mixture of varieties. Pound stated that the cacao seen in the Cametá area was similar to that described for the Belém area. His description strongly refutes the concepts held in certain quarters that the cacao of the region belongs to one variety with rounded fruits having smooth surfaces. With regard to the above extract, several comments are necessary to clarify the statements made:

1. The name 'Para', which he used to denominate the 'central type', has no connection with the variety name 'Para' used in the State of Bahia, which will be referred to in the following section.

2. Presumably, the 'Amelonado' of Surinam refers to the variety that inhabited the Mamaboen Creek, which was described in Chapter 5, and not the variety cultivated on the coast of that country.

3. The word *largarta* (as copied from the original) is the same as that spelled *lagarta* elsewhere in the report, that is, a fruit with a rough surface, presumably with an elongate shape.

As far as can be ascertained cacao was never cultivated in the area lying between the upper reaches of the R. Guamá, in eastern Pará State, and the westernmost of the rivers that flow into the bay of São Luis in the State of Maranhão (see Figs 23 and 38). In the early days of Portuguese occupation the entire area was governed from Maranhão. The first recorded plantings of cacao in the present State of Maranhão were made with the initiative of a Jesuit missionary who took a canoe-load of fruits from Pará to the Order's property in Maranhão. Apparently, from this shipment, 2000 plants were established, probably near the town of Viana on the R. Pindaré. However, correspondence dating from the same period indicates that one of the settlers, who had visited Mexico, also started planting cacao in the area.

Whatever the origins of the attempts at cultivation may have been, the idea spread to other parts of the State that offered suitable conditions. The forested areas that were in the colonized zone are small in extent and limited to near the coast around the capital city and adjoining the State of Pará. In general, environmental conditions are not very favourable for commercial production. The cacao tree has to contend with heavy soils, in which it has to compete with the dense root system of the 'Babaçu' palm. In addition, the area is characterized by well-marked very wet and prolonged very dry seasons. As a result there are no extensive areas of cacao and what survives from the original introduction occurs in small stands in scattered localities.

It has been possible to study some of these areas and collect samples of the genotypes. A comprehensive characterization of the specimens obtained has not yet been obtained but it appears that all the trees concerned belong to the same variety (as a consequence of the limited quantity of material introduced and inbreeding) and that this variety, perhaps, conforms to the 'central type' described by Pound. The fruits, of the type depicted in Fig. 37, are of intermediate size, dark green in colour, the shape being broad elongate with a slight acute apex and low ridges, which are slightly rough. The cotyledon colour is dark purple but of a lighter shade. The ovule number appears to be 55. The flowers are white in colour and the stamens are not pigmented. All the specimens obtained seem to have one common and apparently unique feature in that the plants from vegetative propagation using plagiotropic branches readily produce orthotropic shoots when in the juvenile stage.

Several of the areas investigated are located on the coast of the estuary to the east of the island of São Luis, formed by the rivers Itapecuru and Munim. The location of Icatú, situated nearest the mouth of the estuary, appears to be the most easterly of the areas where cacao cultivation was attempted.

Pound (1938) suggested that the variety he called the 'central type' might be related to the germplasm of the Guianas. Since we know something about the variability that exists in the valley of the R. Jarí, it could be speculated that

Fig. 37. Brazil, Amazon Extension Zone: fruits of the variety cultivated at Icatú at its eastern extremity ($\times 0.27$).

some of the types found in the extended region were introduced from an area comprising this valley and all lands south of the Cayenne–Brazil border. This area would include the island of Gurupá and the banks of the R. Amazon in its vicinity. However, since the cacao populations of this area are not yet known, such a relationship cannot be confirmed. It is likely that the cacao population of the island of Gurupá could provide a key to determining the origin of much of the variability encountered in the zone.

The foregoing account of the cultivated populations of the Amazonian Region demonstrates the extensive diversity that is encountered. That the ample range of diversity had been recognized earlier than Pound's visit can be shown by the classification made by Paul Le Cointe (1934), who was one of the closest observers of the scene in the days when the populations were not contaminated by introduced varieties. The classification referred to, perhaps, was based on the varieties of the Óbidos-Alenquer area but it would also apply to the other areas where varieties possessing similar characteristics occur. The importance of the classification is that it emphasizes the diverse nature of the populations that comprised the cultivated populations of the Amazonian Region. Four variety types were described based mainly on fruit characters. (The names attributed to them were probably adapted from those found in the literature available at the time with the Portuguese equivalents and are not necessarily the locally used names.)

These types are:

1. *Crioulo* [term related to *Criollo*? Not in sense of 'native']. [Fruits] most voluminous, shape usually 'oblong' with well-marked ridges. 'Beautiful fruits'. [Largest seeds – average weight 1.26 g.] However, seed numbers recorded are high. This is sparsely dispersed [confined to certain localities?]. (Perhaps the clone 'CA 4' in Plate 23c belongs to this class.)
2. *Jacaré* ['Alligator', equal to Pound's *lagarta*, which is derived from the name used in Central America?]. Most common type, elongate, sometimes with the apex curved; always has a well-pronounced constriction below the base, which also is somewhat curved [asymmetrical?]; husk thick, very rough with deep furrows; [dimensions] up to 23 cm in length and 8 cm in diameter; average seed weight 0.95 g.
3. *Amelonado* [partly in the sense of the customary use of the term]. [Fruit] shape ovoid; [dimensions] length 18 cm, diameter 9–10 cm; surface smooth; ridges slightly prominent; seeds well developed, more rounded; average seed weight 1.0 g.
4. *Cabacinho* [equivalent to *Calabacillo*]. [Fruit] almost rounded and smooth; ridges almost imperceptible, [dimensions]: length 11.5 cm, diameter 8.5 cm; seeds small and very bitter; tree vigorous; average seed weight 0.6 g.

Le Cointe added that the cacao of the Amazon Basin generally had fruits with thick husks and small seeds in which the colour of the cotyledons was dark purple. The knowledge that has been acquired about the cultivated populations of the Amazon shows that within each of the above classes a significant degree of variability occurs and that the generalizations do not hold true.

Brazil – the Extra-Amazon Cultivation

Not many years had elapsed after cultivation of the species had begun in the Amazonian Region when suggestions were made regarding its introduction to other parts of Brazil, which possessed environmental conditions similar to those of its native habitat. Whether any action was taken at the time (the end of the 17th century) is unclear. The claim is made (as far as the author is aware based on undocumented evidence) that the first introduction of cacao to the coastal provinces was made in 1746. This resulted in the establishment of a plantation on the R. Pardo in the south of the State of Bahia.

Since the most extensive cultivation of cacao in Brazil developed in the State of Bahia, most attention has been focused on its cacao populations. However, during the 18th century the species was experimented with all along the coast from Pernambuco in the north and at least as far as Rio de Janeiro in the south. The States in which some cacao cultivation is practised are shown in Fig. 38 and from which their relationship to the Amazon Extension Zone and the lower part of the Amazon Basin can be appreciated. During the period of experimentation with cacao cultivation the development of these parts of Brazil was largely confined to the coastal areas, penetration into the interior being possible along the rivers that flowed into the sea where colonies had been established. Even then, some rivers had remained untouched until decades later.

The descendants of some of the earlier plantations, as well as those created in the 19th and 20th centuries, may still be encountered. All of these experiments would have involved the introduction of seed from the Amazon Basin. It is probable that these undertakings involved independent introductions from unspecified sources, although the possibility of some exchange among the new locations must be considered. Consequently, the extra-Amazonian enterprises constitute an additional reservoir of the diversity contained in the region and an opportunity for the development of new combinations of genes.

However, except for the research conducted on the genetic bases of the cacao populations of the State of Bahia, and to a minor extent those of the adjacent State of Espírito Santo, these reservoirs of diversity have remained untapped.

What resulted from the putative plantation of cacao on the R. Pardo has never been recorded, in spite of claims made to the effect that trees of the original variety still exist. Zehntner (1914) refers to the oldest tree on the farm supposed to be that on which the 1746 introduction was planted. It would be plausible to imagine that other trees on the same property are descended from the original introduction. Collections have been made of the genotypes concerned and it is hoped that their status will be clarified in due course. In the context of this matter attention would be given to an article published in 1789 on the agricultural possibilities of the district of Ilhéus, which refers to cacao as a potential crop as if it did not already occur in the area. Some cacao trees existed in the area at the time it was visited by Spix and Martius in 1818, but, apparently, only a few trees were seen. Spix and Martius (1828, vol. II)

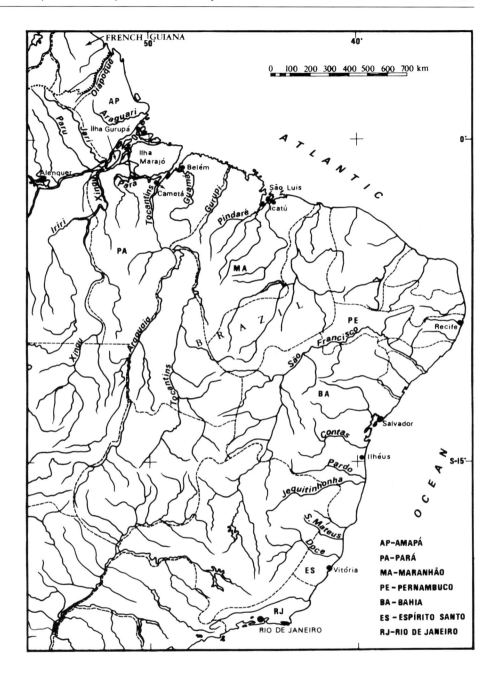

Fig. 38. Map of eastern Brazil with the States in which cacao is cultivated (indicated by their acronyms as defined in the area to the bottom right of the map) and their relationship to the Amazonian Region.

described the flora at the mouth of the R. das Contas (to the north of Ilhéus) and wrote:

> ...Und wir fanden hier, wie in Camamú, nur einige Stämme, deren gesundes Aussehen allerdings die Zweckmässigkeit das Cacaobaumes in diesen Gegenden bewärte. (...and we have found here, such as in Camamú, a few trees of cacao whose healthy appearance shows, of course, that they are well suited to the area.)

The authors would not have been able to describe varieties since it would be several more months before they became familiar with the plant in the Amazon Basin. It would have been just prior to their visit that the German colony had been established in the area of Ilhéus, which appears to have stimulated the cultivation of cacao.

The Ilhéus area was not the only part of the State of Bahia in which the cultivation had been ventured. Planting would have been tried in Salvador as trees may be found in gardens in the city. The areas that encircle the bay, normally cultivated with sugarcane, are, also, with certain limitations on account of the nature of the soils and tendencies for long dry seasons, potentially suitable for the crop. For this reason references to cacao in the area are encountered. The agricultural school at São Bento must have had adult cacao trees in 1882 to be used for instruction on cultivation practices.

One writer (Leal, 1893) mentions seeing attempts at planting cacao in districts to the south of Salvador, where the conditions are considered to be less suitable. In spite of successive attempts results were unsatisfactory since, although the trees 'prospered magnificently' yields were low. This situation could be attributed to the use of self-incompatible genotypes whose flowers would not have been pollinated because of the absence of suitable pollinating agents in the new environments.

The most useful history and description of the varieties of cacao that were cultivated at the beginning of the 20th century in the zone fanning out from the port of Ilhéus is that of Zehntner (1914). This shows the manner in which the original introduction provided planting material for some of the other farms on which the crop was established. Information was included regarding the origins of the cacao plantations south of the Ilhéus zone in the river valleys as far as, and including, the R. Jéquitinhonha. He states that the oldest trees at Una dated from 1809 and that material from these was used for plantings on the other rivers, such as the R. Pardo. The statement raises questions concerning the relationship of these trees to the supposed 1746 planting and whether, if the latter trees existed, they did not provide seed for planting in the neighbouring properties. The plantation system in the south of the State of Bahia had developed by 1849. Export of cocoa from the region was taking place on a small scale by 1835.

From the information presented by Zehntner it could be assumed that a single variety was involved during the first century of the establishment of the crop in the region. It is suggested that this variety was that to which the name 'Comum' (= common) was given. Probably, as seems to have been the practice in other cacao growing regions, the name was applied to distinguish the variety from those introduced at later periods, in addition to implying its extensive distribution. Successive generations of reproduction of the same variety would have resulted in the formation of stands consisting of homozygous individuals that were self-compatible.

Although the variety name 'Comum' is associated with the type of fruit, exceptionally for variety descriptions at the time, Zehntner (1914) described in detail the characteristics of the trees' habits and leaves as well as their cultural requirements. The fruits were described as being large and elliptic in shape, tending to be cylindrical in the middle. The apex was short acute and straight and a tendency for a constriction at the base was evident. The fruit surface had fairly pronounced ridges and varied from being rough (which was more frequent) to smooth. Seed numbers reached 50.

The trees of the 'Comum' variety were distinguished from those introduced later by their vigour, height and size, possessing leaves that were broad at the petiole end, the blades being undulating and thick textured.

The subsequent introductions are attributed to a German colonist named Steiger, who would have brought seed from various trees when he travelled on the R. Amazon. The exact details of these introductions are unknown. In 1859 Steiger had a planting of cacao on his farm. However, Zehntner (1914) supposed that the varieties whose introduction is attributed to Steiger arrived between 1874 and 1876 and seeds from them were used to form a plantation, which eventually numbered 20,000 trees.

The origin of the seeds introduced was given as 'Pará'. Although the name 'Pará' is that of the State at the mouth of the R. Amazon, previously it applied to all of the Brazilian territory falling within the basin. This means that the seeds would have been obtained at various locations in an extremely large area. The description of the descendants of the introduction divides them into two distinct varieties.

One of the varieties was named 'Maranhão' and two forms were recognized. The derivation of the name is somewhat confusing. The name itself is that of the first Portuguese colony, as stated in the previous section, and continues to be the name of the State in which the colony had been established. However, the knowledge regarding the cacao varieties present in the State of Maranhão indicates that it is unlikely that a corresponding variety exists there. The possibility is that a misinterpretation of the name became current and that the actual source of the variety was the R. Marañon in Peru, the population of which was described in Chapter 5.

Zehntner (1914) described two types to which the name 'Maranhão' was attributed. One of these was called 'Maranhão Rugoso' (the adjective means 'rough'). He stated that the fruits of this type were similar to those of the 'Comum' variety but were more variable with regard to their shape. This was slightly more narrow and elongate with an attenuate apex. It would appear that the type was not adapted to some conditions, probably where drainage was poor. When grown in more favourable environmental circumstances it was as good as the 'Comum' type and the characteristics of the surface roughness and the high ridges were most evident.

It is apposite at this juncture to mention that seeds of the 'Maranhão' type (perhaps from the original planting in Bahia) had been introduced into Trinidad in the 1930s. The trees seen at an age of about 25 years were vigorous and tall with elongate rough fruits. The productive capacity appeared to be low and little attention was paid to this sample. It was not possible to locate these trees in 1983 and it could be presumed that this potentially valuable germplasm had been destroyed.

The other type belonging to the 'Maranhão' group is the one that Zehntner referred to as 'Maranhão Liso' (smooth 'Maranhão') but which, evidently, was given various names by the farmers. These suggest that the concept covered several variations in the progeny of the original introduction. In this regard he explained that the fruits varied greatly in terms of length, the range including the longest fruits that he had encountered. The fruits of medium length had the general shape of the 'Maranhão Rugoso' except for the smooth surface with almost imperceptible ridges but maintaining the basal constriction.

On the other hand, the elongate fruits have more extended bases and apices, producing an effect in that the sides of their midsections appear to be parallel. According to Zehntner some farmers gave the variety the name of 'Maranhão Melão' (melon shaped). Such fruits tended to have thicker husks than the other variants. From his samples of this type he obtained maximum seed numbers of 57, which place the type in the group of varieties with high ovule numbers and suggest a relationship to genotypes such as 'MA 12' (Plate 23b). The experiences with trees of this type indicated that productivity levels were low in some situations but were observed to be potentially high in others.

The third main group of varieties that existed in Bahia (and the second group of Steiger's introduction) was given the name of 'Pará'. As stated above, this name could apply to any of the components of the Amazonian Region populations covering the extensive distribution area that was described in Chapter 5. The typical fruit lacks a basal constriction and the shape is relatively wider and more rounded than the 'Comum' variety. The fruit surface is smooth and the ridges insignificant. In general, the fruit size is smaller than that of the other varieties, regardless of whether they are spherical or more elongate in shape. The small, broad and short fruits with their smooth surfaces resemble those of the fruits of the passion fruit (*Passiflora* sp.), of which varieties of a yellow colour are produced in Brazil. On account of this resemblance, Zehntner added the word *Maracujá* (= passion fruit) to the classification of the 'Pará' type.

Among the interesting comparisons made by Zehntner in the description of the 'Pará' group is that with the genotype known in Java with the designation of 'Suriname'. As will be seen later, the cacao cultivation in Surinam appeared to have been based in the early period on a similar variety to which the name 'Porcelaine' had been applied on account of the smooth husked fruit and with which Zehntner probably had become familiar in Java, where he worked prior to going to Brazil.

The 'Maranhão' and 'Pará' types were distinguished from the 'Comum' types, not only by the fruit characters, but also by the habits of the trees. In contrast to the erect habit of 'Comum' Steiger's introductions and their descendants produced wide canopies with long branches that tended to curve towards the ground. The cork on the branches was smoother and the leaves larger with the base narrower than that of the 'Comum' type.

Zehntner provides data on samples of fruits harvested on various properties that are grouped according to the concepts established relating to the variety names. From these data it can be concluded that significant variability occurred among samples to which the same variety names were attributed, the variability depending on the locations at which the samples were produced. Thus, the fruit dimensions of some 'Maranhão' samples do not differ from those of some 'Comum' samples, the distinctions probably being based on other features of the fruits. The main distinctions concern the fruit shapes as measured by the length:diameter ratios. These are, in ascending order: 'Pará' (small) – 1.45; 'Pará' (Special) – 1.55; 'Comum' – 1.64; 'Maranhão Rugoso' – 1.72; 'Maranhão Liso' (average size) – 1.78; 'Maranhão Liso' (large) – 2.00. The largest seeds on a fresh weight basis were recorded in the 'Pará' (Special) category and the smallest in the 'Maranhão Rugoso' group. The first of the varieties listed would be named in Portuguese *Parazinho* (=small Pará).

The classification of the varieties into the above groups shows the scale of variability in the Bahia population and that attempts at classifying the varieties by other authors result in over-simplification of the situation. Admitting that it would be possible to encounter trees that would be descended directly from the original introductions as named, it is more likely that a certain degree of heterogeneity would have been developed, starting from the nature of the parent trees and their populations. If Steiger's original planting of cacao as seen in 1859 was of the 'Comum' type and the later introductions were planted in the same area, hybridization between the three 'varieties' would have occurred. The fruits supplied from Steiger's farm to other properties probably included those resulting from mating among the different varieties in addition to those produced by intra-variety fertilization. Consequently, some of the individual populations that developed were heterogeneous in content but contained individuals that were homozygous. An example of this situation was the farm on which the original 'SIC' selections were made. Other farms that received only a single variety derived from inbreeding would have developed homogeneous populations with the degree of homozygosity being enhanced in the succeeding generations.

This probability would account for the present status of the Bahia cacao population as it evolved since the latter half of the 20th century. On the one hand, the population contains trees that possess characteristics that conform to the descriptions of any one of the variety groups and that are essentially, if not entirely, homozygous. These individual genotypes would be the result of several generations of inbreeding. On the other hand, mere observation of the populations of certain farms indicates that phenotypic variability can be very large. This range of variability includes trees with characteristics that differ considerably from those of the attributes of the named types.

In 1980 a survey of the traditional cacao populations in the R. Jéquitinhonha system revealed the existence of genotypes that did not fall within the accepted range of the variability dispersed in the Ilhéus zone. The existence of other varieties, albeit equally from the Amazonian Region, is attributed to the introduction of other types by the cacao farmers of the valley. These would have been maintained in the original form on account of the poor communications between this area and the Ilhéus area, which existed until about 50 years ago.

The variability expressed in the populations that evolved in Bahia, therefore, is significantly larger than that of the varieties considered by custom to have constituted the basis of the local population. The range of fruit shapes and other characteristics that are expressed in the total population encompasses many of the types that can be encountered in the Amazonian Region. From the data collected by Zehntner we know that genotypes occur with ovule numbers of 65 or more, but the maximum number found in the other types is 55. The seed size does not appear to vary to the extent that definite distinctions can be identified, although certain genotypes do have seeds that are larger than the normal range, these being independent of the fruit shape or of the variety to which they can be classified on the basis of the information given by Zehntner (1914).

It is normally accepted that the genotypes in this population are self-compatible. However, in view of findings elsewhere there is a possibility that, owing to the fact that only a few genotypes have been tested for their compatibility status, other trees may be found to be self-incompatible. The varieties are known to be uniform with regard to the dark purple cotyledon colour, with the exception of those possessing the anthocyanin inhibitor gene, but, again, the entire diversity range has not been analysed for this character.

As indicated above, the wide range of phenotypic expressions does not permit all of the genotypes to be assigned to the three (or four, if the 'Maranhão Rugoso' and 'Maranhão Liso' are treated as separate varieties) groups described. The fruit illustrated in Plate 7d is presently classified as 'Comum' but, in addition to differing from the description given by Zehntner, it is probably closer to the large 'Pará' type. The almost spherical, smooth walled fruit in Plate 7a may be classed as 'Pará' but differs from the normal type of this variety by the very thick husk, which quite likely is the result of low seed numbers. In fact, the variability in husk thickness is noticeable and thick husks were possibly a feature of the large fruits assigned to the 'Pará' group in Zehntner's account with the title of 'Special' as shown by the low ratio of seed weight to total fruit weight.

The knowledge that the cacao populations of the State of Bahia represent a significant range of diversity leads to the elucidation of the origins of the varieties involved and their relationships to the components of the diversity that existed in the Amazonian Region. However, owing to the incomplete knowledge of the Amazonian diversity and the fact that few individuals have personal contacts with the range of variability in the Region it would be presumptuous to imagine that the subject would be adequately treated at the present time.

It would seem that the 'variety' called 'Comum', as described by Zehntner, may be similar to Pound's (1938) 'central type' in the Cametá area. The complex in Steiger's introduction that was designated as 'Pará' contains fruit types that are common in many places of the lower reaches of the R. Amazon. Some phenotypes in the complex resemble those of genotypes that are known from the State of Pará. If it is admitted that the name *Maranhão* may really refer to the R. Marañon, then it will be necessary to seek a relationship with the varieties that are known from that river. For example, the rough fruits could be considered to relate phenotypically to the genotype 'PA 150'. However, if Steiger obtained the fruits of the 'variety' he introduced with the name 'Maranhão' from parental trees belonging to populations in the State of Pará, it is possible that they were descendants of varieties from the R. Marañon that had been introduced to the particular location and were named for their origin on this river. Some of the genotypes to which the title 'Maranhão' could be applied would compare with Le Cointe's *Crioulo*, especially if the ovule numbers were in the same range.

The large scale on which cacao was planted in the southern municipalities of the State of Bahia, which involved various generations of reproduction, provided ample opportunities for the occurrence of mutations. In the course of time several mutations have been observed on account of the ease of identifying them in populations that are essentially homozygous. Most of the mutant genotypes are outstanding on account of the unusual characteristics they possess rather than on economically valuable traits. Reference was made previously in this section to the similarity of the variability expressed in the Bahia population to genotypes that occur in the Amazonian Region. From this it may be inferred that some of the fruits obtained in the region were from trees that were heterozygous for some of the loci in their genotypes.

It was evident that the ability to mutate was recognized in the earliest plantings in Bahia. Zehntner (1914) singled out the appearance of an abnormality that had been named 'Folha de Louro' (bayleaf), characterized by plants that were slow growing and of small stature, the leaves being dark green in colour and very small. Plants with similar characteristics are encountered in various Amazonian populations. Among the better-known mutations are those that possess the anthocyanin inhibitor allele, which, as referred to in Chapter 5, is common throughout the Amazonian Region. The genotypes carrying this allele, which are familiar to students of the genetics of the species, are those named 'Catongo' and 'Almeida', but the allele is present in other components of the Bahia population, which are very different to the two named with regard to their other characteristics.

One of the most remarkable of the mutant genotypes discovered in Bahia is one that is named 'Maracujá'. The genotype's name is not connected with the same name that was applied to Steiger's 'Pará' introduction but to the resemblance of the flower to those of the genus *Passiflora*. In this genotype the sterile staminodes are replaced by normal stamens, the latter producing fertile pollen. The significance of the 'Maracujá' mutant is that it would represent what would have been a wild type ancestor of the genus *Theobroma* that had ten fertile stamens. The flower type is illustrated in Plate 8c from a progeny in which the mutant character is combined with that of 'Catongo'. The gene has other expressions that include fruits with a smooth surface on a glossy bright green colour, a thick husk and abnormally low seed numbers.

Another mutant genotype that is important to mention is the one named 'Jaca' (from the smooth entire broad leaves of the jackfruit tree). This mutant (Plate 4d) shows some relationship to the 'Maracujá' mutant in its flower characters. One of its main features is the predominance of whorls of three plagiotropic branches (instead of the five branches of normal *T. cacao*), as shown in Fig. 50. The possession of this trait shows an interesting affinity with all of the other *Theobroma* species whose architecture is formed by three plagiotropic branches.

Cacao is also cultivated to a minor extent in the State of Espírito Santo, located immediately to the south of the cacao growing region of Bahia, and similarly restricted to the coastal areas. The history of the introduction and cultivation in Espírito Santo is not very well known, which means that it is impossible to determine the genetic constitution of the population that was established. Apparently, the first attempts at cultivation were made at least in the latter half of the 18th century. According to the history of the State, cacao was being exported in 1822 at a time when cultivation in Bahia was insignificant. Production appears to have been carried out initially in the northern part of the State, principally in the valley of the R. São Matéus. Unfortunately, no survey of the cacao population in the valley has been conducted, a situation that, therefore, prohibits a description being made. It would be assumed that the establishment of cultivation in the valley would have taken place independently from that in the valleys in the State of Bahia.

In later decades attempts at cultivation were made in the valleys in the centre and south of Espírito Santo. These included unsuccessful attempts made in the municipality of Cachoeira de Itapemirim prior to 1885. The central area in the drainage system of the R. Doce developed into the most important producing zone in the State. Although publications refer to the commencement of cultivation in 1895 it is probable that plantations existed prior to this year. The same sources also mention that the varieties planted were the same as those introduced into Bahia, which is a probability in certain respects since plantations of trees with similar types can be observed; but the relationships have not been verified. The trees in the field depicted in Fig. 39 may be descendants of one of the components of the Bahia population.

The trees in some of the cacao plantations in the R. Doce valley that derived from these introductions were observed to have fruit types that are not found in the State of Bahia. Some of the genotypes with fruits that differ from those that form the basic population of Bahia are represented among the selections made in the State of Espírito Santo. The fruits are broad in shape with a dark green colour and a more prominently ridged surface. There is a tendency for the apex to end in a short point. It was discovered that some of the selected genotypes were self-incompatible. The self-incompatible genotypes differ from the types usually considered to belong to the Bahia population by producing larger fruits whose surfaces tend to be smoother. They also have larger seeds than the other selections from Espírito Santo and most of the Bahia genotypes. In general, the Espírito Santo selections of the dark green fruit variety tend to be more productive than many of the Bahia selections.

The existence of self-incompatibility in the Espírito Santo population indicates the presence of a variety element that perhaps does not occur in Bahia, as far as is known. The alleles controlling the self-incompatibility reaction have not been determined as yet, a situation that precludes the determination of

Fig. 39. Brazil, State of Espírito Santo: a cacao plantation in a remote area established about 60 years previously, to show the variety cultivated.

the relationship of the genotypes that carry the alleles to other Amazonian genotypes. It is accepted that the Espírito Santo population would have been established from introductions descended from the Amazonian Region germplasm. This germplasm would have included components of the varieties that comprise the Bahia population. However, in view of the fact that the available selections represent only a small part of the potential variability in the State of Espírito Santo, it is impossible to guess at the degree of influence of the Bahia varieties until the variability that existed in the State has been adequately analysed.

Although the cultivation of cacao in the States of Bahia and Espírito Santo is almost entirely based on the germplasm of the Amazonian Region, it is interesting to refer to the presence in the area of germplasm with *Criollo* ancestry. The background to the acquisition of the varieties concerned was described by Zehntner (1914). In his article he stated that he had directed the importation of 'some plants' from Ceylon of the varieties belonging to Venezuelan ('Caracas') and Nicaraguan types. (The Nicaraguan type mentioned would refer to the 'Pentagona' variety that had been introduced into Ceylon a few years earlier from Trinidad.) These plants were established in 1907 in the orchard of the Agricultural Institute (presumably the school in São Bento, referred to earlier). The purpose of the introduction relates to the desire to produce cocoa of high quality in Bahia instead of the product exported from the State, which was considered to belong to the lower quality standards.

The number of plants introduced and established is not recorded. Bondar (1924, 1929) wrote a summary of the history of the introduction in which he indicated that only one plant survived. Some second-generation progenies of this plant were planted in the Botanic Garden in Salvador and their progenies established in the Experimental Station of 'Agua Preta' belonging to the Instituto de Cacau da Bahia. Seeds from this generation were distributed to various farms with the objective of promoting the cultivation of what were presumed to be varieties that would produce cacao of superior quality. However, these varieties were never planted on a large enough scale to make any impact on the genetic constitution of the Bahia cacao population.

The story of the introduction is a remarkable one from the point of view of the genetics of the variety concerned. The study of the diversity exhibited in their third or fourth generation progenies provides an insight into the complex heterozygous nature of the plants in the source population, that of Ceylon (Sri Lanka), the history of which will be discussed in due course.

Some of the progenies at the 'Agua Preta' experimental station were selected and the resulting clones maintained in the germplasm bank of the Centro de Pesquisas do Cacau. The eight genotypes from the fourth generation and two from the third generation exhibit a remarkable degree of variability. They include individuals that are obviously descended from the *Criollo* ancestors but possessing distinctive traits in terms of leaf shape, plant habit, lax branches and fruits of an unusual green colour with a relatively smooth surface. At the other end of the range of variability are two almost identical genotypes with a short height and broad, dark green leaves with a thick and hard texture. The fruits of these types are dark red in colour, small in size and nearly spherical in shape with prominent ridges.

It is possible that the progenies of the second generation of the plant that survived the introduction may have resulted from hybridization with one of the other varieties at the location but this is not evident in the segregation observed in the subsequent generations. If the second-generation progenies resulted from self-fertilization it could be concluded that the parent tree was self-compatible and heterozygous for the red fruit colour. Accordingly, the plant would have been descended from what was classed in the source country as the 'Caracas' variety or Venezuelan *Criollo*. Perhaps the plants obtained from Sri Lanka had resulted from hybridization with any of the other varieties supplied from Trinidad.

An interesting mutation would have occurred within this introduced family. It is entirely sterile but produces a few small, red, parthenocarpic fruits. On account of their resemblance to chilli peppers the mutant is called 'Pimenta'.

Surinam

Cacao cultivation in the coastal belt of the Guianas, the extent of the area is shown in Fig. 23, has a fairly long history and provides a unique situation in which parts of each of the countries in the region fall within the natural distribution range of the Amazonian Region. The coastal belt, however, on account of the lack of evidence of the prehistorical occurrence of the species, can be considered to lie outside of the natural range.

Surinam is an example of such a situation and worth referring to on account of its history of cacao production and its relative importance in this respect. The evolution of the cacao diversity since the first attempts at cultivation would serve as an example of the trends that occurred in similar circumstances. The country has a minor importance as a contributor of the germplasm that exists in some Old World producers through the exportation of a few varieties. It is also interesting from the point of view of variety adaptation to unique growing circumstances and the cultivation methods that were developed to overcome environmental conditions that were, for practical purposes, unsuitable for cultivation.

As far as is known the cacao industry of Surinam was based on a single variety. The origins of cultivation in the coastal plains and the rivers in the hinterland probably date from the late 17th century. It is the opinion of the author that the seeds used to establish the first plantations had been brought by the Jews who settled in Surinam following their expulsion from Pará. As will be described in the relevant section a similar event led to the introduction of cacao to Martinique. Apparently, at the end of the 18th century, the export trade in cacao was in the hands of the Jewish community in Surinam.

The conclusion regarding the variety first cultivated derives from the account of Bartelink (1884), who called it 'the indigenous variety', which contains the following information: 'Surinam cacao is the ordinary yellow sort and is planted everywhere. In a cacao field many different varieties of cacao, sometimes even as many as 20, are found. The best sort, however, is the so-called Porcelain cacao, distinguished by a thin smooth shell and the fullness of its beans.' The variety, also referred to as 'Suriname' in his account, was described by Preuss (1901) as having fruits whose colour, when unripe, was light green, with a shape that was very short and probably tending to be spherical with a rounded apex. The surface of the fruit was essentially smooth. In 1885 fruits of this variety were sent to Guyana on the basis of a very favourable recommendation that referred to it as being 'esteemed for its good qualities'.

The name 'Porcelain' used in the Surinam context should not be confused with the 'Porcelana' of the Maracaibo cacao variety although there may be some similarity regarding the phenotypic appearance, but the Surinam type definitely belongs to the Amazonian Region. Since varieties with similar fruits can be found in Pará it would be possible to determine which type was the ancestor of the Surinam variety. Preuss (1901) indicated that the Surinam variety resembled the 'Forastero' of Trinidad and that its descendants in Venezuela were classed as 'Carúpano'.

The Surinam variety would have been essentially uniform in respect of its genetic constitution. It does not appear to be related to the genotype inhabiting the area of the Mamaboen Creek of the River Coppename that was described in Chapter 5. The only specimen of *Theobroma cacao* in the Carl Linnaeus herbarium was collected by Dahlberg about 1770. Only leaves have been seen, these having a rather thick texture.

Bartelink (1884) stated that the first introductions of 'Caracas' cacao, obviously belonging to the Venezuelan *Criollo* group and containing genotypes with red fruits, had been made in 1845. Presumably the planting material was obtained from Venezuela, possibly from a specific location but could also have been related to similar varieties whose cultivation would have been established during the period of the Dutch colonies of Essequibo and Demerara. These introduced types began to enter into cultivation. Again, we can refer to Preuss' (1901) observations that indicate that two types would have been distinguished.

One of these types was called *Alligator*, the name presumably being given on account of the rough surface. The fruit colour of this type was green when unripe and its shape was very narrow elongate with an attenuate apex.

Preuss related the fruit type to that called *Cundeamor* in Venezuela and Trinidad but this may not have been wholly accurate. Evidently, this variety belonged to the green-fruited component of the Venezuelan *Criollo* group, the separate name indicating an origin distinct to that of the other variety named 'Caracas'.

The main distinguishing feature of the 'Caracas' variety was the preponderance of red fruits. The fruit characters of this variety would have been similar to those of the *Alligator* type, but apparently with more pronounced ridges. The other features of the fruits mentioned were the thick husk and the large placenta. Preuss observed that the trees given this name in Surinam differed from those named 'Caracas' in Venezuela. It would seem that, even at the time of Preuss' visit to the country, the introduced varieties did not comprise a significant proportion of the cacao population as he remarked that their introduction would improve the quality of the produce.

Where the two varieties had been planted together, hybridization would have taken place since Bartelink referred to the degeneration of the 'Caracas' types by fertilization with the 'native' variety. No real study of the cacao population that evolved in Surinam has been made and descriptions of genotypes are non-existent. In 1900 a report was put out about the discovery of a 'new cacao variety' on a farm on the R. Nickerie that had very large fruits, measuring 300 mm in length but nothing more seems to have happened to this type. Even the programme of selection of superior trees conducted in 1914 and 1915 provided no descriptions of the trees that were selected. It is expected that the populations that evolved up to about 1950 were substituted by imported varieties except for a few selections.

In the report of his visit to Ocumare de la Costa, Stahel (1924) stated that he had made arrangements for seed from one of the farms to be sent to Surinam. No mention of the establishment of such an introduction has been found. However, one of the selections made some years later is obviously descended from the Venezuelan *Criollo*, with red fruits and a smooth surface similar to those of the Chuao types.

PART 2: THE CIRCUM-CARIBBEAN REGION AND NEIGHBOURING TERRITORIES – POPULATIONS THAT EVOLVED FROM A *CRIOLLO* GERMPLASM BASE

Since the original populations associated with the Circum-Caribbean Region have existed for a very long time their germplasm was discussed in Chapter 5 as being indigenous to the Region; although, based on the knowledge that we have about them, we may infer that they were entirely cultivated. It was concluded that all of the germplasm involved belonged to a group of varieties with common characteristics and possibly a common origin to which the name *Criollo* would be applied. Since the mid-19th century considerable changes have taken place in the composition of the germplasm cultivated in the Region. In this chapter an attempt will be made to describe the modern status of the populations that have evolved in the areas where the varieties of the *Criollo* group were exclusively grown in the past.

During the past 500 years two lines of population development have occurred in the Circum-Caribbean Region. In the historical sequence the first stage was the expansion of cultivation of the *Criollo* varieties into regions outside the pre-Columbian distribution areas. This process took place over a period of time and would have resulted from the initiatives of European colonizers to satisfy the demand for cocoa to meet their own needs as well as for export. The other aspect of the development of the modern cacao populations in the Region concerns the subsequent alteration (or adulteration) of the original genetic base through the introduction of germplasm belonging to the Amazonian Region diversity.

The areas that are considered for the purpose of this account as belonging to the Circum-Caribbean Region or coming under its influence in terms of the initial germplasm base of their populations are those in which planting material derived from the *Criollo* group was exclusively involved. The geographical nature of the Region differs markedly from that of the Amazonian Region, where the distribution range is essentially contiguous, by the fact that the cacao areas are discontinuous and isolated. This situation resulted in the formation of populations whose genetic composition may be differentiated because they evolved from disparate germplasm bases and by separate processes. However, the common factor is the ancestral *Criollo* diversity base.

Although regional and local differences can be observed between most of the populations it is also necessary to take into consideration the movement of germplasm between the locations. As will be seen some sources of germplasm have contributed to the composition of many of the cultivated populations that were present in the mid-20th century. These contributions would be expected to result in similarities in the genetic composition of these populations. However, the differences in the range of genotypes that are observed occur because the introductions to the individual locations would have involved specific parts of the germplasm base of the source populations. The introductions also occurred at different stages and their effect on the compositions of the local populations would have been determined by the rates of distribution of the introduced varieties and the number of generations involved as well as by selection.

The variability that was created in the Region as it concerns the 'modern' population situation compared with the ancestral germplasm base resulted from the addition of germplasm from the Amazonian Region that commenced in the 17th century. Although the intrusion of Amazonian genes took various forms from more than one source the consequences were a tendency for a shift in the genetic complexes in the producing areas, especially those in the major producing countries. The effects of the introgression of the foreign genes became more pronounced in time as selection for the superior adaptability of these varieties and their hybrids resulted in a tendency for the traditional populations to be substituted by them. This process continues in spite of efforts to conserve the traditional varieties on the basis of their special quality characteristics.

The heterogeneous nature of cacao cultivation in the Region results in difficulties in attempting to describe the formation of their populations. Certainly, no single history can be described for the Region as a whole. Accordingly, it will be necessary to deal with the producing areas individually, as far as circumstances justify their inclusion in this account, in relation to the influences that the germplasm from specific areas has had on the diversity found in other components of the Region. These influences extend beyond the Region to the other areas of the western hemisphere (as described for some of secondary producers in South America) and especially to the Old World.

The Caribbean Islands and Northern South America

Although some claims have been made for the existence of spontaneous populations of cacao on some of the West Indian islands, the early literature concerning their natural history makes it clear that the species was not present on any island in the pre-Columbian era. In view of this situation the Caribbean islands cannot be considered to belong to the ancient distribution area of the Circum-Caribbean Region. However, the development of cacao cultivation on the islands during the past 500 years is integrally linked with that of the mainland. Figure 40 shows the relationships among the islands and with the mainland countries in the Region.

Trinidad and Tobago

The choice of this country to open the description of the diversity that occurs in the Circum-Caribbean Region is based on the contribution that its cacao population has made to the diversity that has evolved in much of the Region. In many respects its influence is a consequence of the active research carried out on the species that commenced in the latter half of the 19th century. Although the two islands have been governed together for over a century they actually have different histories as far as cacao is concerned.

Cacao cultivation was practised only on the island of Trinidad when it was under Spanish rule since the early days of the colonization of the New World. The origins of the species on the island are unknown but, although claims are made that it was indigenous (that is, existing prior to the arrival of the

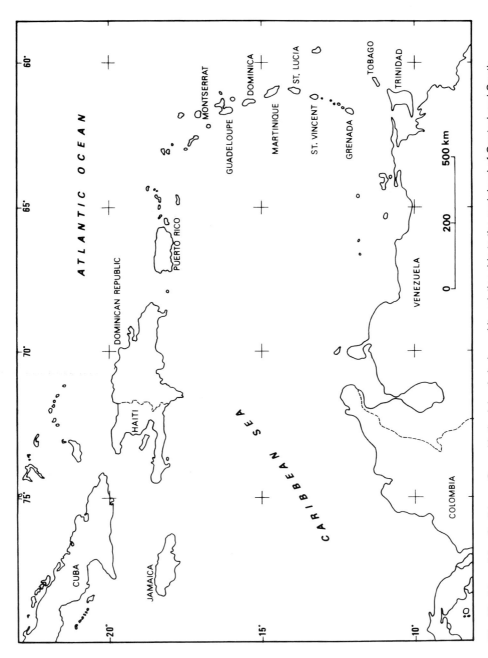

Fig. 40. Map of the Caribbean Sea with the island chain and its relationship to the mainland of Central and South America.

Europeans), no proof of these claims is available. The most likely scenario is the introduction from Venezuela during the early era of the establishment of the Spanish colony of samples of the same population that was found in the west of the country and which was carried eastwards to form the cultivations established along the coast of Venezuela.

There is no doubt that cacao was present early in the history of the island as it already occurred on the hills at the beginning of the 17th century. Presumably, since the principal activity was the production of tobacco, little attention was paid to the species at that time. It was likely that the crop received more attention as the trade of 'Caracas' cacao increased. Trinidad cacao may have entered into the trade under the name of 'Caraquilla', exported from the West Indies towards the end of the 17th century, this being considered in Spain as being of inferior quality to the 'Caracas' product. However, from our standpoint it is not possible to tell exactly what varieties composed the original Trinidad cacao population.

Only in the first decade of the 20th century was any evidence discovered of the existence of types that may have belonged to the original *Criollo* population. The fruit of this type is shown in Fig. 41 and it was described as having seeds with white cotyledons. Apparently, the surface was smooth and the husk thin, characteristics that relate it to some of the types belonging to the diversity present in the coastal valleys of Venezuela. Other trees were described as having rough fruits but all could be related to the *Criollo* population of Venezuela, especially on account of the red pigment.

A phenomenon, generally referred to as a 'blast', but the cause of which is not exactly known, affected the West Indies from Trinidad to Jamaica in 1727. This resulted in the destruction of, or serious damage to, the cacao plantations and a decline in cultivation. It seems that no attempt was made to resuscitate cultivation using the remnants of the original *Criollo* population, probably because of the miserable state the island was in at the time. The situation

Fig. 41. The fruit of the genotype considered to be a survivor of the original *Criollo* population cultivated in Trinidad, which was found in 1908.

continued for many years until priests of the Capuchin order, who had recently arrived in Trinidad, introduced seeds of another variety in 1757. Although some authors have referred to the origin of this variety as being the R. Orinoco, this seems unlikely, as no similar type is known from this valley. The more plausible source of the seed would be the R. Amazon.

Another possible source is the variety with similar characteristics that would have been grown on the island of Hispaniola, which in turn was derived from the type already established on Martinique. Even if this alternative did not apply to the 1757 introduction, the Hispaniola variety may also have arrived, in the same manner as happened in Jamaica, with the French settlers who fled from Haiti prior to the rebellion and settled in Trinidad.

The better adaptability of the new variety to the local conditions led to its adoption as a basis for expanding cultivation, which may have coincided with the arrival of the French. Hybridization between the introduced type and the remnants of the former *Criollo*-based population probably occurred. However, it is likely also that, during the succeeding years, other introductions of *Criollo* from Venezuela were made, since a close relationship between the territories was maintained with cacao farmers owning properties in both places. At least one, subsequent, importation from Venezuela is known to have occurred. This was of the type referred to as 'Ocumare', probably introduced at the beginning of the 20th century.

Regardless of whatever introductions from Venezuela may have taken place, hybridization would have produced the same results owing to similar gene bases being involved. The hybridization gave rise, after various generations of replanting, especially when the newer plantations were established in the latter years of the 19th century, of a large variety of combinations. The extent of segregation from the recombination created a population in which a great array of phenotypes can be found, but in which it is sometimes possible to recognize the contributions of the parental varieties. As mentioned previously the presence of red-fruited trees would indicate that they had *Criollo* ancestors.

Another importation from Venezuela concerns a true *Criollo* that was planted in a progeny row in the Botanic Garden of Tobago. One of the selections made in the 1930s is of one of these progenies. It is heterozygous for the red fruit colour and self-compatible. Seeds from this plot had been distributed for planting in Tobago.

Apparently, a certain degree of selection was practised during the formation of new fields. On the one hand, preference was given to maintaining the *Criollo* variety on account of its quality features. On the other hand, the Amazonian type was favoured because of its superior adaptation to conditions that were less adequate for the *Criollo* variety and also on account of its higher yield potential. As a result the situation arose where the composition of the populations on different farms differed according to the preferences in the choice of the planting material.

The range of phenotypic variability created led to the determination of the manner in which the various types could be identified. Presumably, the variety that was cultivated exclusively prior to 1757 was recognized by the name *Criollo* and this would have applied to the similar varieties in Venezuela. It would be expected that the proper differentiating designation of the Amazonian variety would have been *Forastero*. However, instead, this variety became known as *Calabacillo* on the basis of its fruit shape. It would appear that the hybrid types were named *Forastero*, as described by Hart (in Bailey, 1947), who gave the characteristics as 'the roughly corrugated pod, containing large flattish seeds of a purplish colour.' Hart mentioned that the type he called *Forastero* was that introduced from South America but, in reality, the hybrid form is the one described.

Sometime before 1880 a collection of cacao varieties had been established in Trinidad to which certain names had been attributed. The details of this collection are not entirely known but it would appear that it contained at least 16 types. Descriptions of these have not been encountered but it is known that a set of paintings of the selected varieties was produced with the intention of reproducing them in a book. However, this idea did not reach fruition and it is not known if any of the paintings still exist. The collection of types provided the basis for the seeds supplied to various colonies in the British Empire from 1880. Some idea of the phenotypes of the 'varieties' in the collection can be obtained from the list of the named types included in these shipments.

In Chapter 2 the names of the 'varieties' in this collection were discussed and the descriptions are repeated here in order to complete the picture of the situation. The names used include 'Ochroe' (the Trinidad name for *Hibiscus esculentus*), which would imply a fruit probably of a green colour, with a narrow elongate shape and pronounced ridges. Another 'variety' had the name 'Verdilico', the fruit of which appears to have had a green colour and a smooth surface. The name 'Sangre Toro' suggests that the fruit colour was dark red. Also included was a type named 'Five-grooved Cundeamor', which suggests that the fruit shape may have been similar to what is called the 'Pentagonum' type.

It is not known on what basis these varieties were named. Other writers who described the forms of fruit types in Trinidad gave other names to them. When Preuss (1901) visited Trinidad he observed that no two growers used the same names for trees producing fruits with the same characteristics. He also discussed the meanings of the names *Criollo* and *Forastero*, being inclined to disagree with Hart's definition of these terms. It is evident that the names that were given in Trinidad did not apply to those used in other cacao populations, particularly the Central American varieties.

Another collection of cacao 'varieties' was established in Trinidad after the shipments that were made in the 1880s. The names of the 18 types that comprised this collection mainly contained the word *Forastero* but these were primarily differentiated on the basis of fruit colour. The condition of the surface and husk thickness were used as well; the *Forastero* types had rough surfaces and the list also included red and yellow *Amelonado* types. The fruits of the types to which the term *Creole* (perhaps this refers to *Criollo*) was attached were, with one exception, described as having smooth surfaces.

In 1893, Hart introduced from Nicaragua samples of seed of the 'Alligator' or 'Pentagonum' type and the 'Cacao del Pais'. (These types are discussed in detail in the account of the cacao of the country in Chapter 5.) Considerable efforts were made to promote the growing of the 'Alligator' type on the basis of the concept that it would result in a product with the best quality. However, these efforts met with varying success and it would seem that, in the long term, none of the plantings survived in the original form. The only example of an 'Alligator' type fruit that has survived has a shape that is rather broader than that of the original sample brought to Trinidad and the fruits of its descendants.

Various plantings of descendants of the other types that were introduced from Nicaragua were made. From their greater vigour and the presence of purple cotyledons it may be presumed that the individuals are of hybrid descent but they segregate for the 'Pentagonum' fruit type and the single plagiotropic branch character. On account of their vigour and large seeds several of these genotypes entered into the first group of selections. Although these genotypes comprise a significant proportion of the first selections made in Trinidad the contribution of the descendants of the introductions from Nicaragua to the variability that subsequently evolved is small.

Although the most simplistic scenario for the creation of the large phenotypic diversity that exists in Trinidad is that it derived from two basic types with a minor contribution of the introduction from Nicaragua, germplasm from other sources could also be considered to explain some of the variability that was created. Among the genotypes that are known and studied are ones that have red fruits of a short broad shape, the progenies of which, derived from self-fertilization, do not show any indication of inheritance from a *Criollo* ancestor. On the other hand, genotypes are encountered whose traits are essentially *Criollo* in nature but differ from other known *Criollo* genotypes in possessing specific growth habits and leaves of thick texture.

In addition to the large range of combinations of characters that resulted from the process of hybridization and recombination over many generations, several noteworthy individuals have been identified that have arisen from mutation. Among these is one named the 'Crinkle-leaf Dwarf', in which the leaves are puckered similar to those shown in Plate 6a and the plants are short. Another remarkable type has an almost spherical fruit with a smooth surface without notable ridges and an unusual green colour (Plate 26d). This combination of characters appears to be caused by a single dominant allele. However, the same genotype possesses another inherited feature resulting from the action of a recessive gene, which produces short plants with white leaves and is lethal in its effect.

The history of cacao on the island of Tobago is not well known. It would seem that an attempt was made to establish the species on the island during the time of the Dutch colonies. This experiment may have been the origin of a group of plants found on the island in 1910, of which the type evidently belonged to the *Criollo* group but the fruit had a distinct mamillate apex. Interest in cultivation of cacao would have begun following the British occupation and began to be promoted at the beginning of the 20th century. Accordingly, the composition of the population would have been similar to that of Trinidad with the exception of a few areas on which descendants of the Nicaraguan introductions were planted.

Venezuela (see Fig. 33 for the locations named)

Since the 19th century the cacao population of Venezuela has undergone a profound change that involved the alteration of what was essentially a single variety of the *Criollo* group with a special reputation for the quality of its produce to one in which introduced germplasm has had a significant influence. The source of the introduced germplasm was Trinidad, as the nearest country with relations to Venezuela. Patiño (1963) stated that the first introduction was made in 1831. Evidently the type differed from the variety cultivated at the time and, therefore, it may be assumed that it was the Amazonian or 'Calabacillo' variety.

There are some areas of Venezuela where the 'Calabacillo' type is cultivated almost exclusively. Pittier stated that, in these districts, the name 'Trinitario' was applied to this variety (see Schnee, 1960). An example of the variety concerned from an area where it predominates is depicted in Plate 26c and it may be supposed that it belongs to the same type that was introduced into Trinidad in 1757. The trees seen in Venezuela had flowers with pigment on the stamens. It is not known whether this also applies to the same variety in Trinidad but trees that may be descended from the latter have pigmented stamens.

The first introductions from Trinidad would have been made into the areas of Venezuela nearest to the island, such as the peninsula of Pária. Here the introduced variety would have hybridized with the local type and, in time, supplanted the *Criollo* and also spread to lands to which the *Criollo* was not adapted.

It would also be supposed that later introductions would have included the hybrid forms produced in Trinidad. These introductions spread to other parts of Venezuela so that the original varieties gradually were supplemented or replaced as they died out. The scene in Venezuela is of a situation similar to that which occurred in Trinidad, except that there is a greater presence of *Criollo* germplasm, partly as components of the original populations and partly in the hybrid descendants.

Accordingly, various scenarios may be envisaged concerning the composition of individual regional populations. One example of the situations that may have occurred is that of the Hacienda Guayabita in Ocumare de la Costa as described by Stahel (1924). In addition to a *Criollo* population mentioned in Chapter 5, there were trees with *Amelonado* and *Calabacillo* fruits. Stahel stated that he found the fruits of the former type had larger seeds than he had encountered in other fruits of the same shape. The planting procedure used was to sow five seeds in each hole and leave only the largest of the plants that developed. The consequence of this practice was that there was a tendency for the 'Amelonado' or hybrid progenies to constitute a larger proportion of the new plantings.

Once such populations consisting of mixtures of the *Criollo* and foreign germplasm and the F_1 hybrids had been formed the seed produced by the hybrids used to establish the subsequent generations would include individuals resulting from backcrosses to the parental types as well as the segregant F_2 families. The progenies of the backcrosses to *Criollo* individuals would have phenotypes similar to these parents but differing in terms of traits such as tree vigour. At the same time future generations resulting from inbreeding among the individuals of the first generation hybrid would exhibit an extensive diversity of phenotypes as a consequence of recombination.

When Preuss (1901) visited Venezuela he described the cacao diversity that had evolved in the district of Carúpano, which probably included the peninsula of Pária. Here the cacao was also referred to as '*Trinitario*' and various forms were recognized, to which names were given. All of these forms would have arisen through hybridization with varieties introduced from Trinidad but with different degrees of intermixing and breeding within the local populations.

A survey of the cacao populations of Venezuela was carried out by Ciferri and Ciferri (1949, 1950), perhaps in the 1940s, which provides some idea of the variability that existed in the various regions of the country.

In view of the fact that cacao is grown in the country in several separated geographical regions the degree of influence of the varieties introduced from Trinidad and the survival of the ancestral *Criollo* germplasm varies according to the region concerned. It is also necessary to take into consideration the number of generations involved in the background of the planting material used to establish individual fields. Consequently, there cannot be a generalized concept of the genetic composition of the cacao cultivated in the country. Each region needs to be treated individually (and in some cases this applies to individual farms) according to the history of the evolution of the present day populations. In some regions there is a greater presence of *Criollo* genotypes or hybrids possessing *Criollo*-like characteristics. In other regions the populations consist of individuals that are similar to those found in Trinidad but that may differ from them in certain respects. In general, the gene base in Venezuela would be the same as that in Trinidad.

Colombia

Cultivated cacao in Colombia, in contrast to the native populations in the parts of the country within the Amazonian Region, as shown in Fig. 21, is found mainly in the river valleys and slopes within the Andean ranges. No evidence has been found for the existence of cacao in these valleys in pre-Columbian times and it can be considered that all the cacao found in them has been introduced and cultivated by man.

Although the Circum-Caribbean Region delimited in the map in Fig. 1 includes only the northern areas of Colombia, much of the cacao cultivated in the country occurs in zones outside this Region. For reasons that will be explained subsequently, it is convenient to deal with all of the areas in which the crop has been established as an extension to the Circum-Caribbean Region on the basis of the common genetic background and evolution of the modern populations. In this respect, it should be remembered that the cacao in the foothills of the Sierra Nevada de Santa Marta, whether native or cultivated, was described in Chapter 5 on the assumption that it is a continuation of the cacao of the Lake Maracaibo region.

Patiño (1963) has described some of the history of the establishment of cacao cultivation in the various zones of Colombia. The earliest attempt at cultivation of any importance could be attributed to the region of San Faustino in the east of the country. The area concerned was located in the upper reaches of the R. Catatumbo, just north of Cúcuta, straddling what would become the border between Venezuela and Colombia. It appears that cacao was cultivated here before the end of the 17th century and, in terms of the planting material, would be expected to be a continuation of the population in the foothills of the Venezuelan Andes. The San Faustino area was, during the peak of its era, perhaps the most important contributor to the cocoa exported as 'Caracas' but it was also a large supplier of the demand in the Cartagena area. The forgotten chapter of the era of cacao cultivation around San Faustino was described in a document by Casimiro Isaru, edited by Fernández Duro (1890). Cultivation was abandoned when the town was destroyed by an earthquake in 1784, after which indigo began to be grown.

It is suggested that the cultivation of cacao that started in San Faustino shifted to the west with its establishment in the Department of Santander in the valleys near Bucaramanga, mostly around San Vicente. The first attempts at growing the crop in the area would have been made with seed from the San Faustino region or otherwise derived from the populations in the foothills of the Venezuelan Andes. Some relics of these *Criollo* varieties were still to be found in the region in the 1950s although the population structure had changed during the 20th century.

The varieties from the western slopes of the eastern Cordillera would have been introduced into other regions of Colombia, providing the origins of cultivation in the Antioquia region, where cacao gained a considerable importance as a part of the diet in the form of the beverage made from its seeds. It would be assumed that the cacao grown along the R. Magdalena, reported by various authors, especially in the 19th century, would have owed its origins to the same source as that of the Bucaramanga area but no descriptions of the types cultivated have been encountered. Cacao fields did occur as well along the route from the R. Magdalena to Medellin. The crop was also established in other parts of Antioquia, principally on the lower slopes of the hills bordering the R. Cauca and in places on the banks of the river. These would have constituted the main centres of production. Cultivation would have spread up the river to Cartago, which was an important cacao area at the end of the 19th century, and then to the region of the R. Cauca valley centred on the city of Cali.

In all of these areas types derived from the *Criollo* populations originating in the Venezuelan Andean foothills would have been used to form the first plantations. However, owing to the changes in the composition of the cultivated populations in these areas, which occurred since the end of the 19th century, it is not possible to give an accurate account of the nature of the *Criollo* varieties. Some of the genotypes that can be ascribed to the group do appear to resemble those of the Venezuelan types, especially the trees with green-coloured fruits and the single plagiotropic branch character is also a feature of them.

It has been considered that, in general, the Colombian *Criollo* populations differ from those of the other members of the group in several respects

(Cheesman, 1944). In view of the absence of details regarding the characteristics of the populations it would be imprudent to attempt to define the attributes that distinguish the Colombian populations from others of the *Criollo* group unless all population groupings have been subjected to proper analysis. The distinctive characteristics of the cacao in the upper R. Cauca valley would most probably have resulted from selection that occurred during the process of the movement into these regions from the sources of the planting material.

Although all cacao originally cultivated in the areas mentioned above belonged to the *Criollo* group the genetic composition of the populations began to change in the 1890s through the introduction of other varieties. The history of these introductions is somewhat ill defined but they were probably motivated by the decline in the production in Antioquia that occurred in the late 19th century. The decline in Antioquian production could be attributed to mortality resulting from an unidentified disease which was prevalent at the end of the 19th century. One source of the foreign varieties is attributed to a Señor Patin, who, about 1890, according to Garcia (1954), introduced several types as seed. The name 'Pajarito' is associated with this introduction, being given on account of the small fruits with small seeds. On this basis, it is assumed that actually one variety was concerned that was relatively homogeneous. The variety may have been brought from Trinidad or one of the other West Indian islands and appears to be the 'Calabacillo' type found on these islands.

A few years later Robert Thomson brought several barrels of fruits from Trinidad and established the resulting plants on his farm, located on the left bank of the R. Cauca, in Antioquia. No information was provided as to the origin and nature of the varieties making up the consignment. It is assumed that, in time, seed from the progenies of the fruits was distributed to other farms in the area. The resulting populations are similar in most respects to that created in Trinidad. However, it would be impossible to determine through casual observation which of the trees that exist are the descendants of Thomson's introduction and which derived from hybrids between the surviving *Criollo* trees and the 'Pajarito'.

Although Garcia (1954) indicated that the 'Pajarito' was the most common type grown in the Cauca valley in Antioquia, this did not seem to be the case when the author visited the area in 1953, the hybrid varieties being more prominent. However, the 'Pajarito' appears to have been widely transported to other regions of Colombia. In the Sinú valley to the north of the R. Cauca cacao zone *Criollo* genotypes were grown in the early days when cacao was established and cultivated by the French, who settled on the river. In more recent times these plantations seem to have been substituted by the 'Pajarito' variety, which occurs even in the indigenous settlements. The 'Pajarito' would also have become the principal variety cultivated in the cacao area in the vicinity of Bucaramanga.

The 'Pajarito' also introgressed into the upper R. Cauca area in the Departments of Valle and Cauca, as Garcia (1954) described the hybrids formed with the original *Criollo* genotypes. Apparently, the trees regarded as belonging to the original 'Pajarito' type could be found as rare specimens.

Another important cacao producing area is located in the upper reaches of the R. Magdalena valley in the Departments of Huíla and Tolima. The map in Fig. 21 gives the location of the upper R. Magdalena valley and the

Department of Huíla. The cacao area of the Department of Tolima is situated to the north of Huíla. The proximity of the area to the headwaters of the R. Caguán and R. Guaviare is evident, although they are separated by the Eastern Cordillera. Since the author has no personal knowledge of the situation in this zone, information about the variety situation has to be interpreted from other reports. The brief accounts given by Pound (1938, 1945) following a visit he paid to the area in 1938 probably provide an accurate picture of the situation at the time.

In Pound (1938) he wrote:

> Cacao was first encountered fairly high up in this warm valley and at many places between Guadelupe and Neiva where sufficient soil had accumulated near the streams, fairly large areas of cacao have been planted within the past 20 years. The type was altogether different from that encountered in the valley of the Orteguaza.
>
> It is impossible to tell whether cacao was indigenous to this area or whether all had been introduced. Local people speak of this type as 'criollo' or 'real'. The trees growing in these small plantations were all mixed and at first sight the population was mistaken for the Trinitario complex. More detailed examination showed that the types could be separated into two groups. One group included Venezuelan criollo-type pods, some pigmented and others not, while the other consisted of a miscellaneous mixture of mostly unpigmented forastero types. There was not the continuous series which is found in the Trinitario complex. The 'criollo' types were not all pure in the now accepted sense for many showed tinted beans.

In the second paper, Pound (1945) added the following information about the variety's characteristics:

> ...the salient type is equally divided between unpigmented and partially pigmented pods of a cundeamor strain, i.e., with rather thick shells [husks], corrugated surfaces and no particularly noticeable Criollo point. The beans are large and plump but rarely pure white in colour. In fact, this population has a great deal in common with the Trinitario cocoa of the lower Orinoco except that there are few small Calabacillo type pods. The cocoa is spoken of as 'cacao del pais' and grows readily at altitudes of 3,000–4,000 ft above sea level.

He added the observation that 'This cacao in my opinion forms a link between the Criollo de la Montagne on the southern side of the Cordilleras and the more distinct andean types to the north'.

A single fruit from an individual tree in the Garzon area was described by the Anglo-Colombian Cacao Collecting Expedition, the notes of which stated that it was a *Criollo* type of a large size, the length/diameter ratio being 2.42. The husk was very rough and of a green colour with red streaks on the ridges. The fruit had a straight apex (the *Criollo*-type apex referred to by Pound tends to curve towards the trunk of the tree) and sharp and there was a basal constriction. The husk was thick with a fairly hard mesocarp. The seeds were very large, the estimated average dry seed weight being of the order of 1.5 g which places the variety in the range of the R. Napo varieties, all of the cotyledons being purple (but the shade was not indicated).

The only information about this population available from Colombia was given by Garcia (1954) who stated that true members of the *Criollo* group were rare, these being referred to as *'Real'* (=Royal), but they were not described or compared with other members of the group which inhabited other zones in the country.

The remaining varieties would have been classed as hybrids but these seem to have differed basically from those in the upper R. Cauca valley. These differences were noted as relating to the shape of the fruits and the larger seeds of the R. Magdalena zone. Within the zone, differences were observed regarding the floral characteristics of the types grown. The types that inhabit the upper (southern) part of the zone appear to flower less profusely than those to the north of the zone. It was assumed that the difference was indicative of the predominance of self-compatibility in the south and that of self-incompatibility in the north.

It would appear that the upper R. Magdalena valley was one of the principal suppliers of cocoa to Antioquia before the crop was established in the latter region. This suggests that cultivation in the Huíla region would have been fairly ancient. One of the interesting questions is whether any germplasm from the upper valleys of the R. Caquetá and R. Putumayo, which are fairly close to the Huíla region, had been introduced to it. The answer to this question would be available only if the cacao population had been adequately analysed with regard to the variety situation.

In the lowlands along the Pacific coast, in the southwest of Colombia, there is an area in which cacao has been cultivated since at least the beginning of the 19th century. The plantations appear to occupy the lower reaches of the valleys of the Mira and Patia rivers in the Department of Nariño (Fig. 21). Little is known about the composition of the cacao population except for a classification of the varieties in Garcia (1954), in which the types *Pajarito* and *Antillano* (the latter name probably given to the descendants of the introduction from Trinidad) are featured. The classification also includes a type called locally *Cacao Antigo* (old cacao), which indicates that it would have existed before the introduced varieties arrived. The fruit type is described as *Amelonado*, that is, a short broad green fruit. If it happened that this variety was that first cultivated the implication is that its origin was distinct from the other non-Amazonian varieties in Colombia.

A sample of the germplasm from the R. Patia has been available for analysis. The sample probably derived from a plantation made at the beginning of the 20th century. A large range of variability was encountered in the progenies of the fruits collected with segregation for red and green fruits occurring as well as for such characteristics as plant habit. These progenies showed that *Criollo* group characters were prominent with some individuals perhaps being direct descendants of *Criollo* trees in the upper R. Cauca valley.

To the south of the Colombian Department of Nariño lies the Province of Esmeraldas in Ecuador (Figs 5 and 21). Most of the Esmeraldas coastal zone where cacao occurs belongs to the same ecological and cultural region as the Tumaco area and, therefore, will be considered as an integral part of this specific region. Until fairly recently the Esmeraldas coastal region was effectively isolated from the remainder of the cacao growing zones of Ecuador and the knowledge about its cacao population has become imperfectly known only during the past few decades. It is known that cacao occurred in Esmeraldas in the mid-18th century. The variety originally grown in Esmeraldas belongs to the *Criollo* group. Red and green-coloured fruits occur that are elongate in shape with a rough surface, as shown in Plate 30b and c. In common with many types of the *Criollo* group the trees tend to grow very tall with relatively sparse canopies. Since it has not been possible to maintain the variety in other environments no comparative studies have been carried out to determine its position in the *Criollo* group. The origin of the variety is unknown. It may have been introduced from Mexico or Central America but appears to differ from the types known from these regions. The name *Marascumbo* is associated with cacao of the Esmeraldas Province but it is not certain whether the name refers to the *Criollo* variety or to *Theobroma gileri*, a species that is endemic to the area.

The Esmeraldas *Criollo* is found occasionally in mixed plantations with a variety having green fruits that are fairly small, short broad in shape with a relatively smooth surface. This type appears to be homozygous and perhaps is the same *Cacao Antigo* or the *Pajarito* of the Tumaco region. In the mixed fields trees that are hybrids between the two types can be recognized on account of exhibiting the characteristics of the parental types.

Guyana

The development of cacao cultivation in Guyana (formerly British Guiana) is interesting and complicated although the country never achieved any importance as a producer. Although the country technically lies outside the Circum-Caribbean Region (the situation may be appreciated from Figs 1 and 33), it is discussed in this section on account of the origin of the varieties originally cultivated in the country. Apparently the country was not colonized until the Dutch colony of Essequibo was founded near the end of the 17th century. Presumably the first attempts at cultivation took place about that time or early in the 18th century. Later cultivation spread to the R. Demerara. Part of the history of cultivation was described by Storm van s'Gravesand (Harris and de Villiers, 1911), who played an important role in promoting its cultivation. At the time the colony of Berbice was under a different administration and although cacao was also planted there details of the establishment of cultivation in the valley are lacking. Cultivation was confined to the river banks in an area not far from the sea.

No descriptions or illustrations of the varieties that may be assigned to this period are available to provide a reasonable idea of the types grown. It may be presumed that the varieties had been introduced from Venezuela and, therefore, belonged to the *Criollo* group. However, the descriptions of the cacao trees that survived in the regenerated forests after cultivation had been abandoned near the mouth of the R. Berbice raise doubts as to whether they belonged to a pure *Criollo*, similar to the types that had been cultivated in the valleys of Aragua. Among the varieties mentioned is one called 'Golden Caracas'. What this refers to is unclear. It is possible that the variety was a green-fruited *Criollo*. The Dutch colony maintained fairly close contacts with settlements of the R. Orinoco and material from the plantations in Barinas and along the R. Apure may have been introduced. In the Orinoco Delta, which is the area in Venezuela nearest Guyana, green-fruited *Criollo* types are encountered. These are somewhat variable in their characteristics but they possess seeds with white cotyledons.

The cacao that was found in the grounds of an abandoned mission station on the R. Berbice, located a short distance from its mouth, was described by Dash (1929). The trees were very large and appeared to be fairly homogeneous. They were identified as belonging to the variety 'formerly known as Golden Caracas, probably the best cacao in the world'. All of the trees had fruits of a yellow (green when unripe) colour. However, there was some variation in the surface characteristics, some being smooth, but the husks were thin. The fruits appeared to be of intermediate size and broad ovate in shape with definite acute apices. The variety would belong to the *Criollo* group but differing from the normal concept of Venezuelan *Criollo*. This would be especially important in view of the fact that the cotyledon colour was purple.

During a phase of resurgence of interest in the cultivation of cacao in the country, during the last two decades of the 19th century, varieties of cacao were introduced from Trinidad. Also introduced were quantities of seed of the Surinam 'Porcelaine'. However, at the beginning of the 20th century a significant proportion of the trees on the farms were identifiable as *Criollo*. At this time a set of paintings of different types based on their fruit phenotypes was produced. Some of these paintings were taken to the UK in the expectation that names would be applied to the types illustrated but no further reference has been found and their present location has not been discovered.

The cacao population grown in the North-west District seems to have evolved separately from that of the R. Essequibo and R. Demerara. Schomburgk indicated that the fruits were different by giving them the same 'species' title that he gave to the varieties that he found in the south of Guyana. The variety he knew was perhaps related to the varieties cultivated in the region of the R. Orinoco Delta, which borders the North-west District. Later introductions from Grenada were reported. It has not been possible to obtain details of the variety originally grown in the district. It appears to be affected to a lesser degree by the pathogen *Crinipellis perniciosa* than the other varieties in the country.

The Lesser Antilles

All of the major islands (see Fig. 40 for their locations and relationships) that make up the chain on the windward side of the Caribbean Sea have cultivated cacao during the past 200 or 300 years. In most cases the history of the establishment of the crop on each of the islands is vague and this leads to conflicting accounts regarding the origins and varieties of the cacao populations on the individual islands. The history and evolution of the crop on the island of Trinidad has been described and this island has had an important influence on the creation of the present populations on some of the other islands.

The general features of the islands as regards the development of the cacao populations include the fact that their areas are generally small and the growing of cacao is limited to certain areas with suitable environmental conditions or lands available after utilization for other crops. Since the areas occupied by cacao are limited in size individual populations are numerically small. Most of the islands are characterized by their hilly terrain, which provides suitable conditions for the cultivation of cacao, especially because of the protection afforded by the hills.

All islands suffer from the effects of periodical hurricanes and other natural disasters, which have had an important effect on the longevity and profitability of cacao growing. Consequently, cacao growing on the islands has been affected by cycles of decline and regeneration. The latter has often prompted the introduction of new planting material, in which case there has been a stimulus for growing what were considered to be the varieties, such as that of 'Caracas', that were considered to produce a higher quality product. The populations that have resulted from the waves of introduction have a large component of hybrid types.

The most important key to what may be the germplasm on which cultivation on most islands is based is Martinique. Although reports exist that mention cacao being found in the forests at an earlier date the main effort of cultivation could be attributed to a Benjamin Da Costa, who is said to have brought seeds with him around 1660 when the Portuguese Jews were expelled from Pará. This introduction would have occurred just about the time that interest in the cultivation of cacao in Pará was beginning. The variety is unmistakably of Amazonian origin and if the quantity of seeds (such as from a single fruit) was small, the resulting plantings would have been homogeneous in constitution and given rise to homozygous individuals in the succeeding generations. It was reported that Martinique experienced the same 'blast' that affected Trinidad in 1727. This resulted in a decline in the cacao population but the Amazonian type probably withstood the effects of the phenomenon.

The variety situation on Martinique was referred to by Guérin (1896), who indicated that the varieties were similar to those described for Guadeloupe. The principal variety, presumably descended from the introduction attributed to Da Costa, was called *Créole* and, evidently, this variety and similar types form the bulk of the cacao on Martinique. A later report on the situation in Martinique stated that *Criollo*, 'Amelonado' and 'Calabacillo' varieties could be found on the island.

In some ways the situation on the island of Guadeloupe mirrors that found on Martinique. It would appear that, at the beginning of the 18th century, a small number of cacao trees could be found on the island. The history includes a claim for the existence of cacao when the French arrived in 1635. The preparation of a cacao beverage by the aboriginal inhabitants similar to that used in Central America could be attributed to their origin on the mainland or contacts with the mainland tribes. Alternatively, the early establishment of cacao on the island could be attributed to Spaniards who may have settled on the island prior to the arrival of the French. However, during the early years of French settlement cacao does not appear to have been cultivated and the history of subsequent exploitation appears to have been marked by considerable changes in interest and activity.

The variety composition of the cacao cultivated up to and during the 19th century is not known but, presumably, the original plantings may have been based on types belonging to the *Criollo* group. The Martinique 'Créole' variety would have been introduced at some stage, perhaps before 1857.

A classification of the cacao varieties that were to be found on Guadeloupe at the end of the 19th century was given by Guérin (1896). This account is of interest because of the illustrations of some of the varieties involved. From these it can be confirmed that the 'Créole' variety has a short broad fruit and is the same as the 'Calabacillo' in other countries. The *Criollo* element was also represented by the variety called 'Margariteno'. Another variation of the Venezuelan *Criollo* differed from the former type, which had large leaves, in having small leaves. Both types had red-coloured fruits. It would be expected that the other varieties described were hybrids of the *Criollo* types with the 'Créole', unless they were derived from some seed imported from Trinidad. The hybrid types were described as having large fruits and segregating for red and green colours. During the 1890s a further consignment of seed was introduced from Venezuela. It is not known whether any of the *Criollo* or hybrid varieties still exist but it is evident that the 'Créole' predominated in the cultivations that existed in the early 20th century. From the descriptions it could be concluded that some variability existed in the 'Créole' variety with fruit apices being rounded or having a slight point. From the description, one of the fruit types of green colour and a rounded shape would appear to have had a slightly rough surface.

Although cacao has played an important role in the economic life of Grenada its history does not seem to have been recorded. During the period of French occupation of the island the presence of the species is not mentioned. From reports of the occurrence of old trees on abandoned farms in the 1880s it could be assumed that some cacao had been planted before the island was ceded to the British in 1763. Cultivation on Grenada appears to have started about the middle of the 19th century. As with all of the islands already described the population structure is based on the presence of the Amazonian variety or 'Calabacillo' and *Criollo* from Venezuela.

The 'Calabacillo' is found in some circumstances in virtually pure populations. It is likely to have been introduced from Martinique. Alternatively, these cultivations may be based on the 'Calabacillo' variety of Trinidad. The *Criollo* of Venezuela is evident in some areas and its influence is marked by the presence of red fruits. Fields have been seen that comprised trees whose characteristics showed a strong affinity to *Criollo*, in which variability for phenotypic characters was so large that it appeared that no two trees were alike. In the development of plantations at the end of the 19th century the planting of *Criollo* was discouraged and the planting of 'Forastero' or hybrid varieties was emphasized. Trinidad probably was the source of the seed for these plantings but some natural hybrids between the 'Calabacillo' and *Criollo* would have been produced in Grenada since genotypes occur with red fruits that do not resemble the Trinidad 'Forastero' descendants. Other varieties can be encountered, occurring in small groups, whose distinctive characteristics indicate that they may derive from introductions from other sources. Also, some genotypes with elongate green fruits classifiable as *Criollo* are found in a few locations that may be progenies of the varieties introduced into Trinidad from Nicaragua.

On the Diamond Estate, trees with the type of fruit illustrated in Fig. 42 occur in sufficient numbers to form a readily recognizable element of the population. The light green colour of the immature fruits distinguish the type from other varieties in Grenada and elsewhere. The ripe fruits are attractive on account of their rich yellow colour.

St Lucia has a small cacao industry but of relative importance in terms of the area available for its cultivation. The early period of settlement of the island was marked by changes in possession between the French and the British. From the distribution of the species on the island it would be possible to conclude that the initial attempts at growing the crop occurred during the period of French occupation, perhaps during the 18th century. The present cacao population exhibits a great range of variability. This is distributed in isolated areas and the individual populations often possess a specific portion of the genetic variability.

Fig. 42. Grenada: the fruit of a variety grown on the Diamond Estate (×0.429).

In some farms the 'Calabacillo' types are frequent and they are also found scattered in other areas. These types probably were introduced from Martinique, being the first variety planted on St Lucia. In other locations trees related to the *Criollo* group occur. Plate 29d shows the fruit type in a group of trees that is unmistakably descended from *Criollo* but the trees are notable for their compact canopies and the fairly small leaves of hard texture (Plate 29e). At one location a large range of variability was encountered, encompassing many different combinations of characters. Several of the trees had fruit types similar to those of the more recent introductions to Costa Rica.

In other locations the trees were derived from seed introduced from Trinidad, possibly from the collection made prior to the 1880s. As a result of the several introductions of planting material the present population generally comprises a range of hybrids, either between the 'Calabacillo' and *Criollo* types from the earliest introductions or belonging to later importation of seed. On account of the various circumstances in which the individual plantations were formed, the variability exhibited in St Lucia is extremely interesting from the point of view of analysing the evolution of the various varietal complexes.

The island of Dominica, like most of the other islands, was the scene of changes of ownership during the 17th and 18th centuries between the British and French. From about the mid-19th century cacao began to be an important element of the economy of the island. Various periods of activity resulted from cycles of interest that were accompanied by resuscitation of old farms or by expansion. At the end of the 19th century a research and development programme, which was established at the Botanic Gardens, contributed greatly to the present genetic base of the cacao population through the introduction of planting material from other countries.

In common with the other islands the first attempts at growing cacao on Dominica probably were made during the period of French occupation in the 18th century but were not of any great significance. A report made in 1886 describes the finding of a few old trees in an abandoned settlement in the centre of the island. These had perhaps survived the hurricanes that scourged the islands and that also have had an influence on the prosperity of cacao cultivation. The oldest trees on the island belong to the variety named 'Calabacillo', which would be the same as the 'Créole' of Martinique and planted during the French occupation of Dominica as the islands are neighbours. Although various introductions of other varieties were made from about the mid-19th century the 'Calabacillo' has survived the longest. The variety occurs in some localities in an almost pure state, covering fairly large areas. The fruits depicted in Plate 26b were harvested in one of these areas; the fruit on the right is an example of the principal type. Only one tree was seen to have fruits of a more elongate shape. This is shown on the left but decidedly belongs to the same variety.

About 1846 planting material was imported from Trinidad, comprising samples of the *Criollo* and 'Forastero' hybrid types on that island. A further introduction of Trinidad varieties was made at the beginning of the 20th century, which involved seed of the 18 types maintained at the experiment station in Trinidad. Seed from these introduced varieties was distributed to various properties as part of an effort to improve the quality of the island's produce. Consequently, the cacao populations on some farms comprise hybrid types similar to those in Trinidad. It is probable that some genotypes having characteristics similar to those associated with the *Criollo* group may be found but it is more likely that pure *Criollo* trees would not have survived on account of their delicate nature and susceptibility to diseases.

Near the end of the 19th century Dominica received seeds of the collections made in Nicaragua by Hart. The 'Alligator' or 'Pentagonum' type progenies were used in experiments on propagation by budding and the resulting plants were fruiting by 1912. However, no trace of any trees of this type has been recorded and it is presumed that the variety was lost.

In 1892 seeds and fruits were introduced from the island of Montserrat of what was considered to be a special variety with exceptional characteristics. It has not been possible to obtain details about the origin and attributes of the Montserrat variety. The lack of this information precludes the determination of the trees on Dominica that may be descended from the variety. The variety may be related to a *Criollo* type or hybrid. Some plantations on Dominica have vigorous trees that have short fruits with a rough surface, which differ from the hybrid 'Forastero' types but would also be descended from *Criollo* because of segregation for red-coloured fruits. The robustness of the trees and their yielding capacity would have made them suitable material for the few selections made in Dominica.

The Greater Antilles

The largest of the islands is that of Hispaniola, which is divided between Haiti and the Dominican Republic. In the past the island was totally under Spanish domination and the origins of the cacao population would have applied to the whole island. Cultivation probably started soon after the island was discovered and the planting material introduced from the Atlantic coast cacao areas of Mesoamerica. Nowadays it is impossible to tell if any remnants of the original introductions may be found, since similar types would have been imported at various subsequent periods.

Later, the island came under the domination of the French and this would have been the cause of the introduction of the Martinique 'Créole' to the island. This variety became the most prominent element in the composition of the population as is to be seen in the Dominican Republic. Some introductions of *Criollo* types from Venezuela were made at various times by several people. The Suchard family was involved in a cacao farm in the region of Samaná, in which the intention was, apparently, to grow the Venezuelan *Criollo* but this was later abandoned. Among other examples of *Criollo* are progenies of a variety from Mérida, brought to the island at the beginning of the 20th century. The variety has green-coloured fruits that are illustrated in Plate 29c. Trees with fruits of the type in Plate 29b occur on the same farm and possibly are hybrids with the *Criollo* or represent another variety of the group, but neither is distributed elsewhere in the country.

It is not known to what extent other *Criollo* types either from Venezuela or Mesoamerica entered into the composition of the cacao population but they are found on other farms and are represented among the selections made in the Dominican Republic, mainly on account of the larger seeds. In addition, it is reported that introductions from Trinidad and Ecuador have been made as well as the hybrid types from Venezuela, identified as Carúpano. As a consequence of the several introductions of different varieties a wide range of fruit and tree types is encountered. Among this variety of types genotypes have been seen that differ markedly from those of the hybrid populations in other countries.

Two examples from the extensive variation which may be encountered on individual farms are depicted in Fig. 43. The two contrasting fruit types represent only a part of the variability that exists on one small farm.

Ciferri (1929) published an extensive classification of the fruit types in the Dominican Republic and the varieties they represented. The classification is based on the dimensions of the fruits, husk thickness and seed numbers. The fruit types were divided according to fruit surface colour. Various local names for the fruit types are given. This is the most comprehensive description of the fruit types that would exist in populations derived from the *Criollo* and 'Calabacillo' origins and their hybrids. The classification is too complicated to undertake an analysis in terms of the origins and relationships among the phenotypes and to relate the names of those used in other countries. The same author (Ciferri, 1933) published another classification that is accompanied by illustrations of some of the varieties distinguished on the basis of their characteristics. The classification includes a variety that appears to be restricted to a certain locality, which he found difficult to name. The trees of this variety were vigorous and the fruits of a green colour and of a large size. The description indicates that these trees may belong to the cacao 'Nacional' of

Fig. 43. Dominican Republic: fruits of two trees on a farm exemplifying the variability of the population cultivated.

Ecuador. Evidently, Ciferri found a range of genotypes that he attributed to the *Criollo* group, principally related to the Venezuelan types, that exhibited a diversity of characteristics, but they are usually related on the basis of possessing red-coloured fruits. The cacao population of the Dominican Republic deserves to be better known but the lack of a research programme and interest in the subject has resulted in little attention being paid to the determination of the country's genetic resources.

During recent years some trees belonging to the population in the Republic have been selected and cloned. The criteria appeared to be based on the possession of *Criollo* characteristics. Some selections produce very weak trees but there are several with remarkably large seeds.

In the case of Jamaica it is known that cacao was planted on the island when it was ruled by Spain. The first plantings would have been established soon after the island was occupied by the Spanish with planting material brought from the Atlantic coast of Mesoamerica. These plantings would have remained when the island was captured by the British and were those seen by Sir Hans Sloane, who recorded his observations on the situation during his stay in Jamaica (Sloane, 1707). Unfortunately, the varieties were not described but the specimen collected by him is the oldest herbarium specimen of the species. The fruit that is depicted in this specimen has a broad but rather short shape. Sloane saw the cacao after the island had suffered a 'blast', probably a hurricane, that destroyed most of the existing plantations and only a few trees remained.

In the early days of the British occupation of the island, cultivation of cacao was neglected but, later, various attempts were made to continue growing the crop. Seed would have been imported from Venezuela since Barham (cited by Lunan, 1814) referred to plantations being formed by trees of the 'Caracas' variety. Later, some of the French who fled from Hispaniola after the uprising in 1804 went to Jamaica and took fruits of the 'Créole' variety with them. Although few of these plantations seemed to have survived at the end of the 19th century the names attributed to various varieties (Morris, 1887) that existed at that time attest to the diversity that was available. The 'Créole' was known as 'Santo Domingo' and the word 'Spanish' forming part of the names of other varieties is an indication of their origin and use in the first cultivation practised.

An article on the growing of *Criollo* cacao in Jamaica (Anon, 1904) classifies the varieties at that time into three groups. The *Criollo* was described as having narrow elongate fruits with a 'wrinkled' surface of a red, yellow or grey colour. A type called 'Forastero' had fruits that were large with a smooth surface and thick husks; their seeds were said to have been larger than those of the *Criollo* and had pink cotyledons. The author suggests that this variety had come from the Amazon Basin. The colour of the fruits was not described but they were presumably red. The third type was called 'Calabacillo', the fruit of which was smaller and had a smooth surface, the cotyledons being dark purple. Trees of the last type can be seen in sub-spontaneous situations and their survival could be attributed to their greater general adaptability to most conditions than the other groups. Hybrids between these types also occurred.

According to Morris (1887) a large quantity of fruits was imported from Trinidad that produced between 20,000 and 30,000 plants. The manner of distribution of these plants was not recorded. One property in the centre of the island had planted what appeared to be a variety of uniform characteristics, probably a hybrid type. Possibly there were other, unrecorded, introductions from Trinidad in later years.

Cuba has a long history of cacao cultivation that would have been started soon after the settlement of the island by the Spaniards and their contacts with the species on the mainland but no records of this are available to mention here. Cultivation appears to be restricted to an area in the east of the island. It would be expected that the first plants derived from trees on the mainland, either on the coast of Mexico or that of the Gulf of Honduras. However, no descriptions have been found of the varieties that were grown in the past or those that are grown at the present time.

Although Puerto Rico is not known as a cacao producer, attempts at cultivation had been made at various times. According to Barrett (1925) the plant was first introduced to the island in 1626. These attempts failed to establish any permanent cultivation, mainly on account of competition from other crops. Most of the attempts at growing cacao in Puerto Rico appear to have been based on seed from Venezuela. At the end of the 17th century seed was imported from Trujillo in Venezuela. Barrett (1925) referred to the existence of a few plantings that seem to have been composed of introduced hybrid and 'Calabacillo' types with red and green fruits. Among the descendants of previous introductions is a variety that belongs to the *Criollo* group and appears to be relatively homozygous, the trees being self-compatible with green fruits and white cotyledons. The basic characteristics of the fruits of this variety are similar to those of the type in Plate 26c that occurs in the Dominican Republic.

Mesoamerica

The Caribbean coast and its hinterland

The only significant addition to the germplasm represented by the varieties that were grown before the arrival of the Spanish is a genotype of Amazonian origin. This is probably the same as the Martinique 'Créole'. It appears to have been distributed along almost the entire length of the coast from Panama northwards and in the interior of some countries. The introduction of this type was perhaps due to the pirates who were based in the Caribbean islands and to the workers from those islands who were brought to the mainland. In some countries, such as Nicaragua, *Criollo* types seem to be absent from the coastal regions so that the 'Créole' is the only variety that was established in the area.

Costa Rica is one country where the development of the cultivated cacao population differs from that which took place in other countries. Whether the species existed in Costa Rica in pre-Spanish times appears to be a matter to be decided since the records of the first contacts of the Spanish are not clear on this point. Stahel (1924) referred to the growing of a green-fruited *Criollo* that was decimated by disease and mentioned that trees of a similar type were to be found growing 'wild' on the peninsula of Nicoya. It is likely that he was referring to the extension of the Nicaraguan cacao varieties as described in Chapter 5 but some cacao varieties may have been introduced from other Central American countries. These varieties would have formed the first plantations along the Atlantic coast.

Later, the Martinique 'Créole' had been introduced. It is the practice to identify this variety with the name 'Matina', the place where it was first grown (Fig. 34). The variety has been maintained in its original form and has the same characteristics as the 'Créole'. The fruit colour is green and the surface smooth or slightly rough, the trees being self-compatible. Figure 44 depicts the fruit of one of the progenies identified as 'Matina' from fruits introduced into Trinidad as representatives of this variety. The tree depicted appears to be homozygous. It should be compared with the fruits in Plate 26b. It would appear that some variability in phenotypic expressions exists within the concept of 'Matina' although homozygosity seems to be the rule.

In 1898 a quantity of fruits was imported from Trinidad by a grower and seed from the progenies raised distributed to other farms (Pittier, 1902a). The type of material involved was not recorded and neither was the location of the farm or

Fig. 44. A specimen of the progeny grown in Trinidad from fruits introduced as the variety 'Matina'. The tree depicted is homozygous.

those of the farms that received seed from it. The result of this introduction would be the existence in some areas of genotypes similar to those in Trinidad.

Most of our current knowledge of the variety situation in Costa Rica is connected with the undertakings of the UFCo. When the company began to cultivate cacao on a large scale the material used in the plantations was the 'Matina' variety. Apparently, later plantings were made using a variety that had fruits of an elongate shape and having a green colour with a relatively smooth surface. However, it would seem that some hybrids with *Criollo* types were also planted.

In the 1930s the company introduced several types from its property in Ecuador, presumably from the collection of varieties at the La Buseta section mentioned in the description of the Ecuador germplasm. A part of the material comprising this introduction was also established at the UFCo's cacao plantations in the Bocas del Toro area of Panama, just south of the border with Costa Rica. Some of the genotypes that are known to derive from these plantations resemble varieties seen at La Buseta as well as others with clear phenotypic similarities to the hybrids descended from the 'Nacional' variety. From its nature the introduction from Ecuador embraces a significant diversity that includes genotypes that would be genetically similar to those in Trinidad.

The 'Créole' type appears to have been established along the coast of Honduras at a fairly early date, such as during the 18th century. As far as can be determined this variety is known in Honduras as '*Cacao del Indio*'. It is in this area that genotypes have been located that possess the anthocyanin inhibitor gene. The production of white cotyledons by the genotypes that carry this gene has led to the mistaken notion that they belong to the *Criollo* group. Hybridization between the Amazonian variety and the original *Criollo* varieties must have occurred soon after the introduction of the former as is shown by the existence of types having red fruits that have a short, broad shape and smooth surfaces. These types apparently are quite common in the neighbouring areas of Guatemala, where they have replaced the *Criollo* varieties.

In spite of the size of the cacao population on the Atlantic side of Mexico the situation regarding the constitution of the present population is rather obscure. As mentioned in Chapter 5 the information available is scanty and the author lacks personal knowledge of the scene. From the work of Martínez (1912) it can be assumed that during that period most, if not all, of the cacao population was constituted of types belonging to the *Criollo* group. The variant types described by this author, which he described as *Amelonado* and *Calabacillo*, would have been variants of the *Criollo* with different fruit characters, since they apparently had red fruits.

In his report of a brief visit Soria (1961) found that the situation had changed since the period described by Martínez (1912). By 1961, the traditional *Criollo* varieties had been replaced to a large extent by an introduced type, or hybrids between it and *Criollo*. The name given to the introduced type was 'Ceylán' and it had entered into the composition of most of the populations at that time. The populations were described as being a mixture of the native *Criollo* with an *Amelonado* of high productivity, the fruits of which were of a small size and light green in colour.

In various reports the variety is identified by several names. One of these is 'Costa Rica' and it may be inferred from this and the description of the fruits that the variety is the same as the 'Matina'.

According to Soria the story of the introduced variety involved the introduction into Tabasco of seeds from Soconusco in 1930. The astonishing performance of the progenies resulted in great interest being taken in the new variety and this led to it being widely distributed. Since the plants of the new variety had been planted within fields of *Criollo*, hybridization occurred between the two types with the result that the subsequent generations would have been composed of mixtures of mating systems involving hybridization and inbreeding. The procedure would have mirrored that described in the case of Venezuela following the introduction of the Trinidad varieties. Consequently, the populations contain individuals that exhibit most of the *Criollo* characteristics as a result of various generations of backcrossing.

Soria (1961) also referred to the presence in the region of trees with 'Amelonado' type fruits with seeds having white cotyledons. Apart from the probability of segregation these trees would most probably be progenies of the 'Matina' variety, which carry the anthocyanin inhibitor gene. They would be similar to the trees of this type that are found in Guatemala and Honduras, indicating a connection between the two areas.

The Pacific coastal zones

Cacao cultivation in Nicaragua would have received the first alteration to the traditional *Criollo*-based population through the introduction, prior to 1893, of planting material from Trinidad. In that year J.H. Hart (1893) took seeds and plants to Nicaragua but discovered that material from Trinidad had been imported prior to his visit. The details of the germplasm involved were not indicated but it could be assumed that the material taken by Hart derived from the collection of types that had been established in Trinidad. These introductions would have resulted in populations containing individuals with the same characteristics as the Trinidad population, as well as exhibiting a large range of phenotypes, as has been reported to occur in farms near the border with Costa Rica. If the germplasm introduced into Nicaragua spread into the neighbouring areas of Costa Rica, especially hybrids of the Trinidad types with *Criollo*, the variability in this country would have been amplified. Also, there would have been a tendency for the genetic constitutions of the populations in the two countries to be similar.

After decades of attempting to prevent cacao from Ecuador entering the country, either to supply local demand or for trans-shipment to Europe, the importation of seed into Guatemala from Ecuador was sanctioned in 1861 in order to resuscitate the declining production in the former country. In time the influence of the Ecuador germplasm resulted in the almost total extinction of the pure forms of the *Criollo* varieties that had been grown during the previous centuries. The result of the introduction of planting material from Ecuador, which would have been entirely of the 'Nacional' type, was the development of a hybrid population. The population that developed would have been basically similar to the hybrids that were formed in Ecuador after 1890 following the introduction of the 'Venezolano' varieties.

In the absence of personal knowledge of the Guatemalan population or of any analysis made of the situation resulting from the introduction of non-*Criollo* genes, it is not possible to describe the population that had developed. Among the questions that are pertinent in this case is whether any 'Nacional' individuals remain from the introduction. The few genotypes that are known outside the country are mainly distinctive in having fruits with relatively smooth surfaces, a situation that would not be expected from the two groups of parents involved, unless the *Criollos* with smooth fruits were the parents of these genotypes.

In Mexico the story appears to be somewhat different. Although it may be suspected that the 'Nacional' type would have spread to the western coast of Mexico, the report of Soria (1961) gave a different picture of events. In this account he gave the names of introduced varieties in Soconusco as *Guayaquil* and *Patastillo*. The last name is applied to typical *Amelonados* that are possibly nearer to *Calabacillos*. Such types may derive from the variability within the concept of 'Nacional', such as the Balao type. In addition, there are types called *Sánchez* and *Costa Rica*. Such types would be descendants of introductions of the Martinique 'Créole' and its congener in Costa Rica, the 'Matina' type. This type would have hybridized with the *Criollo* forming hybrid populations of various types known as *Injerto*.

From the above accounts it can be assumed that the populations that have developed in the Circum-Caribbean Region have the same genetic bases but, on account of the isolation of the individual units, different results have been produced in terms of the phenotypic characteristics of the populations that are encountered.

PART 3: CACAO BEYOND THE AMERICAS – THE EXPORT OF DIVERSITY TO THE OLD WORLD

In the introduction to this chapter it was indicated that there had been a considerable movement of cacao germplasm from the western hemisphere to the Old World as various attempts were made to establish the crop in other regions. This movement took various forms as cultivation was taken up in different countries. Figure 35 shows some of the principal routes by which cacao arrived in various countries and gives an idea of the scale and nature of the movement of germplasm that took place in the establishment of new areas of cultivation.

The first attempt to plant cacao outside the American continent is attributed to the introduction of a plant from Mexico into the Philippines in 1670. From that period interest in growing the crop developed and various countries commenced planting cacao. The growth of the plantings in the Old World was gradual until the end of the 19th century when the process of expansion of cultivation became more rapid, especially in Africa. As will be apparent from the following descriptions of the development of cultivation in individual areas, the germplasm introduced was generally derived from a few sources and with limited genetic content. Some variability developed as a result of mutation and during the later periods from hybridization following the addition of other germplasm.

South Asian and Pacific Region

The oldest record of cacao being introduced outside the Americas is that described by Blanco (1837), which deals with the arrival of the species in the Philippines through the Manila galleons that made the crossing of the Pacific from the port of Acapulco in Mexico. The date of the introduction is given as 1670, although other authors put the date at a few years earlier. According to this account a single plant was brought to the Philippines. This plant probably came from the area then called Tecuanapa, to the southeast of Acapulco. Blanco stated that this plant was the origin of most of the cacao found in the country during the first half of the 19th century. The crews of the Manila galleons carried cacao seeds with them regularly to make the cacao beverage or chocolate. It is likely that attempts had been made to bring fruits and seeds to the Philippines on previous voyages but the seeds did not survive the journey across the Pacific.

The cacao that was cultivated in the Philippines during the first two centuries following the first introduction belonged entirely to the Mesoamerican *Criollo* group. If it was true that a single plant was introduced then it would have been heterozygous for fruit colour as red and green fruits occur in the present population and the plant would have been self-compatible. However, it is probable that further introductions were made from Mexico (and Guatemala) on other occasions. These would explain the presence of self-incompatibility in the present population. Owing to the absence of personal knowledge of the Philippine cacao varieties and the scanty information available in the literature, a description of their characteristics will not be attempted. It appears that within the elements that can be considered as belonging to the *Criollo* group some phenotypic variation occurs. The Philippine cacao diversity was also augmented by additional introductions as reference is made in more recent publications of the presence of *Amelonado* genotypes, and this may include hybrid types. As a result, a degree of hybridization would have taken place.

Figure 45 shows the connection between the Philippine group of islands and those of the Indonesian Archipelago. Thus, it is probable that the introduction of cacao into some of the latter group would have been the second stage in the dispersal of the *Criollo* variety into the region. It seems that the crop was first adopted in some of the small islands to the east of the archipelago and the transported to the larger islands to the west, the movement taking place in stages. As in other situations, smaller amounts of planting material would have been involved which represented parts of whatever variability that existed in the islands that supplied the planting materials.

By the mid-19th century a sufficient quantity of cacao was being produced on the island of Ambon to provide for local consumption. The connection with

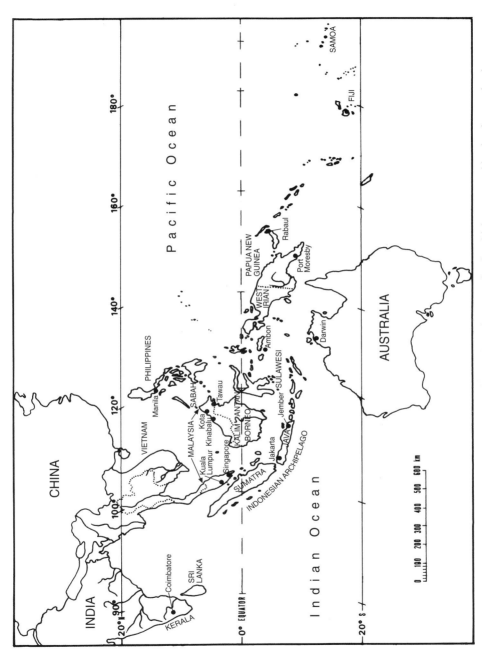

Fig. 45. Map of South Asia and the Western Pacific Region: the countries in which cacao cultivation is practised are identified, from which a comprehension of their relationships can be obtained.

the Philippines (or going back to Mesoamerica) is evident in the preparation of cakes of the cocoa mass in a manner similar to that in the Philippines, called *champorado*, a term that originated in Guatemala.

The *Criollo* from the Philippines and Indonesia probably is responsible for the first cacao varieties that were present in Ceylon (Sri Lanka) at the beginning of the 19th century. Cacao appears to have arrived on the island of Mauritius fairly early, perhaps at the beginning of the 18th century. In turn, cacao from Mauritius was involved in the implantation of the crop on Madagascar and probably on the island of Réunion in the west of the Indian Ocean. By the end of the 19th century cacao had been planted in Thailand and India and was beginning to be cultivated in the islands of the Pacific in addition to as far south as northern Australia.

The most important producer in the Southeast Asia region at the beginning of the 20th century was Java. According to Toxopeus and Giesberger (1983) the first encouragement to plant cacao on the island was made in 1778 and plantings were made in the following year. Although the origin of the planting material is not known it may be assumed that it came from the Philippines. The authors suggest that cacao existed in Java at an earlier date, perhaps established between 1750 and 1778. Development of cultivation appears to have been slow and did not reach any significant area.

The same authors summarize descriptions of the variety that was grown to which the name 'Java Criollo' was given. The type described was stated to be very constant in its characteristics, probably on account of its descent from a single relatively homozygous source. The predominant fruit colour was red and the main feature of the shape was its broad shoulder. The fruit surface was very rough. The above description agrees with that given by van Nooten (1863) and the accompanying illustration. The latter shows a fruit with a thick husk. Toxopeus and Giesberger (1983) mentioned that some trees had red fruits with smooth surfaces and relatively broader shapes. These trees perhaps belonged to the variation referred to in some accounts of the cacao in Java as 'Java Porcelaine'.

During the mid-19th century a variety was identified that had green-coloured fruits and unpigmented flowers and flush leaves. This was known as 'Witte Cacao' (white cacao), also referred to as 'variety *alba*', and would have resulted from a mutation to the anthocyanin inhibitor gene. Toxopeus and Giesberger (1983) referred to trees with green fruits that had smooth surfaces and low ridges. Presumably all of the trees with green fruits (owing to the inhibition of the synthesis of the anthocyanin pigment) possessed this gene. The description of 'Witte Cacao' was published by van Nooten (1863) and the accompanying illustration is reproduced in Plate 32a; it is the first example of the action of the anthocyanin inhibitor gene.

Two plants were introduced from Venezuela in 1888, presumably of the 'Caracas' type, and one of these survived to be multiplied through its seeds. The fruits of this tree were light green in colour and had a smooth surface. They would have differed from the variety being grown in Java at that time by possessing a conspicuous basal constriction. The progenies of the tree produced a wide range of variability in which the presence of red fruits indicated that hybridization with the original *Criollo* had taken place. Several abnormal types apparently were also produced. The introduced genotype itself would have been a hybrid between the Venezuelan *Criollo* and the Trinidad varieties as indicated by the segregation for white and purple cotyledons.

It appears to the author that the 'Caracas' type introduced in 1888 came from Ceylon. The origins of the varieties cultivated on that island, prior to those sent from Trinidad from 1880, will be discussed later and it will be seen that material identified as 'Caracas *Criollo*' had been introduced into Ceylon. However, since the plants that had been introduced into Ceylon from Trinidad would have reached the adult stage the possibility should be considered that the seeds introduced into Java came from one of these types, identified as *Forastero*, which had *Criollo* characteristics.

Another source of information, from an earlier date, is that of Roepke (1917). In general, the information given above coincides with Roepke's descriptions. The photographs in both publications give an idea of the principal fruit types that existed. Roepke (1917) gave the following classification of the cacao population as it existed at that time, which on the basis of fruit characteristics, could be divided into four groups, i.e.

Fruits of red colour, rough surfaces – 'Java Criollo'.
Fruits of green colour, rough surfaces – 'Witte Criollo'.
Fruits of red colour, smooth surfaces – 'Java Porcelaine'.
Fruits of green colour, smooth surfaces – 'Witte Porcelaine'.

The 'Witte Criollo' appeared infrequently. It seems that the term 'Witte' was applied to genotypes with green fruits rather than on the basis of producing seeds with white cotyledons.

It appears that some introductions of what was called 'Forastero' were made about 1886, both on an official basis and as private initiatives. Therefore, it may be concluded that more than one introduction had been made. No details of the types have been found but it would appear that the origin of the planting material would have been Ceylon. In this case it is probable that the types introduced would have been the progenies of the material sent from Trinidad and named 'Forastero'.

The literature on Java cacao refers to the introduction of varieties from Surinam but it is not known to what extent they influenced the cultivated germplasm. The Surinam 'Porcelaine' may have been included among these varieties. References to 'variety' names related to several agricultural properties indicate the degree of variability that may have existed in the island's cacao population.

The next most important cacao producer in Asia would have been Sri Lanka, which gained a certain reputation for the quality of the product during the period when the island was named Ceylon. According to Wright (1907) cacao trees existed in the Botanic Garden in 1819. These probably were descendants of the variety established in the Philippines and perhaps introduced from those islands or from Java.

Planting material was obtained from Trinidad in 1834–1835. It could be assumed that this material formed the basis of the plantations that were established about 1840. Wright (1907) stated that seed was introduced from Venezuela during the period between 1866 and 1876; presumably several introductions were made from this source but no details as to the specific origins were given. One of these introductions was made in 1873 and involved what were supposed to have belonged to the 'Caracas' variety but it was found that they did not possess the characteristics expected. A Mr Tyler is said to have introduced seeds from Trinidad in 1866 and, on account of the successful establishment and productivity of the trees that were established, seeds from them were obtained by other farms, from 1877 onwards.

In 1880 and 1881 plants of 11 of the 'varieties' in the collection established in Trinidad were sent to Ceylon. The types were made up of those named 'Cundeamor', 'Forastero', 'Criollo', 'Cayenne', 'Verdilico' and 'Sangre Toro', which are described in the section relating to Trinidad. The shipment represented the range of variability among the hybrid population that had evolved in Trinidad. It was made clear that since the plants had been derived from seed from naturally pollinated fruits no claim could be made that any of the plants belonged to a specific variety. In fact, it was concluded that all of the material established in Ceylon appeared to belong to a single variety, which was identified as 'Forastero'. In 1895 plants of the Nicaraguan varieties, including the 'Pentagonum' type, introduced into Trinidad by Hart, were received. Since only plants were involved in the shipments sent from Trinidad at the end of the 19th century, it could be taken for granted that those established in Ceylon may provide an understanding of the varieties in Trinidad that bore the names by which the plants are identified.

A few years before 1894 fruits were received from Grenada. The details of this shipment with regard to the type involved and the results after establishment are not known. Apparently, this introduction was made to a certain property and the trees established perhaps did not become part of the population structure that developed elsewhere on the island.

Among the varieties mentioned by Wright (1907) is an 'Amelonado' with green fruits. This type probably represents the same variety introduced from Trinidad, where it would have been named 'Calabacillo'.

An interesting contribution to the knowledge of the varieties that existed in Ceylon at the time was that of Stockdale (1928). In this report he reproduced the results of a study of the fruit types produced in progenies (from natural pollination) of a single tree. The paper is important in that the names given to these types are those of the varieties sent from Trinidad and, therefore, it provides a basis for relating the fruit types to the names given in Trinidad. One type illustrated shows a 'Trinidad Amelonado', while another is named 'Porcelaine Amelonado', the fruits of which appear to have a red colour and which would probably be called 'Calabacillo' in the present day nomenclature system. However, this suggests a relationship with the use of the name in Java for a similar fruit type. The fairly high percentage of seeds with pale pigmentation in the trees whose fruits were classified as 'Porcelaine' indicates the degree of descent from the *Criollo* varieties that were present.

For a long time Ceylon was renowned for its product and it was considered that this was due to cultivation on a large scale of a variety that came to be known as 'Old Red Ceylon'. One correspondent considered that the variety was descended from the introduction from Trinidad in 1834–1835 and represented a form of the Venezuelan *Criollo*.

Ratnam (1961) reviewed the introduction of cacao into South India and its subsequent history. This stated that the first consignment of cacao seeds came from Ambon Island and was received in Madras in 1798. By the mid-19th century the descendants of the few trees raised from the original introduction had been distributed to various other localities in South India. Perhaps the best known of these plantations is one near Coimbatore, which had been described as consisting of a pure *Criollo* type. This type undoubtedly is related to the variety introduced into the Philippines and later to Ambon. The article also refers to seedling plants of the 'Pentagonum' type having been sent from Guatemala in 1873 but it was not clear whether any results were obtained but, in any case, there was no long-term survival.

Cacao seems to have been planted in Peninsular Malaysia at least before 1777 but the crop did not gain importance in cultivation until recent times. The first record of a cacao plant in Peninsular Malaysia refers to a tree in Malacca. The material that comprises the herbarium specimen from the town, attributed to Chris Smith and dated 1796, may be from this tree. A survey of the occurrences of cacao carried out in the late 1940s (E.A. Rosenquist, personal communication; Rosenquist, 1950) showed that almost all of the trees planted in scattered small areas belonged to the *Criollo* group. Although it was stated that some of the planting material had been obtained from Ceylon the presence of the 'white cacao' mutant indicates that a proportion of the trees would have been derived from the Java population, or another island in the archipelago.

Moving to the east of the Philippines, the most important producers of cacao were on the islands of the Samoa group and in Papua New Guinea, cultivation having been introduced to these territories when they were German colonies. Cultivation started in Samoa in 1883 with plants introduced from Ceylon. Presumably, these belonged to the *Criollo* group, as the material sent from Trinidad in 1880 would not have been able to produce plants for the purpose. In 1884 plants were introduced from Java, again of the *Criollo* group. Wohltmann (1904) gave an interesting account of one of the varieties whose descendants were being planted, presumably from one of the above sources. This was introduced in 1883 and 1884 and the following extract from the report is reproduced as follows:

> It is always known in Samoa as 'Criollo', but it quite unlike the 'Criollo' of Trinidad, surpassing it in quality; it is also unlike the 'Forastero' cacao. In fruit, nibs and form of tree it most nearly resembles the cacao of Guatemala and Colombia.

The description is accompanied by illustrations of two fruits, which probably represent the variation in the *Criollo* that was cultivated at the beginning of the 20th century. One fruit is described as being dark red in colour and had a larger size. The other fruit was smaller and of a lighter red colour, its apex being more pointed. It is likely that the relationship of the Samoan *Criollo* to the Guatemalan (Soconusco) and Colombian types was made by Preuss. This would apply to the cacao that had been introduced into the Philippines from Mexico and to the cacao introduced from the former to the other countries in Asia and the Pacific, of which Samoa was about the last in the line. When he referred to the *Trinidad Criollo* he may have mistaken the variety for the Venezuelan *Criollo*.

It would appear that the cacao that was planted in the early years of development entirely consisted of red-fruited trees. Wohltmann (1904) referred to the introduction of varieties with green fruits. The characteristics and origin of these types are unknown. Possibly, they were introduced from Ceylon and, in that case, were progenies of some of the varieties sent from Trinidad.

During the next decade the situation changed. When Demandt (1914) wrote about the cacao industry in Samoa, the *Forastero* had been firmly established and, apparently, was known as 'Ceylon-Cacao'. Demandt described the fruits of the *Forastero* as having an elongate shape and surfaces that were smoother than those of the *Criollo* types. These would have been similar to the 'Verdilico' of the Trinidad collection and probably included the green-fruited type mentioned by Wohltmann. By the time Demandt (1914) wrote, the *Criollo* variety had been pushed into the background (as he described the situation) and was found only in the older fields. Hybrids had been produced between the *Criollo* and *Forastero* and preference was given for using planting material of these.

After Demandt wrote, World War I resulted in the cessation of the German colony and the Samoan archipelago was split into two administrations. The development of cacao cultivation was practically unnoticed since 1914. It was stated that in Western Samoa the policy was adopted of planting only green-fruited types, thus selecting against the types belonging to the *Criollo* component. The only genotype from Samoa known outside the island has a green fruit and possesses a distinctive leaf form that may indicate an origin in Ecuador. However, the genotype possesses characteristics that relate it to the *Criollo* but it is very different to any of the plants of hybrid origin that are known elsewhere.

Cacao cultivation in Papua New Guinea commenced at the beginning of the 20th century. As this territory was also a German colony at the time the most probable sources of planting material would have been the same as those of the types introduced into Samoa. Therefore, the first plantings would have consisted of the same *Criollo* types and this is evident in the genotypes that are known outside the country. It would have been expected that later attempts would have involved the *Forastero* varieties that had been established in Samoa. Other introductions probably were made but not recorded. Ceylon, once again, is a likely source of other types but the *Criollo* continued to be the predominant element in the constitution of the cultivated population. It is known that a variety with green fruits that are short and broad in shape and apparently homozygous occurs on the island of Bougainville. The variety may be ascribed to an introduction of the 'West African Amelonado' but it differs from it in certain respects.

Fiji is a group of islands in the Pacific Ocean where experiments with cacao cultivation were undertaken during the last two decades of the 19th century. It would appear that the first consignment of seed was received in 1883 and consisted of some of the same varieties that had been sent from Trinidad to Ceylon through the Royal Botanic Gardens in England. Although only 20 years had passed Anon (1903) reported that the trees were growing wild in the forest and that the records of the plantation had been lost. The report stated that most of the seeds of these trees had purple cotyledons. It is probable that mostly the hybrid types survived except for rare individuals with light coloured cotyledons. The report described one of these trees in terms that are worth including here:

> we have one dwarf tree the leaves of which are very much smaller than any others. This tree carries a small yellow pod deeply furrowed and pointed, the bean when cut is much lighter than any others, in fact is a pink colour.

Harvey and Harwood (1958) reported that the plantation consisted of two types, namely, 'Five-grooved Yellow Prolific' and 'Sangre Toro', from which seeds were distributed for planting. They stated, 'Excellent drawings of these original introductions enable them to be classified as Forastero, type Amelonado, and Criollo, respectively.' Whether these were correctly classified is a matter of conjecture, especially since the original list has 'Five-grooved Cundeamor' as described in the section on Trinidad and Tobago. Later the authors refer to the presence of 'a *Cundeamor* tree', which is classified as *Forastero*. However, it should be noted and emphasized that, if the 'Amelonado' in this case had been sent from Trinidad, it is not the same variety as the 'West African Amelonado'.

During the period after 1948 clones were introduced from Trinidad, Papua New Guinea, Western Samoa and Grenada and seeds from Papua New Guinea and Western Samoa. However, it was decided to concentrate on growing the 'Amelonado' and 'Cundeamor' types. It was found that there was a strong element of 'Criollo' types on the island of Vanua Levu.

A later account by Vernon (1971) gave a different picture of the variety situation in the Fiji archipelago. In it he stated:

> It appears that the planting material used was largely either pure Criollo or
> Trinitario (using this term to mean any cocoa of part Criollo ancestry).
> Contemporary and subsequent references to 'Forastero' material were almost
> certainly based on pod-character classification... The main sources of this material
> were Ceylon, Java and Samoa.

The *Criollo* and 'Trinitario' material would have belonged to the selections referred to by Harvey and Harwood (1958) as being introduced from neighbouring countries, whose populations are described above. This account highlights the differences in the concepts regarding the classification of varieties by various persons. Later introductions included the 'West African Amelonado', which would have introduced a complicating factor in the identification of types denominated 'Amelonado'.

Africa

While the Asiatic and Oceanic attempts at cultivating cacao were based on the initial introduction of *Criollo* varieties that were distributed widely, the history of the origins of cacao in West Africa is quite different. In this region the first material introduced was of Amazonian origin. Although there are indications that previous attempts at introducing cacao to the colony of São Tomé and Príncipe had been made it is generally conceded that the cacao industry of these islands is based on plants established on the island of Príncipe in 1822. The source of the seeds that constituted this introduction is generally taken to be the State of Bahia in Brazil. If this had been the case the most likely source of the seeds would have been in the vicinity of the city of Salvador. However, there are two other possible areas from which the fruit transported to Príncipe could have come but the exact place has yet to be determined. These areas are the northern part of the State of Espírito Santo and the State of Rio de Janeiro, cacao having been present in both States at the time as described under the extra-Amazonian cultivated areas.

A single fruit probably was concerned in the first introduction since it has been stated that the original planting consisted of 30 plants. After some time the progenies of these trees provided seed for planting in other areas of the island and by 1840 small quantities of cacao were being exported. In the 1850s the cultivation spread to São Tomé, the main island of the group. Even at the present time homogeneous plantings of this variety are still to be found on the island. These would be the result of several generations of inbreeding. The variety as it is known is self-compatible and homozygous. It could be related to the variety in Bahia called 'Comum', described under the cacao varieties cultivated in the State but it has not been possible to match the São Tomé 'Creoulo' to any other variety in Bahia.

The São Tomé 'Creoulo', as depicted in Fig. 46 was taken to other West African countries, either directly or through the island of Fernando Pó, in subsequent years and became the basis for the cacao produced in most of the countries on the continent. Because of its important position in these countries it became known as 'West African Amelonado'. From West Africa this variety has been introduced into most of the cacao growing countries in Africa, South Asia and Oceania.

About 1880 (as far as can be determined) other varieties were introduced into São Tomé from Ecuador, Trinidad and Venezuela. The varieties that comprised this introduction were established in progeny rows and some records were kept of their performance. Unfortunately, it has not been possible to see the original planting or find the records of the data.

Seed from this collection of introduced varieties was distributed to several farms on the islands during a period in which active expansion of cultivation was taking place. Since the seed used would have resulted from natural fertilization most of the plantings that were established would have been formed by hybrids among the varieties and later between these and the original 'Creoulo'. From these beginnings the São Tomé and Príncipe cacao population developed a wide range of phenotypic variability. Basically, the variability is similar to that found in Trinidad except for the possible differences resulting

Fig. 46. An early photograph of cacao trees on the island of São Tomé, presumably depecting the São Tomé 'Creoulo'.

from the effect of the 'Creoulo' as compared with the 'Calabacillo' and the influence of the Ecuador 'Nacional'.

Johnson (1912) described the situation regarding the São Tomé varieties at the beginning of the 20th century in a book that includes photographs of the fruits of the more outstanding types. The conception of the characteristics of these varieties is enhanced by the existence of collections of dried specimens of the fruits. Johnson described a type with red fruits that had a smooth surface. This type would represent the *Criollo* introduced from Venezuela as 'Caracas'. However, other specimens have fruits whose shape is narrow elongate but the ridges are very pronounced. The objective of producing a product similar in quality to that of the Venezuelan 'Caracas' prompted some farms to prefer the *Criollo* types, but it is likely that most of the plants for these derived from hybridization. It has not been possible to identify any of the existing trees as belonging to a genuine 'Caracas' variety but the connections with it are to be found in trees that possess most of the *Criollo* traits.

The presence of any representatives of the Ecuador 'Nacional' variety would be difficult to distinguish from some of the hybrid combinations that occur. However, trees can be found that possess characteristics that differ from those of the Amazonian × *Criollo* hybrids and have a stronger resemblance to types found in Ecuador, especially in those genotypes that have fruits of a green colour. The principal confirmation of the existence of cacao of the 'Guayaquil' variety is its inclusion in the sketches in Chalot (1901).

The São Tomé diversity includes one exceptional type that, according to information provided, had been introduced especially and established in an isolated area. The fruits of this type are depicted in Plate 7b. In some respects these fruits do not differ phenotypically from those of the São Tomé 'Creoulo'. Although it appears that segregation occurred for fruit colour the type is homozygous for other traits and it is possible to produce families of trees that are homozygous but differ only with regard to fruit colour. The origin of this variety is a mystery since it is unlikely to have been developed from the *Criollo* genotypes even after several generations of inbreeding and no other similar variety is known, which would be the source of the seeds imported.

The São Tomé population contains several genotypes that possess characteristics that are probably unknown in other countries. These include individuals with very large fruits, the surfaces of which are slightly ridged and smooth. Since the fruit colour is red in these cases the trees are obviously descended from the Venezuelan *Criollo* that was introduced. Other genotypes with green fruits appear to be hybrids but their progenies from self-fertilization reveal that they are homozygous for their visible characteristics. Also, among the individual genotypes with fruits of the 'Amelonado' shape a certain degree of variability can be discerned when they have been carefully examined.

The production of mutant genotypes is also observable in the population. The most outstanding of these is the genotype locally called 'Laranja', which is depicted in Plate 7c. This mutant became universally known when Chevalier (1908) gave it the specific epithet of *Theobroma sphaerocarpa*. The mutation probably arises in a variety of types in São Tomé but no similar genotype has been reported in other countries. Another mutation produced trees with thick trunks and short broad green fruits that have a distinctive short pointed apex. Another feature of this genotype is the peduncle that is so short that the fruits on the trunk are kept erect in contrast to the hanging fruits of the normal cacao trees (see Fig. 4). Female sterility also occurs. Another genotype encountered carries a gene that, when the recessive allele is in the homozygous state, segregates for progenies with yellow leaves but its effect is lethal.

The map of West Africa in Fig. 47 shows the position of the islands of São Tomé and Príncipe. From this it is possible to recognize the role played by these islands as the central point from which the other territories along the coast obtained the planting material for the establishment of their cacao cultivation. All of the countries in the area have cultivated cacao to a greater or lesser extent and the evolution of the germplasm in most of these countries will be described in the following pages.

As mentioned above, the first African territory to benefit from the cacao that had been introduced to São Tomé would have been the island of Fernando Pó (Malabo – Equatorial Guinea). The first date mentioned for the establishment of cacao on Fernando Pó is 1854 using seeds from São Tomé or Príncipe. A few years later one of the persons to whom a concession had been granted brought 400 fruits from São Tomé. The variety exclusively cultivated in the early years would have been the São Tomé 'Creoulo'. Other shipments would have followed, eventually including the varieties imported into São Tomé about 1880. There was a close relationship between the agricultural enterprises on the two islands.

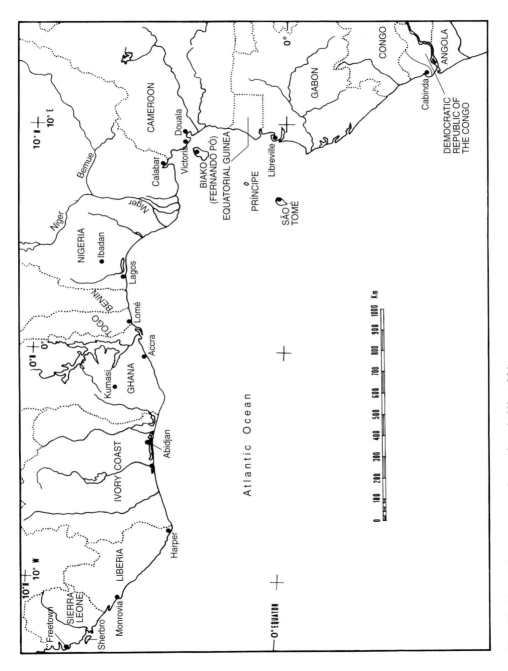

Fig. 47. Map of the cacao growing regions in West Africa.

The first recorded attempt to grow cacao in Cameroon is connected with a shipment made in 1876 containing 13 plants from the Royal Botanic Gardens (probably from Trinidad) to a British missionary on the Cameroon Mountain. Presumably these plants were established in the country but the location of the first planting is unknown and so, too, is the role they may have played in the development of cultivation in Cameroon. The varieties received are also unknown but their connection with the collection that had been established In Trinidad may be assumed, although the date is earlier than the first shipments from the variety collection.

When Cameroon became a colony of Germany cacao cultivation already existed and was planted by the natives. As in other parts of West Africa, planting material had been obtained from São Tomé or from Fernando Pó by the Baptist missionaries, according to the information given by Preuss (1901). The plantations formed by the natives or the missionaries are the likely sources of the fruits taken to Ghana by the Basel missionaries in 1889. After he took over the Botanic Garden at Victoria, Preuss introduced some of the 'better' varieties that had been established in São Tomé. According to a report written about 1904 he also set out to acquire the varieties cultivated in other countries and established them in the Victoria Botanic Garden.

Gosselin (1895) referred to the introduction of 332 plants from Trinidad. Presumably these included the 'Forastero' of which Preuss (1901) wrote that it gave the best product.

It was the failure to obtain a satisfactory product from the initial plantings that resulted in the journey Preuss undertook to South and Central America in 1899–1900. Hence, the progenies of the material provided from São Tomé would have been in production prior to 1899. The journey of Preuss (1901), in addition to the study of the agricultural systems employed in the countries he visited, involved the acquisition of planting material for establishment in Cameroon. This was collected as fruits and plants at the various places he visited and shipped to the Botanic Garden in Victoria or to Berlin. Preuss (1901) summarized the various shipments involved with the sources of the varieties and the quantities concerned.

The information regarding the condition of the plants after establishment at Victoria and the use to which they had been put is scanty. The 1904 report mentioned above refers to the collection of varieties occupying an area of 15 ha and containing specimens of 'Forastero', 'Criollo', 'Puerto-Cabello', 'Soconusco', 'Venezuela', 'La Guaira', 'Maracaibo', 'Guayaquil', 'Nueva Granada' (Colombia) and 'Suriname' besides the cacaos from São Tomé and other varieties from Trinidad. Some of these were planted separately in blocks. The other varieties, for example, 'Soconusco' and 'Guayaquil', were planted mixed with the type known in Cameroon as 'Victoria-Kakao'. Presumably this was the 'Caracas' variety introduced from São Tomé since it was stated that the 'Maracaibo', 'La Guaira' and 'Puerto-Cabello' were identical with the 'Victoria-Kakao'.

However, regarding these varieties, Preuss (1901) wrote that several of the types acquired did not correspond with the true varieties. For example, he found that the specimens identified as 'La Guaira' belonged to the 'Cojón de Toro' instead of the *Criollo* expected. The misidentifications especially concerned the types named 'Criollo' and 'Soconusco'. In some cases the identifications of the varieties were incorrect but it is possible that he may have referred to the quality expected of the cocoa produced. The 1904 report stated that although the best variety should have been the 'Criollo' it grew slowly and was unproductive.

In this way Cameroon had established the most diverse collection of cacao varieties that existed at the beginning of the 20th century.

The outbreak of World War I probably put a stop to the publication of results concerning the development and production of the different varieties so it is not known what planting materials were supplied from the Victoria Garden. It may be imagined that seeds from naturally pollinated fruits from the mixed planting of the 'Victoria-Kakao' were distributed.

Another West African country in which the initial efforts at cacao cultivation gave promising results was Gabon. The variety cultivated in the early years was the São Tomé 'Creoulo', from which a satisfactory product was obtained. In order to have varieties of superior quality the preferred types were imported from Cameroon in 1898. Chalot (1901) listed the names of several of these varieties, such as 'Soconusco', 'Guayaquil', 'Forastero', 'Criollo', 'Surinam', 'Venezuela', 'Caracas' and 'Victoria'. Also mentioned were nine 'species' from Trinidad; perhaps these were the plants that were introduced in 1893. The list gives an idea of the range of variability and diversity of origins of the germplasm that existed in Cameroon prior to Preuss' journey and probably includes the types that had been imported from São Tomé.

A later account of cacao in Gabon (Congo Française) by Chalot and Luc (1906) listed the varieties that were present in the experimental garden of Libreville with data on fruit and seed characters, the list containing 22 items but apparently the number of distinct varieties and origins was 19. The fruits and seeds were also described and accompanied by outline drawings of the fruits. Regrettably, the existence and knowledge of these varieties at the present time is unknown on account of the absence of an evaluation of them. The information provided in the article provides an excellent manner by which the basic differences among the varieties that were the most sought after at the end of the 19th century may be appreciated. It is interesting to note that among the types identified as originating in Trinidad several had short broad fruits but these were usually inferior in performance to the São Tomé 'Creoulo' or 'Amelonado'.

Further south from Gabon the former Portuguese colony of Angola made its first experiments with cacao about 1860, utilizing seeds from São Tomé. These would have been of the 'Creoulo' variety, which continues to be the most common in the areas where the species was successfully established.

Cacao planting was also attempted in what became the Belgian Congo. A satisfactory history of cacao cultivation in the colony and an analysis of the varieties grown are not available. According to de Wildeman (1908) a German agronomist made the first planting in 1884. With regard to the varieties used de Wildeman stated that seeds were obtained from São Tomé, Venezuela and Colombia. The inclusion of Colombia is curious, since this country has not been mentioned as a supplier of planting material and it would be interesting to know from where such seeds were obtained. Possibly, this refers to the variety

'Peñon', that is, the *Criollo* from the foothills of the Sierra Nevada de Santa Marta. De Wildeman mentions this variety in another part of his book, with regard to the trade with its product.

The history of the establishment of cacao cultivation in the British West African colonies follows a somewhat different line to that of the territories described above. In both the Gold Coast (Ghana) and Nigeria the origins of cultivation were based on the importation of seed from São Tomé, which, at the time, would have been exclusively of the 'Creoulo' variety. At least this variety was the one predominantly grown during a period stretching over more than six decades. Various versions regarding the first introduction of the species into Ghana have been made. According to an account published by the Basel missionaries there was an unsuccessful attempt to grow cacao made in 1857 with seed brought from Surinam. From material, whose date of introduction was stated to have been 1861, only one tree was alive in 1864. The story was that this tree descended from seed that had been brought from Liberia but there is no record of cacao existing in Liberia at the time. If cacao in production did exist at Harper its establishment there would date from before 1850. There was the possibility, of course, of a planting at Harper being made from seed brought from Príncipe since labour from Liberia used to be contracted to work in São Tomé and Príncipe.

Unlike Cameroon no attempts appear to have been made to introduce other varieties until about 1900, with the exception of a single plant whose origin is unknown, sent in 1890 and which died in transit. The introductions made at the beginning of the 20th century seem to have had the purpose of providing varieties that were considered to give the higher quality products. The first variety introduced was the 'Pentagonum' or 'Alligator' (that had been introduced into Trinidad from Nicaragua), which consisted of plants grown from seeds in Trinidad sent to Ghana in 1900. On account of its inferior development the variety did not prosper in the new environment. However, one plant differed from the others and survived. This was named 'Cundeamor' on account of its resemblance of its fruits to those of the trees with this name shipped from Trinidad to Ceylon in 1880. The genotype was a result of hybridization with an unknown parent in Trinidad, this perhaps being another member of the sample of Nicaraguan varieties. While the 'Alligator' plants disappeared during the following years the 'Cundeamor' was maintained in the form of its progenies.

In 1900 or 1901 small numbers of plants were sent to Ghana through the Royal Botanic Gardens in England of which no details are available about their establishment. The shipments included plants identified as a 'Red Forastero' and a 'Criollo' as well as a 'White Variety', the origin of which was recorded as Jamaica. A later report refers to a 'White Variety' of the 'Caracas' type that may have been successfully established.

The next dispatch of plants from Trinidad to the Aburi Botanic Garden in Ghana was made in 1903 and comprised a variety called 'Ocumare', the origin of which is described in the account relating to Trinidad. Although the records sometimes refer only to *Criollo* it appears that the shipment comprised a 'Trinidad *Criollo*' and a 'Nicaraguan *Criollo*'. Trees were seen at the Aburi

Garden in 1990 that resembled the descendants of the Nicaraguan *Criollo*, which are conserved in Trinidad. Later records refer to 'Red Criollo' and 'Yellow Criollo', which would agree with the Trinidad (perhaps a sample of the 'Caracas' variety introduced into Trinidad) and Nicaraguan types, respectively. The progenies of the 'Ocumare' variety were all red-fruited although there was, apparently, some segregation for the intensity of pigmentation. These trees and the 'Red Criollo' from Trinidad would be the ancestors of most of the red-fruited trees that occur in the country.

During the subsequent years planting material, probably mostly as seed from natural pollination, was distributed to the agricultural stations and private farmers. The populations derived from the original introductions would most likely have been produced by hybridization, either within the varieties or between them. It was later found that most of the trees from these introductions were self-incompatible. This situation would have resulted in hybridization taking place only with the self-compatible individuals in the introduction plots or with the self-compatible 'Amelonado'.

The large scale on which the 'West African Amelonado' was planted in Ghana should have resulted in the appearance of mutants but few have been reported and conserved. The reason for this situation could be ascribed to the absence of searches for mutant individuals. The level of knowledge among the farmers and the system of cultivation practised would have prevented the occurrence of unusual trees from being identified. Among the few mutant types that have been reported is one that carries a gene for what was termed an 'albino leaf' character. In the literature reference is found of the occurrence of a genotype that produced seeds with white cotyledons. These seeds may have been the expression of the anthocyanin inhibitor gene but it seems that the genotype was not preserved.

Cultivation in Nigeria developed independently from that in the Gold Coast and, from the point of view of the germplasm, presents a different picture. The situation is more complex especially since, at the time that the crop was being introduced about 1880, colonization was taking place under separate administrations, each of which pursued its own policies. Consequently, regional differences in the genetic composition of the populations created may have evolved. Verification of this hypothesis is impossible owing to the absence of descriptions of how the varieties were distributed and an analysis of the structure of the populations.

In general, the common basis of the germplasm in Nigeria in the three regions was the São Tomé 'Creoulo' that was imported in the form of fairly large quantities of seed from São Tomé or from Fernando Pó. This variety constituted virtually all of the planting material cultivated at the middle of the 20th century.

The most active area as regards the introduction of genetic types seems to have been in the Niger Delta region. This was administered, from 1884, under the Niger Protectorate, which included the States of Benin (mid-Western) and Cross Rivers. The economic development of the region was entrusted to the Royal Niger Company, which fostered the introduction of plantation crops and established agricultural stations for this purpose. The first experiments with cacao would have been established at the beginning of the administration of the Niger Protectorate. In 1899, one, or two, shipments of cacao plants were made by the Royal Botanic Gardens in England. These probably were plants that had been kept from the shipments made to Ceylon in previous years or progenies of fruits produced by them. No details were given of the varieties represented.

In 1900 six plants of the 'Pentagonum' type were received at the Old Calabar Station. They would have been part of the same shipment that had been supplied to the Aburi Garden in the Gold Coast at that time. No information was provided about the outcome of this shipment, which could have also included plants from outcrossing in the same way as occurred in the Gold Coast. In 1905, six plants of other types were sent to the Old Calabar Station from Trinidad. Later, a request from the Old Calabar Station to Trinidad for the supply of improved varieties led to the sending in 1909 of a barrel containing 60 fruits of the *Forastero* type. With regard to this shipment the information was given that fruits from the Old Calabar Station had been identified 10 years previously as *Forastero* from trees bearing this type of fruit that occurred in quantity on the Station. It is likely that genotypes with similar fruits, belonging to the more recent introductions to that island, had been introduced from São Tomé with one of the imports of seed from that source. No information has been discovered regarding the outcome of the above introductions and the extent to which material that survived from them had been distributed to farmers in the region and which may be found at the present time.

The first planting of cacao in the Lagos Colony is reported to have taken place in 1882 and consisted of 1500 plants. The seed from which the plants were raised was probably introduced from São Tomé and there were regular imports from that source at least up to 1889. By 1893 a plantation of cacao whose origin was given as plants 'introduced from Fernando Pó some years ago' was exporting several tons of cocoa. One or more shipments of plants to Lagos were made from the Royal Botanic Gardens in England in 1887 and 1888. Again, these may have been derived from the plants that had been sent from Trinidad in 1880 and 1881 destined for Ceylon. One of these shipments contained three plants. A single plant was sent to Forcados in 1905.

With regard to the Lagos Colony no indication is given as to the location where the introductions were established or the results obtained and their utilization in terms of distribution to farmers.

Later, it was reported that varieties of cacao had been introduced from other West African countries, the probable sources being Ghana and Cameroon. The research station that had been established at Moor Plantation near Ibadan started a programme of selection in the cacao populations in the southern States of Nigeria. The first activity in this regard concerned the detailed study of 49 selected trees of the yellow-fruited 'Forastero' variety. It is likely that these trees belonged to the 'West African Amelonado' as they were self-compatible. Trees producing seeds with light coloured cotyledons were identified among the other introductions but these were found to be mostly self- and cross-incompatible, a situation which would have indicated the presence of *Criollo* germplasm. Reports of the results of experimental work carried out prior to 1938 describe the varieties used as 'Red Forastero' and 'Yellow Forastero' as well as red- and yellow-podded (fruited) 'Amelonado'. This indicates that red-fruited varieties had been introduced at some time, the presence of the red 'Amelonado' suggesting a similarity to the same type in São Tomé.

Sierra Leone is a country where a small amount of cacao is cultivated. It can be seen from Fig. 47 that Sierra Leone is situated on the extreme west of the West African cacao belt. This is an area of transition to drier climates and, consequently, the area suitable for cacao is limited. The history of the introduction of cacao into Sierra Leone is rather different from that of the other British colonies and, in fact, it may be the first territory on the African mainland in which cacao was established (Howes, 1946). A shipment of young cacao plants was sent from the West Indies to the Glasnevin Botanic Garden in Dublin, Republic of Ireland, for shipment to West Africa. It is suggested that the destination of these was Sierra Leone. Apparently, cacao plants had existed in the garden for several years previously but it has not been possible to determine their origin or the variety. Also, the location of the establishment of the plants in Sierra Leone was not recorded. It has been reported that a few very old trees are to be found in Sherbro.

Sixty plants were sent in 1902 from Trinidad to Sierra Leone of varieties named as 'Ceylon Red', 'Nicaraguan Criollo', 'Forastero' and *T. pentagona* (the first had been introduced into Trinidad a few years earlier) and the shipment also contained 1000 seeds of a unspecified variety that had been requested. The outcome of this shipment also was not recorded, but it is probable that survival of the plants was poor under the unsatisfactory conditions.

Newland (1916) had a photograph of a plantation of very large trees that appeared to belong to a single variety, this resembling the São Tomé 'Creoulo'. The same book has an appendix containing advice for farmers by H. Hamel-Smith, in which it is stated that the growing of 'fancy varieties' along the West African coast should be avoided. The type that was grown in São Tomé and Cameroon was recommended as the best for the purpose.

It is interesting to note that several Sierra Leonians were among the principal cacao farmers in Fernando Pó at the end of the 19th century. Hence, this island could well have been the source of some of the plants in Sierra Leone.

A certain amount of cacao has been cultivated in Liberia although the quantity appears to be small. If the report that the Basel missionaries had brought fruits from Cape Palmas in 1861 could be verified it would show that the origins of cultivation in Liberia were fairly old and possibly predated the introduction into Sierra Leone. It has proved impossible to obtain any information about the varieties cultivated in Liberia but the possibility occurs that some planting material may have been introduced from São Tomé during the period when Liberians were contracted to work on the plantations on the islands.

In East Africa conditions would seem to be generally unfavourable for cacao and several attempts to establish the crop did not lead to any great success. The oldest planting appears to have been in Zanzibar where in 1891 it was possible to obtain ripe fruits. These trees probably would have been obtained from a place such as Mauritius and were descendants of the *Criollo* introduced into the Philippines from Mexico. However, this plantation in Zanzibar stimulated the idea of establishing the crop in the, then, British Central Africa, now Malawi, and 100 fruits were imported from Grenada in 1895. However, no further information about the outcome of this importation has been found but it probably did not have any success.

The German East Africa colony started experimenting with cacao towards the end of the 19th century. Plantations were established in Usumbara at an altitude of about 2000 m (in what is now probably Burundi). The planting material perhaps came from Cameroon and it was recorded that the *Forastero* cacao had been planted.

In 1901 the British colony of Uganda made its first experimental planting of cacao at the Botanic Station of Entebbe. This experiment was made with young plants obtained from the Royal Botanic Gardens in England. These would have been the same varieties that had been distributed to other British colonies during the previous decade, descended from the plants sent from Trinidad in 1880/1881. In 1911 yield data were given for the first planting from red and yellow 'varieties' of *Forastero* and red and yellow 'varieties' of *Calabacillo*. The *Forastero*, in this case, would apply to the progenies of hybrids with *Criollo*. Segregation would have occurred among their progenies so that as reported two years later, 'We have in Uganda only "Forastero" types ranging from a high type approaching "Criollo" to a low type approaching "Calabacillo".' (The same segregation would have occurred in the other countries that had received material of the same type from the same source.)

On a final note, the confusion regarding the concepts that concerned the application of 'variety' names is very evident in the above account.

7 The Genetics of the Diversity

An outline of the extent to which the diversity of the species can be recognized on the basis of the large number of morphological markers that are exhibited by the various populations was presented in Chapter 3. The knowledge of the inheritance of the attributes of the populations and varieties would be an important tool in the determination of the genetic structure of these populations and, consequently, the relationships between them. However, at present, this knowledge is limited to a few characters that have been analysed or observed in some depth.

Various causes are responsible for this situation. One is that many of the character expressions have become known only during the past few years. The specimens in which these traits are expressed are still in the development stage; accordingly only observations made on a casual basis provide information regarding their characteristics. Although breeding programmes have generated a large number of families through inbreeding and hybridization, analysis of the segregation that occurs in the many characters in these families has not been undertaken or, if at all, in an incomplete manner. Among the factors responsible for this situation are that the families are represented by small numbers of individuals, a situation that does not permit reliable conclusions to be arrived at from the analyses of the data. Also, as a consequence of the nature of the species, it may take a long time to carry out the necessary observations in order to obtain a reliable basis for analysis. One of the principal causes of the absence of adequate knowledge of the inheritance of the species is the fact that most breeding programmes are conducted by persons with limited concepts of the subject or lacking an interest in the determination of the inheritance of the character expressions.

In view of the inadequate knowledge of the inheritance in the species this account will be restricted to describing the genetic mechanisms that control the expressions of the characters about which some knowledge is available. It should be emphasized that most of the studies on inheritance that have been conducted so far have been based on a limited number of individual genotypes. As a result it is not always possible to extrapolate the results obtained from the characters of a given population to predict the likely situation in another, perhaps unrelated, population whose individuals may exhibit similar phenotypic expressions.

The phenotypic expressions of several characters are subject to the influence of non-genetic factors, some of which could be of significant importance. These effects have not been studied to any degree and the absence of knowledge about the reactions of some characters to various environments precludes an analysis of the extent to which non-genetic factors are involved in the expressions of the phenotypes observed or their stability.

Plant Colours

The coloration of the various organs of the cacao plants is often the most visible indicator of variability. Therefore, since these colour expressions are the easiest to observe the knowledge of the mechanisms of their inheritance is the most developed. Although the phenotypic expressions of colours of specific organs may be controlled by genes that are specific for the organ concerned, the situation is complicated by the action of genes with pleiotropic effects or linkages, as will be explained below.

Fruit colour

The basic colour of cacao fruits is green. This character exhibits an ample range of variability, with shades of green varying from very light green (called *blanco* = white, by some authors) to dark green, the fruits in Plate 18a being an example of the latter. In the families in which segregation has been observed the darker shades of green are dominant to the lighter ones, but it is not known whether this applies universally to all combinations involving parental genotypes that differ with regard to the shade of green.

Superimposed on the base green colour are various degrees of red pigment. The usual source of red pigment is that found in members of the *Criollo* group. In this case a single gene appears to be responsible for the phenotypes descended from the group with the red pigment being dominant to the state in which it is absent. However, when this gene is present a wide range of phenotypic expressions is observed in terms of the intensity of the pigment and the manner in which it is distributed in the epidermis of the fruits. These phenotypic expressions are partially related to the genotypes of the individuals concerned. When the individuals have similar genetic backgrounds the homozygous genotypes would be expected to express a greater intensity of pigment than the heterozygous genotypes. The intensity of pigment would be influenced by the status of the base green colour. In some cases a dark red surface pigment may result from enhancement of the pigment effect when the base colour is dark green while in other cases the dark green base may inhibit the full expression of the pigment gene.

In individuals descended from the *Criollo* group the pigmentation could vary from a light red or pink, or with an effect that is barely visible, to a very dark red that is almost purple in intensity. The latter degree of expression would probably be more intense than that found in the *Criollo* genotypes in which the gene is intrinsic. Such situations lead to the possibility that other genes may be involved in the range of red pigment observed through a modification of the expression of the pigment gene in the *Criollo* genotypes or by the presence of factors that limit the synthesis of the anthocyanin pigment. Detailed studies of such situations have not been made to elucidate these hypotheses. It may also be conceived that two loci are involved. One locus would carry alleles that determine the intensity of the green colour. The other locus would be involved in the expression of the red pigmentation with various alleles determining the expressions of pigmentation in a sequence of dominance, the recessive allele being an amorph.

Various genotypes from different populations of the Amazonian Region produce fruits with varying degrees of surface pigmentation and varying expressions of it. The low frequency with which such genotypes occur, pigmentation often being expressed in progenies of genotypes with green fruits, indicates that the pigment expression may result from a combination of various factors, the inheritance of which has not been studied. Nearly all of the genotypes with superficial pigment do not express the same effects on other plant characters as the red-fruited gene.

Hypocotyl pigment

One of the pleiotropic effects of the fruit pigment gene is the development of pigment in the hypocotyls of the emerging radicles following germination of the seeds, as observed *in vitro*. In this case the dominant expression of the gene is well marked and provides an easy procedure for determining the fruit colours of the individuals that will develop in segregating families.

Flush leaf colour

This character exhibits considerable variability in respect of its expression and is, in many cases, related to the fruit colour. At one end of the scale are the genotypes in which no anthocyanin pigment is developed in the young leaves. Descriptions of the populations that contain genotypes with such phenotypes (as found in the chapters that summarize the nature of the diversity) indicate that differences exist with regard to the manner in which the phenotypes are expressed. Such a situation suggests that different genetic mechanisms may be involved according to the populations concerned. As will be seen below there could be a relationship between cotyledon colours and flush leaf colours in such populations but each of them has to be considered individually as explained in the examples that are described. However, in general, it may be concluded that the expression of non-pigmented flush leaves is due to recessive alleles when the genotypes concerned are homozygous for these alleles.

In the populations derived from the Equatorial Oriental Piedmont there is a clear association of the non-pigmented cotyledons with the non-pigmented flush leaves, from which it may be inferred that the character is due to a single recessive allele. However, the mode of inheritance of the character may not be as simple as it appears. When individuals with non-pigmented flush leaves are crossed with those having pigmented flush leaves their F_1 progenies produce pigmented flush leaves. This result would be expected if the non-pigmentation was due to the action of a single recessive allele. However, combinations of the F_1 progenies with other pigmented genotypes segregate for non-pigmented flush leaves as though the character is inherited from a dominant allele.

Similar situations may occur with genotypes in the Amazon Extension Zone, which produce pale green flush leaves. However, other genotypes from the zone exhibit other expressions in which the pale green flush leaves are produced from seeds with purple or pink cotyledons. Hence, in these cases the absence of pigment in the flush leaves is inherited independently of the cotyledon colours. Genotypes from the R. Envira population also produce non-pigmented flush leaves, the character similarly being determined by a single recessive allele. No seeds with white cotyledons were reported in the collection data and none was found in the fruits of the progenies with non-pigmented flush leaves. It may be concluded that, in this case, the cotyledon and flush leaf pigment expressions are inherited independently.

Other genotypes with green fruits have flush leaves containing varying degrees of anthocyanin pigment. The gene that confers the red pigment to the fruit surface also confers pigment to the flush leaves. The genotypes involved would show enhanced intensity of pigment in comparison with the green-fruited genotypes belonging to the same families. In such situations the red-fruited gene would be dominant. The action of this gene also results as pigmentation being expressed in the pedicels of the young leaves and in the adjacent axillary node, which persists for a few days after the flush leaves have hardened. This character has been called 'axil-spot' and is useful for identification purposes and, when observed in young plants, for determining the fruit colours of the plants that will develop. The character is, of course, the result of the action of the same gene that produces pigment in the hypocotyls and is supplementary to it. However, it should be pointed out that, as depicted in Plate 4b, some green-fruited genotypes also show pigment in the leaf pedicels.

Pigmentation of the flush leaves is a normal occurrence in genotypes with green fruits and varies in intensity. Hence the character is inherited independently of the fruit colour. Even in the Equatorial Oriental Piedmont genotypes the flush leaves of the plants that develop from the pigmented cotyledons are intensely pigmented. Variation in the degree of intensity of pigment occurs and it would be necessary to study the differences in the expressions resulting from the influence of the gene that produced pigment on the fruit surfaces.

Although it would be expected that all genotypes possessing the gene producing pigment on the fruit surface would develop pigmented flush leaves, a case has been recently observed where plants with red fruits have unpigmented flush leaves. The occurrence of this combination needs to be investigated as well as the system of inheritance involved, which may be due to genes whose effects are, as yet, unknown.

Cotyledon colours

The cotyledons present a complex situation regarding the nature of their pigmentation. This varies from white (lack, or inhibition of the synthesis of anthocyanin pigment) to varying shades of pink, violet or purple. In some cotyledons various degrees of mottling are encountered, resulting from mixing of pigment in the cells of the cotyledons. Also involved in the determination of the specific expressions of this character are the presence of the gene producing pigment on the fruit surfaces and the association of factors linking cotyledon pigment and flush leaf pigment.

In the situations in which there is a simple effect of pigmented cotyledons versus those in which anthocyanin is absent the gene producing the pigment is dominant to the non-pigmented genotypes. Wellensiek (1931) observed the cotyledon colours in open-pollinated fruits of trees that were heterozygous for the trait and found that, in most cases, there was a segregation ratio of three purple cotyledons to one white cotyledon. On this basis he concluded that a single gene was involved. However, in fact, a range of shades of purple pigment was observed and the pigmented class was the total of the cotyledons with pigment.

The expression of a range of shades of pigment is common in hybrid families descended from genotypes belonging to the Criollo group, such as that found by Wellensiek. In most cases this situation applies to the progenies of red-fruited Criollo genotypes, where, as referred to in Chapter 5, the allele controlling fruit colour is expressed as a slight pigmentation in the cotyledons.

The genetic factors responsible for the various levels of pigment expression have still to be determined in those situations where the expression of cotyledon colour is not confused by the factors influencing fruit pigmentation; that is, when only green-fruited genotypes are concerned. One of the problems involved concerns the accuracy with which the pigment expression can be determined, which is particularly important in view of the gradations of pigment expressions within fruits. Another matter refers to the effects that non-genetic factors have in determining the degree of pigment in a given cotyledon.

Bartley (1964b) discussed some of the findings from a diallel mating system involving inter-crossing genotypes homozygous for various levels of pigment expression. The results showed that genotypes with intermediate levels of pigment produced the same result as genotypes with intense pigmentation when crossed with genotypes having non-pigmented cotyledons.

When green-fruited 'Criollo' genotypes were crossed with the 'Catongo' genotype (carrying the anthocyanin inhibitor gene) the progeny seeds had cotyledons with some degree of pigment. In contrast, the progeny of crosses between 'Catongo' and other genotypes with purple cotyledons produced the same intensity of pigment in all of the seeds as was exhibited in crosses between 'Catongo' and *'Criollo'* genotypes with white cotyledons.

The cross between 'Catongo' and the clone 'SPA 17', which is presumed to belong to the lower R. Napo population, segregated into equal proportions of light purple and dark purple cotyledons. The plants of the progeny from the lightly pigmented cotyledons produced light green flush leaves while those from the dark purple cotyledons produced pigmented flush leaves. From this result it may be concluded that 'SPA 17' is heterozygous for the recessive gene of the R. Napo population that is responsible for the non-pigmented cotyledons and flush leaves.

The combination of 'BE 2', which produces pink cotyledons (Plate 9d), and 'Catongo' resulted in seeds that had the same pigment expression as 'BE 2'. In this case it appears that the pink cotyledon expression of 'BE 2' is produced by a dominant gene. However, the results of hybridization of 'BE 2' with trees homozygous for dark purple cotyledons are not available to determine the relationships among these genotypes.

Flower colours

The flower is a complex organ that shows different colour expressions in its component parts and also a variety of combinations where two or more parts are concerned. Consequently, it is necessary to consider both the pigment expressions in the individual parts as well as the possible genetic relationships among the parts, in addition to the relationships with the expressions of pigment factors in other plant organs. So far, very few studies of the inheritance of pigment characters in the flowers have been conducted. The results of studies carried out to date are difficult to interpret on account of the complexity of the subject as well as the fact that not all of the expressions have been adequately identified. An analysis of these observations will not be undertaken at this time owing to the complexity of flower expressions and the inter-relationships of the characteristics of the various flower parts.

Many genotypes with green fruits have sepals that lack anthocyanin pigment. Other varieties possessing the anthocyanin-free allele show various expressions of pigment in this organ. These expressions are usually specific to given varieties and populations indicating the possible existence of different genes responsible for the expressions in individual cases. One of these cases concerns varieties that show conspicuous red pigment on the external edges of the sepals (observed where they join in the unopened buds). Genotypes with red fruits produce pigment on the inner surfaces of the sepals, but the intensity of pigment varies according to the homozygosity status of the genotype. Hence, the pigmentation of the inner surface of the sepals would be the result of action of a dominant allele linked to fruit colour. However, there are green-fruited genotypes that also have sepals pigmented on the inner surface, which indicates the existence of different genetic mechanisms acting independently.

The pigmentation of the staminodes and the guide-lines on the inner surface of the petal pouch presents some interesting situations. Plates 8a and 9a contain examples of flowers where both of these organs have the same shade of dark purple pigment and Plate 9c is an example of a case where the colour of both organs is pink. However, in other genotypes the shades in the individual parts may appear in various combinations. Examples of the combinations that occur are dark staminodes with pink guide-lines or light coloured staminodes with dark purple guide-lines. Evidently, the genetics of pigmentation in these cases is somewhat complicated but not enough data are available to reach any conclusions as to the mechanisms involved.

The pigment of the stamen filaments is a sufficiently distinctive character to be of considerable value in the identification of genotypes belonging to distinct populations. The differences among the populations are quite marked, some of them showing distinctive pigment intensities and patterns. At one end of the scale are populations in which no pigment is visible and at the other end are genotypes, such as those from the R. Purús system, in which the pigment is intense and very evident. Here, again, although efforts have been made to study the segregation of the character in the progenies of some genotypes not enough data have been collected to obtain a definitive idea of the mode of inheritance. Bartley (1964b) summarized some of the observations made of the segregation that occurred in several crosses. It would seem that, in the group of clones identified as 'IMC', the distinctive pigment of the stamens is inherited as a partial dominant. In the absence of observations on the parental genotypes, interpretation of the incomplete analysis of the segregation occurring in the hybrids derived from its progenies would be premature.

In most examples the flowers of red-fruited genotypes, even those resulting from hybridization with genotypes with pigmented stamens, do not exhibit colour in the stamens. In such cases it would appear that the allele that confers the red pigment to the fruits (and other organs) inhibits the expression of the stamen pigment. However, at least one genotype is known with red fruit surface colour and stamens that are distinctly pigmented.

The anthocyanin inhibitor gene

The presence of this gene in the homozygous condition is recognized by the absence of anthocyanin pigment in all organs of the genotypes that carry this allele. Hence, when the allele is present in the homozygous state the genotypes that possess the allele for fruit surface colour produce fruits that are green. This would be the case of the genotype illustrated in Plate 32a, which is the oldest known example of the effect of the gene, the illustration having been published by van Nooten (1863). Although there are some well-known genotypes from the Amazonian Region that carry the allele it occurs fairly frequently. It is found in heterozygous individuals belonging to a wide range of varieties and also occurs in Central America.

The allele responsible for the mutation is recessive and, consequently, only detectable in the cotyledons and progenies of the heterozygous individuals whose fruits have been self-fertilized. The penetrance of the allele may not be complete since a few progenies have been observed with slight surface pigment of the fruits and also is observed as slight coloration of the cotyledons. While combinations of plants carrying the gene with those producing dark purple cotyledons have seeds with cotyledons of the same shade, the seeds produced in crosses between the mutant type and green-fruited *Criollo* genotypes with white cotyledons are light purple in colour. This result indicates that one of the parents has a cotyledon colour expression that is partially dominant (Bartley, 1964b).

It would appear that the absence of anthocyanin pigment in any plant organ, or combination of organs, would be the result of genes that block the action of the enzymes responsible for the synthesis of anthocyanins. Although the analyses of segregation of pigmented/non-pigmented plants may indicate that single genes are involved in any particular process, it is probable that the expressions observed are the result of the involvement of the whole genotype of an individual. So far no studies appear to have been undertaken to determine whether there is a relationship between the shade of purple pigment in the cotyledon and the shade of flush leaf colour; excepting, of course, the association between non-pigmented cotyledons and flush leaves in certain populations.

Fruit Shape

The overall fruit shape is a composition of various effects that influence the appearance of a given fruit. Accordingly, the fruit shape needs to be broken down into its various components. In particular the length of the fruit is determined to a large degree by the nature of its apex. Longer fruits usually possess long acute or attenuate apices whereas short fruits are generally the result of having short acute or rounded apices. However, there are exceptions to these relationships.

Progenies of crosses between genotypes of the *Criollo* group with elongate fruits and Amazonian types with short fruits produce elongate fruits. This explains why the hybrid descendants of *Criollo* genotypes tend to resemble the parents belonging to the *Criollo* group and are, therefore, difficult to distinguish from the *Criollo* parents. The progenies of combinations of *Criollo* hybrid genotypes with those having short broad fruits segregate for a range of fruit types including short broad fruits and elongate fruits. The inheritance of the elongate fruits of the *Criollo* genotypes appears to be the result of the action of a single dominant allele.

It would appear that the greater frequency of elongate fruits in some populations is due to the dominance of alleles producing longer fruits. The hybrids of the elongate-fruited genotype 'MA 15' with homozygous genotypes of the variety shown in Plate 7d produce fruits that are uniformly of intermediate length. However, there are other exceptions to the possible generalized dominance of elongate fruits.

In Fig. 48 two examples are presented that illustrate the influence of different parents on the characteristics of the fruits of their F_1 progenies. The

Fig. 48. Demonstration of the influence of the parents on the characteristics of the fruits of their hybrid progenies. (Top) 'MA 15' × 'Mocorongo'; (bottom) 'Mocorongo' × 'CA 6'. The parents are depicted in Plate 23.

fruits of 'MA 15' × 'Mocorongo' are shorter than those of 'MA 15'. It may be inferred that the fruit length of 'Mocorongo' is dominant to that of 'MA 15'. On the other hand, the smooth surface of the F_1 is probably inherited from 'MA 15'. In contrast, the short, broad shape of the F_1 of 'Mocorongo' and 'CA 6' indicates that the fruit shape of the latter is dominant. The prominent ridges with a tendency for surface pigmentation are probably inherited from 'Mocorongo'. Some surface pigment is developed in progenies of 'Mocorongo', presumably in the individuals lacking the anthocyanin inhibitor allele.

The segregation for fruit shapes in the F_1 generation of the combination 'PA 16' and 'PA 148', shown in Fig. 11, presents a different picture with the fruits of the progenies tending to be more elongate than those of the parents.

Fruit Husk Characters

The hardness of the mesocarp layer of the fruit husk is one character that is intriguing for the geneticist. The very hard mesocarp of fruits of the R. Guaviare variety (which is homozygous for the character) appears to be dominant to the less woody mesocarps of other varieties with which it has been hybridized. At least the dominance is partial if not always complete, indicating that a single gene would be involved.

On the other hand, the very soft husks of the R. Juruá system's specimens (in which the mesocarp is absent) give every indication of being dominant. Hybridization between the two extreme types has not been attempted so far.

Some genotypes with thick husks tend to transmit this character to their progeny. The subject of the inheritance of husk thickness has not been studied in detail although a quantity of data would be available for analysis. The character presents some difficulties in measuring owing to the variability with regard to the characteristics of the ridges and fruit diameter; these leading to the need to define the quantitative aspects of husk thickness in order to estimate it objectively.

The 'Pentagonum' Fruit Shape

The low rate of incidence of fruits with the 'Pentagonum' shape among progenies of heterozygous members of the *Criollo* group indicates that the fruit shape is the result of the combination of action of several recessive factors when they occur in the homozygous state. On account of this situation it would be expected that families could be produced from genotypes bearing the character that are homogeneous for it. This would have occurred in the progenies of the 'Pentagonum' trees introduced into Trinidad as well as their descendants from seed exported to other countries.

Leaf Characters

The extensive variability encountered in the various expressions of the leaf characteristics provides excellent opportunities for studies of inheritance of the individual characteristics. However, the leaf is another complex organ and insufficient attention has been paid to the studies of the genetics of its components. An attempt was made to study the inheritance of leaf shape in crosses between genotypes with narrow elongate leaf blades and those with broad leaves. The data obtained were not completely analysed but indicated that, in the parent genotypes used, the narrow shape was dominant. In the case of the narrow-leaf genotype 'CA 3', whose inbred progenies are shown in Plate 31b, there is some segregation for shape but narrow leaves predominate.

Leaf texture and thickness is another outstanding character worthy of study. The thick, hard leaves of 'ICS 89' were shown to be dominant over the finer leaves of the other parents with which it was hybridized on the basis of subjective observations but, again, data are not available from the range of parents used in combinations with this genotype.

Dwarfs in the 'Scavina' Family

Progenies of crosses between some of the members of the family denominated 'Scavina' ('SCA') originating on the R. Ucayali in Peru (described in Chapter 5) segregate for dwarf plants in varying proportions (Fig. 7). Similar genotypes were obtained in crosses between F_1 individuals in hybrids with 'SCA 6' as a parent and in backcrosses of the resulting F_2 families to a 'SCA' clone.

Bartley (1967) reported the proportions of dwarf segregants obtained in some of these combinations. The proportions of the dwarf progenies obtained were as high as 25% in some combinations but the low frequency in most combinations indicated that perhaps the character was the result of the combination of recessive alleles at two loci when both were in the homozygous state. No studies were made of the behaviour of progenies obtained from the dwarf genotypes. The lack of this information prevented definite conclusions from being reached regarding the genetic mechanism involved. The dwarf character persists during the juvenile phase of plant development but in the course of time the plants reach a more normal stature but are still distinguishable from the normal genotypes. The appearance of this character is one example of the existence in the genetic composition of cacao genotypes of traits that are of little practical use but may have an effect on the breeding ability of the parent genotypes.

Leaf Abnormalities of Lethal Effect

Among the deleterious characters encountered in the production of advanced generations from cacao genotypes are various manifestations of abnormalities in their leaves. These abnormalities are identifiable in the early stages of development of the seedling progenies and take the form of alterations of leaf size and shape and various colour effects. The progenies that exhibit these characters die during the first months of growth. Some examples of these characters are shown in Plate 31a and c. The fact that these abnormalities occur in various forms in genotypes belonging to different populations indicates that they arise from individual mutations specific to the populations or genotypes. Most of these abnormalities are due to the action of single alleles that are recessive.

The appearance of segregants with yellow leaves in self-fertilized progenies of clones from the Parinari region of the R. Marañon was mentioned in Chapter 5. The allele was also lethal in its effect. The character was named 'Luteus-PA' and the results obtained in the progenies were reported by Bartley et al. (1983). The ratios obtained indicate that, in this case, more than a single locus was involved. Since various genotypes from this population carry the factors responsible for the expression of the character it could be used as a tool for determining the relationships among the genotypes that belong to the population.

Compatibility and Reproduction

The discovery of the occurrence of self-incompatibility in the species was followed by intensive investigations to determine the causes of the condition and the genetic mechanism involved. The condition has important consequences for the improvement of the species as well as an influence on the reproductive capacity of the self-incompatible individuals. The present knowledge of the genetic mechanism is based on the work of Cope (1962) and the consequences of the various situations involving the possible mating systems were summarized by Bartley and Cope (1973).

The mechanism by which incompatibility occurs in cacao was found to take place in the ovary of the flower during fertilization. At the time that the mechanism was discovered the species was unique in this respect. The incompatibility could be described as a rejection by the egg nucleus of a male gamete carrying the same incompatibility allele as the female gamete. When non-fusion of gametes occurs in a proportion of the ovules the development of the ovary is halted, causing the flower to abort.

The studies carried out showed that a series of alleles was involved of which one had a neutral effect, in which case the rejection mechanism did not function in the egg cells when the gametes were identical. This allele is the self-compatibility allele. When a particular flower is fertilized by pollen gametes carrying the neutral allele all egg nuclei are compatibly fertilized and normal fruit development occurs.

The alleles that are involved in incompatible fertilizations were found to exist in several states according to the dominance relationships between pairs of alleles. This situation is illustrated by the following sequence of dominance, the alleles being identified by letters:

a > b > c = d ... 0

The signs > and = indicate complete dominance between the pairs of alleles and co-dominance, or independent action, respectively, and 0 the self-compatibility allele.

The genotypes obtained from various combinations of alleles can be determined by cytological examination of the ovules after fertilization or by the segregation that occurs in the progenies of the families obtained in subsequent generations.

The various situations that can occur in individual self-incompatible genotypes are:

1. In the case of a genotype such as *a0* self-fertilization results in the occurrence of non-fusion in 25% of the ovules. When the genotype is fertilized by a *00* genotype 50% of the progenies will be self-compatible and 50% self-incompatible.

2. When the genotype possesses two incompatibility alleles of which one is dominant, for example, the type *ab*, the action of the recessive allele is cancelled and a *b* male gamete will fuse with a *b* female gamete. In this case the proportion of non-fusion alleles will be 25% but when the genotype is fertilized by a *00* genotype all progenies will be self-incompatible, 50% of these being compatible with the *ab* parent.

3. In the genotypes carrying two co-dominant alleles, for example, *cd*, the male gametes will not fuse with the corresponding female gametes, leading to a proportion of 50% of the ovules in which no fusion occurs. The product of fertilization of this genotype with the *00* genotype will result in 100% of the progeny being self-incompatible and also incompatible with the *cd* parent.

The above examples show how the genotypes that determine whether an individual is self-compatible or self-incompatible produce various situations that determine the ability of self-incompatible genotypes to inter-cross. In the populations in which only one incompatibility allele is present all self-incompatible genotypes are cross-incompatible. Where more than one self-incompatible allele is present genotypes possessing the same dominant allele are cross-incompatible but are cross-compatible with genotypes carrying other incompatibility alleles. This situation provides for the possibility of genotypes occurring that have various combinations of alleles and of hybridization among self-incompatible genotypes even if the self-compatible allele is absent.

Genotypes that are homozygous for the self-compatibility allele are self-compatible and cross-compatible with all other genotypes regardless of whether they carry any self-incompatibility alleles.

Among the populations studied a number of alleles have been identified and many of these are specific to the populations in which they occur. Because of this situation the compatibility/incompatibility relationships between populations and also within them are useful tools for differentiating among populations and identifying individual genotypes. An example of a study in which the incompatibility relationships have been used to assign genotypes to related groups in a particular set is that reported by Yamada *et al.* (1982, 1996). The next step would be the determination of the genotypes through the analysis of the progeny of inter-crosses. The use of the determination of incompatibility genotypes and their relationships for establishing inter- and intra-population diversity has not received the attention it deserves on account of the effort required.

The incompatibility mechanism in cacao is not an insurmountable barrier to obtaining inbred progenies from self-incompatible genotypes. Since in any mating system fusion occurs in at least 50% of the ovules any technique that overcomes the abscission mechanism can be utilized to produce seeds resulting from self-fertilization. One technique involves mixing compatible pollen and pollen of the genotypes that it is intended to self-fertilize. The value of obtaining self-compatible progenies in this manner depends on whether the genotype concerned is heterozygous for the self-compatibility allele.

In spite of the apparently numerous alleles that have been identified in the populations in which the incompatibility status has been studied, the concepts related to the subject are fairly easy to understand. However, the matter is not so simple since other possibilities of determination of self-incompatibility have been identified. In the case of the genotype 'IMC 67' the segregation of self-compatible and self-incompatible progenies of crosses of the genotype (which is self-incompatible) with self-compatible genotypes was found not to produce the expected ratios. The genotype appears to combine with the genotype that is understood to be its female parent and it also is compatible with some of its F_1 progeny. This situation indicates that 'IMC 67' possesses a compatibility allele that behaves differently to the allele of this type in other genotypes.

Another deviation from the expected norm was discovered when a self-compatible genotype of the *Criollo* group was hybridized with a self-compatible genotype of another population or origin. It was found that, in this cross, all of the F_1 progenies were self-incompatible. On the basis of studies of the zygote reactions of the progenies and those of other generations of the combination an explanation of this behaviour was proposed that involved the action of two modifying genes (which we may name *A* and *B*). In this case one of the parents should be homozygous for one dominant gene, with the other gene in the recessive state (that is *AAbb*). The genotype of the other parent would have the dominant and recessive genes in a reversed state (*aaBB*). Consequently, when both genes are in the heterozygous state, as occurs in the progeny, the individuals are self-incompatible. The same results are obtained when other genotypes of the *Criollo* group that are self-compatible are fertilized by self-compatible genotypes from the Amazonian Region populations.

Situations of this nature have significant implications for the determination of the incompatibility/compatibility relationships of varieties that have been derived from hybridization with members of the *Criollo* group. The exact effects of this situation have not been adequately investigated, especially on account of the small scale on which the *Criollo* group has been involved in breeding. All genotypes in which this situation has been encountered belong to the Central American *Criollo* populations, since the situation in pure *Criollo* genotypes from northern South America has not been investigated. On the basis of the findings observed it may be supposed that the Central American *Criollo* genotypes possess a gene that distinguishes them from the Amazonian Region populations.

Besides the incompatibility situation several other mechanisms appear to act in cacao that influence fruit development and productivity. One such mechanism concerns the production of what are known as 'male trees', that is, individuals whose gynoeceum is sterile and, consequently, unable to form fruits. Since the androeceum is fertile the pollen from the flowers can fertilize other genotypes and the progenies obtained seem to behave normally. Although the actual genetic mechanism has not been fully investigated it would be assumed that the female sterility is caused by the action of a homozygous recessive gene. This condition is infrequent but has been observed to occur in several populations.

Another situation is one that involves individuals that produce fruits with low numbers of seeds. In some of these individuals the number of seeds rarely exceeds half of that of the ovules. The same results are obtained from cross-fertilization as well as from self-fertilization. The likely explanation for the incomplete seed content is the action of factors during the development of the zygotes which causes the abortion of a proportion of them. Studies were initiated on one of the genotypes that show this abnormality to determine the cause of the abortion of the zygotes and the genetic mechanisms responsible. The procedure included the establishment of progenies resulting from self-fertilization and crosses with normal genotypes. However, these studies have not been concluded, for which reason it is premature to discuss the mechanisms involved. It seems that, from the nature of the abnormality, cytoplasmic factors should be considered. The individuals known to behave in this manner belong to the Amazonian Region populations.

A special case concerns the clone 'CA 3', which is distinguished by its narrow elongate leaves and also possesses a relatively high productive capacity. The clone produces a low proportion of fruits when self-pollinated. The fruits produced contain few normal seeds; the highest number obtained per fruit is seven. The resulting plants develop slowly and present a dwarf appearance when young, as shown in Plate 31b, in which the self-pollinated seedlings are compared with normal plants derived from cross-pollination of 'CA 3'. However, the selfed progeny produce the elongate leaves of the parent. They eventually develop into normal plants with a short stature but their productive capacity is similar to that of the parent. Although investigations into the mechanisms concerned were initiated in the progenies they were not concluded.

In all of the cases where barriers to the normal mechanisms of fruit or seed production occur the genotypes depend on cross-fertilization to ensure the formation of normal fruits. The identification of genotypes with similar abnormal behaviour is subject to uncertainties owing to the vagaries of the pollination events and the absence of information regarding their ovule numbers.

Ovule Number

The descriptions of the various populations refer to the extent to which they vary with regard to the number of ovules in the ovaries of their flowers. The highest number encountered is of the order of 80 and it appears that numbers as low as 30 may be found, especially in some mutant genotypes. The data obtained from progenies of hybrids between individuals belonging to populations that are distinguished by having different ovule numbers indicate that the higher numbers are dominant to the lower numbers in each combination. These results may be interpreted on the basis of a series of multiple alleles being responsible for the differences and that each allele may add one number to the number of ovules in each loculus in a sequence of dominance. There are suggestions from some data that heterosis for this character may occur but these need to be verified. The ovule number character is a distinguishing feature of several populations. The higher numbers are found in genotypes from the Amazonian Region and the lowest numbers (with the exception of some mutants) are encountered in the *Criollo* group.

Polyembryony

Occasionally, seeds are encountered that contain more than one embryo. Where two embryos occur they will give rise to twin plants. It is shown by cytological examination that some ovaries have one or more ovules containing two egg sacs. The egg sacs are fertilized by separate male gametes, often producing two distinct genotypes, an example of which is illustrated in Plate 32c. In this case the parent genotype was homozygous for the anthocyanin inhibitor gene, one egg sac being self-fertilized while the other was fertilized with pollen from a tree with the dominant cotyledon colour. While the occurrence of twin embryos has been noted to be more common in some genotypes the lack of information resulting from inadequate observations prevents the determination of the genetic situation involved.

At the present time there is little evidence to link the occurrence of polyembryony to any particular populations or any set of gene action. The only examples of seeds containing multiple embryos were found in trees belonging to the Ecuador 'Nacional' variety or descendants of it, several seeds in each fruit being polyembryonic. Plate 32d shows the developing plants from some of these seeds. The occurrence of polyembryony in the population in which the seeds were produced appeared to be influenced by seasonal factors, which means that the determination of genetic effects may be difficult to establish. Double embryos have been found mainly in genotypes belonging to the Amazonian Region populations. One of the more consistent producers of twin embryos and one in which more than two twin embryos have been found in each fruit is the variety from the R. Coppename valley in Surinam.

Seed Size

Although seed size is an attribute that is very important in improvement since it is a component of fruit size and, therefore, of the yielding ability of cultivars, its inheritance is poorly known. The inheritance of the seed size of the large array of genotypes used in improvement programmes could be determined from the analysis of the large number of hybrid combinations and inbred lines that have been produced. However, only one serious study of the inheritance of this attribute has been carried out (Bartley, 1965). The study concerned the determination of seed size in progenies of the clones 'SCA 6' and 'P 12' combined with 'ICS 6', the self-compatible genotype possessing the largest seeds known at the time. The determinations were based on fruits obtained by artificial fertilization to the large seeded parent (that is, a backcross mating system was used). The comparative frequency distributions for the variable, average dry seed weight, for the two crosses, are presented in Fig. 49. These results may be interpreted as showing that the small seed size of 'SCA 6' is

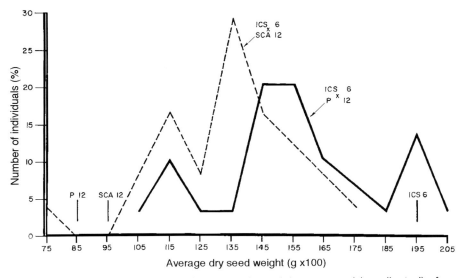

Fig. 49. Frequency distributions of average dry weights per seed (unadjusted) of the offpring of the crosses 'ICS 6' × 'SCA 12' and 'ICS 6' × 'P 12' to show the inheritance of the seed sizes of these parents. Graph based on data by Bartley (unpublished).

inherited as a dominant to the large seeds of 'ICS 6' and that segregation in the cross is transgressive. In the case of 'P 12' the dominance of the smaller seed size is less noticeable although a tendency for this can be detected. In another study, the seed sizes of the progenies of the F_1 of 'SCA 6' and 'ICS 6' were compared to those of one of the F_2 families derived from this cross. In the latter there was a general increase in seed size but the mean of the family was lower than the median value of the seed sizes of the parents.

Mutations

The inheritance systems of a few of the mutations that have been identified in cacao have been worked out. The majority of the mutations known result from single recessive genes and become apparent when they are present in the homozygous state. As a result, although the mutation rate may be fairly high, the presence of such mutations may remain undetected in normal genotypes owing to their existence in the heterozygous condition.

The genotype known as 'Crinkle-leaf Dwarf' (identified as 'M 253'), the leaves of which are similar in appearance to that of the genotype illustrated in Plate 6a, was found to be the result of a single recessive gene with pleiotropic effects with which a low level of seed formation and a high rate of zygote abortion were associated (Bartley, 1964b). In the third generation progenies of hybrids with normal plants the 'Crinkle Leaf' character is combined in plants with normal growth. Hence, although the dwarfing character appears to be produced by the same gene as the leaf character, the two characters would be inherited from two genes with close linkage, giving the effect of being produced by a single gene.

The 'Jaca' genotype, named from its leaf shape (Plate 4d), is the result of the action of a recessive gene. The characteristics of the genotype include the tendency to produce three plagiotropic branches instead of five, as may be noted in Fig. 50, and a possible abnormality in the flower. The inheritance of these traits has not been investigated.

Occasionally, various abnormalities occur in cacao plants, which may be the result of non-genetic factors and could be of a transitory nature. Plate 6c shows a chimaera that developed from the base of a plant. It was impossible to reproduce this chimaera and since no flowers and fruits have been formed the determination of the origin of the shoot has not been possible. Another abnormality is the appearance of fasciated stems, sometimes of large proportions, that are localized in their occurrence and probably owe their origin to imbalances produced by non-genetic factors. The development of the fasciation would take place during the formation of the zygote in the form of multiple-fused terminal buds. Some plants, apparently belonging to certain populations, produce abnormal development of the whorl of plagiotropic branches in the form of repeated branching and multiple branches.

Fig. 50. Tree of the 'Jaca' mutation showing the leaf shape and three-branched habit.

Under certain conditions parthenocarpic fruits are produced, the numbers of which range from a few to many fruits per tree. Parthenocarpy occurs mainly in self-incompatible genotypes and varieties, some being particularly prone to this disorder. However, except for the association with self-incompatibility no genetic causes of parthenocarpy can be identified since the production of such fruits cannot be considered to be a permanent feature of the varieties in which it occurs.

8 The Relationships among Populations

The distinguishing features of the naturally distributed populations described in the previous chapters were based on the observed differences in the morphological characteristics of their component genotypes. However, an analysis of the relationships among the populations depends on the availability of an integral picture of the genetic composition of each population based on the genotypes of the individuals that belong to it. Since an individual genotype comprises many genes, some of which may occur in other populations, the individuals that appear to belong to distinct populations may have common DNA compositions. The determination of the common and different genetic elements in the individual populations is important with regard to understanding their origins and the manner in which they diverge.

Taking into consideration the nature of the species the analysis of the relationships among the populations would be an impossible task if we had to depend on the study of the observed variation of the morphological markers. During the past two decades the molecular biology techniques that have been developed have been applied to the determination of the genetic constitution of the genotypes by means of the analysis of their DNA profiles. The analytical procedures have evolved from the identification of the allelic differences using isozyme markers. Since then attention has shifted to the use of other genetic markers such as restriction fragment length polymorphic DNA, random amplified polymorphic DNA and satellite markers.

A considerable number of genotypes have been assayed using these techniques and this has provided us with an idea of how some of the genotypes are related or can be placed into groups. A summary of the results obtained from earlier analyses using molecular markers is given by Lanaud *et al.* (1999). However, other groups have undertaken similar analyses with genotypes belonging to several sources from natural and cultivated populations.

It is not the intention at this time to analyse all of these results in detail in terms of the techniques used but it would be pertinent to point out some aspects of this work as it relates to the subject of this chapter. One difficulty is that several authors have reported on the relationships between the genotypes used in their studies without naming the genotypes. This prevents a serious analysis of the results being made. For example, we may cite the work of Ronning and Schnell (1994) and Whitkus *et al.* (1998) in which the unnamed genotypes were allocated to hypothetical 'population' groups. The authors of these studies lacked the knowledge of the origins and relationships of the genotypes analysed. Results of this nature are found in reports of other investigations and, for our purposes, have no application whatever in unravelling the inter-population relationships.

Another aspect to be considered is that the sets of genotypes analysed vary according to the study conducted. The results obtained in a given set of genotypes analysed, in terms of the relationships among them, are determined by the genotypes that are included in the set. This leads to the situation in which the relationships between two genotypes or between two populations may be presented in different ways, making it difficult to interpret the results for this purpose. Similarly, it is difficult to extrapolate the relationships between populations analysed in various studies in which two or more genotypes may be common to more than one analysis but a different set of other genotypes were included in the individual studies. For example, the studies of Marita (1998) and Sounigo (2001) included the same members of the series 'PA' and 'IMC'. The position of these genotypes in the plots depended on their relationships to the genotypes from other populations, of which a different set was analysed in each case. The situation in these studies is such that comparison of the relationships within and between the respective series is difficult because of the manner in which the results are presented. Hence, the determination of genotypic relationships is hampered by the lack of uniformity in the presentation of the results.

It is probable that some degree of divergence in the results obtained occurs owing to the differences in the numbers of molecular markers used in various analyses. There may also be a lack of uniformity in the markers used for a given set of genotypes. Other sources of variation in the degree of association among genotypes are the differences in techniques and the possible misidentification of the samples as well as of the plants from which they were obtained.

By and large the genotypes analysed in most of the studies conducted so far belong to the selections made in cultivated populations. As demonstrated in Chapter 6, the majority of the selections were made among the progenies resulting from hybridization between a restricted range of Amazonian germplasm and members of the *Criollo* group from various populations. It is necessary to emphasize that in no case do the selections cover the range of variability present in the hybrid populations in the regions in which the selections were made. Since the ancestral populations of most of these regions were attributed to the *Criollo* group some genotypes belonging to the group may have entered unintentionally, through misclassification of their variety status, into the sets of genotypes analysed. The relationships among the selections would be confused since they are individually products of various degrees of hybridization and recombination and, consequently, have genetic constitutions that are specific to the individual genotypes.

A second group of genotypes that have been subjected to analysis relate to the Amazonian populations constituted by the progenies of the collections made in Peru. The range of the variability contained in each of the groups of progenies that has been analysed in these studies has been limited to a few genotypes. Presumably, genotypes from this origin were classified by Cruz *et al.* (1995) as South American 'wild' varieties, but as the genotypes attributed to this class were not named, verification of this claim is not possible.

The above is one example of the situation that applies to several of the earlier analyses that were undertaken by institutions in countries outside the areas where the crop is distributed. Since the specimens analysed would have been supplied to the laboratories in which the analyses were conducted the persons who carried out the analyses had no knowledge of the veracity of the identifications of the plants from which the samples were obtained. In addition, judging from the interpretations of the results, the information supplied regarding the origins, classifications and relationships of the genotypes was often of dubious quality. In the cases of analyses undertaken by persons who have no knowledge of the diversity of the species and no contact with the plants in the field, conditions do not exist for relating the genotypes sampled to their characteristics. In some ways, such a situation may be beneficial since it could lead to unbiased conclusions when the current concepts of variety classification are ignored.

There is a tendency for some authors to interpret their results in a way to prove certain pre-established theories regarding the position of the material analysed in the scheme of the diversity. In view of this situation it is necessary to adopt unbiased interpretations according to the indications revealed by the morphological differences among and within populations.

Among the information that can be obtained from the analysis of the molecular markers is that relating to the degree of heterozygosity of each genotype, based on the numbers of polymorphic loci that are determined in its genetic composition. Such information is a valuable aid in determining whether a genotype derives from hybridization and the extent of genetic variability that a population possesses. Some disparities in the values obtained from genotypes that have been analysed several times may be attributed to the differences in the numbers of molecular markers and the techniques used. The degree of heterozygosity is presented as the proportion of polymorphic loci to the total number of loci determined in the analysis of the genotype. However, the results do not tell us which loci are heterozygous nor the characters involved. Although progress has been made in mapping the cacao genome with regard to assigning the molecular markers to their relative positions on the chromosomes, the relationships of the loci related to the expressions of the morphological markers have not been determined.

For our purposes the most useful study is that of Marita (1998) since it includes several populations of the Amazonian Region that are not discussed in other studies. In order to interpret the results it is necessary to know which samples were misidentified and adjust the population groupings accordingly.

The genotypes derived from artificially produced hybrids should be eliminated in order to remove their influence on the distribution patterns of the genotypes that are relevant to our objectives. Also, it is essential to have a conception of the geographical origins of the populations in order to understand the patterns of the distribution of the variability within the populations sampled and the relationships among them. It is necessary to take into account the fact that, with regard to the Amazonian Region's populations, only genotypes from the parent trees were analysed. As explained in Chapter 4, the collections made in the populations sampled included progenies from other trees (as well as from some of the same parental genotypes) and these progenies would represent a much wider range of the variability contained in a population. Consequently, it should be appreciated that the variability shown for the individual populations may be only a fraction of the total variability exhibited by their components.

We may start our analysis of the relationships between the populations of the Amazonian Region with the drainage systems that radiate from the Province of Caravaya. The study of Marita (1998) is important in that it analyses various populations on these drainage systems. The most extreme of these populations is that of the valley of the R. Urubamba and its source tributaries. Unfortunately, we do not have any information about the primary varieties in the area. The primary germplasm may be connected with that occurring in the upper R. Bení valley and forming part of a dispersal pattern along the eastern foothills of the Andes.

It would seem that the variety that could be considered to be indigenous to the R. Urubamba does not extend through the valley of the R. Ucayali. Here, the principal type known is that identified as 'Scavina', which is represented by a few progenies. The multi-dimensional scaling (MDS) plot shows the 'Scavina' genotypes forming a dispersed group at one corner of the plot (other misidentified members of the family need to be included). The uniqueness of the type is confirmed by other results, such as those of Charters and Wilkinson (2000), who also placed this group at the extreme end of the dendrogram. Two of the three genotypes resulting from more recent collecting activities (analysed by Marita, 1998), from the same area, could be considered as belonging to the same range of variability as the 'Scavina' type. These few genotypes perhaps do not represent the entire diversity of the area and, certainly, the phenotypes of the newer accessions differ from those of the 'Scavina' type. The remaining genotype was located downstream from the 'Scavina' location and appears to belong to another variety.

As speculated in the account of the genotypes from the R. Juruá valley, they are related to the 'Scavina' type but some of them show a degree of divergence from the 'Scavina' type indicating that they contain other factors. Pires *et al.* (2001) stated that 'SCA 6' and 'SCA 12' possess ten unique bands and found that some of the R. Juruá genotypes possess some of the same unique bands. The number of bands found in each genotype varied in the same way as those encountered in the 'SCA' genotypes. It is not stated whether this situation applies to the newer genotypes from the R. Ucayali.

The statements made by Chandless (1868) regarding the connections between the R. Juruá and the R. Ucayali leave no doubt as to the probability that the same variety was exchanged across these river systems as well as along their tributaries by which the exchange would have taken place.

If this exchange did take place the pertinent question is, 'on which river did the variety originate?' Judging from the enormous size of some of the trees seen on the R. Juruá, it may be supposed that these trees are older than the trees on the R. Ucayali. However, the small sizes of the samples from these rivers do not provide an adequate basis for answering the question. Further investigations would be necessary which would include a survey of all the cacao occurrences along the Rivers Moa and Tapiche.

The study shows that the populations from the other drainage systems arising in the Province of Caravaya form discrete groups in which the genetic distance from the R. Ucayali types increases as they progress towards the east. Although the samples analysed by Marita represent only a portion of the variability that exists in the populations of the R. Purús and R. Madeira drainage systems, nevertheless, the results indicate that a significant degree of diversity may be found in them. In addition, in this analysis, no attempt has been made to determine the extent to which sub-populations, or distinct varieties, occur along the course of the rivers or on their various tributaries.

The only specimen from the R. Envira system was located within the range of the R. Purús population. This situation may be expected on account of the proximity of the R. Envira to the upper reaches of the R. Purús system although the phenotypes of the R. Envira population are very different to those of the R. Purús. It is suspected that the sample analysed is of a misidentified genotype from the R. Purús. It would be important to determine the degree of relationship between the genotypes occurring near the headwaters of the R. Acre and those of the western tributaries of the R. Madeira, the courses of which run close to the upper reaches of the R. Acre and sometimes parallel with it.

The R. Jamarí/R. Ji-Paraná population, on the basis of the morphological characteristics and the results of the analyses of samples used in the study of Marita (1998), is distinct from the diversity of the R. Purús. The types known from the R. Guaporé also appear to be unrelated to those of the R. Jamarí/R. Ji-Paraná populations. However, comparisons are unreliable since the number of samples analysed from the R. Jamarí/R. Ji-Paraná population is small relative to the number of genotypes which represent a sample of its diversity. In this sense no valid conclusions can be reached about its diversity until samples from the various localities have been analysed. No analysis has been attempted of the diversity that has been collected in the southeastern extremity of the R. Ji-Paraná valley that constitutes the boundary of the natural range of distribution of the species.

At this point it is in order to refer to some analyses undertaken in France on genotypes from the Brazilian Amazon, some of which are summarized in Lanaud *et al.* (1999). The samples involved were from single genotypes representing different drainage systems. As such they do not offer any information regarding the variability within each drainage system. Few of these specimens are actually mentioned in the results and they are usually identified as a group with the title of 'Forastero du Brésil'. For a person unfamiliar with the geographical background the results would be difficult to follow since many of the location names are erroneously spelled. It can be concluded that these samples show a significant range of variability. The most outstanding is the sample from the R. Japurá. For our purposes the results need to be studied in greater detail than that presented in the publications.

As explained in Chapter 5, the region defined as the Equatorial Oriental Piedmont is an area from which river systems diverge towards the south and east towards the R. Marañon, and others belonging to the R. Napo drainage system flow towards the east to join the R. Amazon. It was postulated, on the basis of the characteristics observed in genotypes from both drainage systems, that there were distinct shifts in the diversity that evolved in either system. However, as far as can be judged from the available information, molecular analyses have been performed on only a few specimens from these systems.

Some of the collections made in Ecuador were analysed by Sounigo *et al.* (2001) and the principal component analysis (PCA) presented indicates that the total diversity could be fairly wide, although the small number of samples is insufficient to permit any conclusions on the matter to be made. If the possible exchanges in the identifications of the samples are considered it would appear that some of the R. Marañon genotypes differ from the R. Napo genotypes. The few specimens from the R. Sucumbios diverge slightly from the R. Napo genotypes and are at a further distance from the R. Marañon specimens.

Sounigo *et al.* (2001) also compared the progenies collected by Pound (1938) in the area of the R. Morona. As might be expected most of the genotypes coincide with those from the samples from the collection made in Ecuador on the R. Marañon tributaries, of which the R. Morona is one.

The phenotypes of some of the R. Morona progenies resemble the cacao 'Nacional' of Ecuador. However, no comparative analyses of samples from the upper R. Marañon valley with those of the traditional variety of the Ecuador littoral have been made. According to the results obtained by Crouzillat *et al.* (2001) the samples of the cacao 'Nacional' analysed by them formed a distinct group from the other genotypes studied. The latter included the hybrids with the 'Nacional' and the comparison was also reported by Lerceteau *et al.* (1997). If a comparison is made of the relative positions of the 'Parinari' genotypes and those from other locations in the R. Marañon system, as reported in Sounigo *et al.* (2001), with those of the relationship of the few 'Parinari' specimens and the 'Nacional' given in Crouzillat *et al.* (2001), it may be possible to conceive that the cacao 'Nacional' falls within the range of the variability of the R. Marañon genotypes.

Several studies have been carried out on genotypes belonging to the group of progenies identified as 'PA' but different genotypes were analysed in the individual studies. The two studies mentioned above indicate that some of these genotypes are related to the populations from the upper R. Marañon. However, since these genotypes represent progenies of seven, perhaps genotypically distinct parents, significant variability may be expected in the group. The analysis of the variability within the group would include assigning the individual genotypes to their parents.

In Chapter 5 it was explained how the compatibility relationships could be used to group the genotypes according to their respective families. The analysis of the DNA would be superimposed on the classifications obtained. In order to do this it is necessary to examine the general results of the analysis of the diversity. It would appear that the genotypes exhibit a large degree of variability among them but, since few genotypes are common to all studies, direct comparisons would be based on insufficient data. The PCA presented in Sounigo et al. (2001) show that the 'PA' genotypes analysed are dispersed in the PCA but they fall in areas separated from the other groups analysed. In the results obtained by Marita (1998) the distribution of the genotypes is more extensive.

What is common to all studies is that a small group of genotypes, perhaps related to the clone 'PA 13', are distinct and bear a closer relationship to the genotypes from the upper R. Marañon than to the other 'PA' genotypes. On the basis of their phenotypes the relationship of these genotypes to the type Pound (1938) referred to as 'Criollo de la Montagne' may be considered. In this case it would be expected that their genetic constitution included elements of the diversity from the Equatorial Oriental Piedmont. Most of the remaining genotypes analysed by Sounigo et al. (2001) could be considered as belonging to four groups on the basis of similarities among them. It may be possible to assign the compatibility alleles to these groups but this depends on the analysis being undertaken of a larger number of genotypes.

The analyses of Marita (1998) involved a set of these genotypes that differed in most cases from those of the work of Sounigo et al. (2001). Several of the genotypes were found to fall within the diversity pattern of the genotypes from the Solimões/R. Japurá area. This situation suggests that some of the 'Parinari' population would be derived from the same source as the diversity occurring in the Solimões area.

The few specimens from the Solimões/Japurá area analysed by Marita (1998) were placed into a fairly homogeneous group, which would be distinct from the other populations previously described except for the relation to some of the 'Parinari' genotypes. The size of the sample probably does not reflect the genetic variability that occurs in the area, as evidenced by the phenotypic variability observed in the parents as well as in the families of their progenies. Besides the 'Parinari' genotypes referred to above several genotypes belonging to the variety groups that originated near the town of Iquitos in Peru also show affinities to the diversity of the Solimões/R. Japurá area.

The population of the R. Javarí system with its unique characteristics was represented by a single genotype (Marita, 1998). This appeared to be distinct from the Solimões/Japurá populations and forms a separate entity with some affinities to the R. Juruá types, although there is a wide difference between them on the basis of their phenotypic expressions.

Of special interest are the genotypes from the island of Careiro. As would be expected they are widely distributed in the MDS plot. Some of these genotypes are associated with those of the Solimões/Japurá specimens, which may indicate that the seed used for establishing the fields on the island came from further up the river. Whether the other genotypes included derive from the same area would have to be determined by the analysis of a larger sample of the Solimões population.

Sounigo *et al.* (2001) included in the populations sampled some genotypes from the R. Oiapoque population. A gradient of diversity was shown to occur among genotypes collected along the river, with the R. Camopi specimens at one end of the distribution range. Although it would appear that the R. Oiapoque specimens belong to a distinct population it is surprising to note that those from the more easterly portion of the river were found to correspond with the specimens from the R. Napo and R. Marañon populations. This situation may be construed as indicating a link between the populations from the extremities of the Amazonian Region, but the real situation may be otherwise.

However, a few of the genotypes from the R. Camopi area were included in the study of Marita (1998). The results confirmed that they appear to form a homogeneous group. Lanaud *et al.* (1999) found that all of the individuals analysed were homozygous, on which basis it can be concluded that the population has a narrow genetic base. Marita's study showed that, in comparison with the entire range of diversity, the R. Camopi specimens are not really a uniquely distinct group but that they are related to the R. Jarí specimens, illustrated in Plate 26a, as well as to the germplasm that exists in the State of Bahia.

Charters and Wilkinson (2000) also found that the R. Oiapoque specimens that they analysed fell into a distinct group but that this appeared to be most related to the 'Parinari' specimens they analysed. In fact, the position of the R. Oiapoque specimens may be interpreted as lying between those of the 'Nanay' and 'Parinari' groups.

A significant number of genotypes that had been selected in the cultivated population in the States of Bahia and Espírito Santo in Brazil were included in the analyses of Marita (1998). As expected from the knowledge of the consanguinity between these genotypes, virtually no differences were found between most of them. However, the distribution patterns within this group indicate that some genotypes may derive from distinctly different parents when this group is subjected to a more detailed analysis. The specimens in this group were not individually identified on the MDS plot but they include two genotypes from the Alenquer–Óbidos district. This result indicates a possible origin of the Bahia population from the latter area. The specimens included in Marita's analysis do not include any descendants attributable to the 'Maranhão' origin, for which reason the position of this 'variety' remains undetermined.

The various sets of genotypes that were analysed in the several studies summarized in Lanaud *et al.* (1999) included a few specimens from the upper R. Orinoco. Although it was stated that this population was considered to be one of four 'poles' on which the diversity is structured, the actual position of the population is ambiguous. The results could be interpreted in various ways since it is variously combined with other genotypes depending on the method of analysis used and the sets of genotypes in which they are compared. It would be expected that the population of the upper R. Orinoco would be homogeneous with, perhaps, a low level of heterozygosity. However, the small sample size analysed does not provide a sufficient basis to reach a convincing conclusion as to the position of these genotypes in the scheme of diversity.

Since the 'West African Amelonado' is derived from a genotype of the Amazonian Region, whose source is unknown, the determination of its relationship to the putative populations in the areas of Brazil where the crop is cultivated would be of considerable interest. However, the only comparisons that have been made involve a few trees selected in the Ivory Coast from populations that presumably belong to the 'Amelonado' and an insufficient number of genotypes from the Bahia population. While a degree of association among the two groups is indicated, the results do not enable a conclusion on the matter to be reached.

The determination of the relationships among the various populations that belong to the *Criollo* group have been ignored until recently, when it has been the object of some research (Motamayor *et al.*, 2002). In this case the genotypes that form the basis of the study were chosen as representatives of the 'ancient' varieties found in the various areas of occurrence; that is, varieties suspected of being contaminated by foreign germplasm were excluded. The principal comparisons were made between varieties from the foothills of the Venezuelan Andean range, the foothills of the Sierra Nevada de Santa Marta and genotypes from southern Mexico. The latter were supposed to represent the Mesoamerican cultivated populations. It was concluded that all of the specimens were related on the basis of being homozygous or nearly so. In this sense the analyses confirm the conclusion reached nearly a century ago that there was a relationship between the northern South American *Criollo* and those cultivated in Mesoamerica, a conclusion verified by Cheesman (1944).

However, in spite of the title of the paper, no Mesoamerican genotypes are specifically mentioned (neither are those from the Sierra Nevada de Santa Marta). It is difficult to accept the conclusion that no genetic differences occur among the genotypes from the two areas. Although it was found that most of the genotypes were homozygous, no evidence was provided as to whether they were homozygous for the same alleles. The fact that some of the genotypes exhibited a small degree of heterozygosity indicates that some loci would have been polymorphic. In substance, the complete information regarding the genetic composition of each genotype according to its location is required in order to provide a convincing analysis of the constitution of the genotypes belonging to each population.

The genotypes from southern Mexico included a sample of the diversity from the Lacandon forest. Here, the suggestion of close affinities within the population based on the homozygosity of the genotypes does not agree with the existence of variability as shown by the illustrations in Plate 30a. The conclusions of Motamayor *et al.* (2002) in this respect differ from the results of Whitkus *et al.* (1998), where a higher level of polymorphic bands (12%) was found in the specimens from the area.

Crouzillat *et al.* (2001) analysed a fairly large number of genotypes from the Mexican populations as well as a few from Guatemala. It would appear that no definite attempt was made to determine whether the genotypes were genuine members of the *Criollo* group. However, a few specimens were found to have zero, or a low level, of heterozygosity and it could be assumed that these might be considered to be *Criollo* genotypes, although the trees were cultivated. The remaining genotypes showed a high level of heterozygosity and they could be considered as being derived from hybridization. In these cases, it would be interesting to differentiate between the hybrids that are descended from the Ecuadorian introduction and the genotypes whose parentage involves the introduction that we may consider to derive from the 'Matina' variety. Many of these hybrid genotypes were considered to possess the same genetic constitution and, since the Guatemalan selections were found to be similar, it may be considered probable that they contain genes from the Ecuador introduction.

The set of genotypes analysed by Marita (1998) included several of the genotypes selected in Mexico that were analysed by Crouzillat *et al.* (2001) and it was also found that these genotypes formed a group in which it was difficult to distinguish between them. However, other genotypes, particularly those descended from the Nicaraguan *Criollo* population, were included in the group as well as others from Costa Rica and one *Criollo* genotype from Colombia. Several of the genotypes with Venezuelan *Criollo* ancestry occurred in scattered positions around the Mexican genotypes. There is not enough evidence to draw a definite conclusion from the genotypes analysed but it would appear that the descendants of the hybrids between the Venezuelan *Criollo* and the 'Calabacillo' introduced into Trinidad differ from the hybrids formed in the Mexican population.

The presence of non-*Criollo* elements in the composition of individuals possessing *Criollo*-like phenotypes, but which are suspected of being descended from hybrids, should be capable of detection. Similarly, it should be possible to identify the individuals belonging to the *Criollo* group that do not exhibit all of the characteristics that are commonly used to define which genotypes can be attributed to the group.

The relationship between the 'Matina' variety and the genotypes from the Amazonian Region cultivated in Brazil is suggested by the results of Marita (1998) but this is subject to confirmation. Several genotypes of the variety named 'Martinique Créole' (whose fruit type is depicted in Plate 26b) from Dominica, Martinique and Guadeloupe were analysed by Charters and Wilkinson (2000). They were found to belong to the same grouping as the selections from the Bahia, Brazil, cultivated population, thus indicating a similar origin.

As mentioned in the introduction to the chapter a large proportion of the genotypes that have been subjected to the analysis of their genetic content by the use of molecular markers consists of those selected in cultivated populations, mainly in South and Central America. These selected clones are usually identified by acronyms that indicate the country or institution in which the selection programmes were conducted.

To undertake an analysis of the relationships among these genotypes it is necessary to appreciate that the selections that are identified by a given acronym do not belong to any specific group since they derive from various origins and backgrounds. There is a general lack of understanding of this fact among the authors of the studies in which such genotypes were analysed. It is a fallacy to consider that all of the selections bearing the same acronym are related and to classify them as belonging to a 'Group' called 'Trinitario' which is anything but the truth. This tendency results in difficulties in interpreting the results of the studies in terms of the relationships among the selections. To do this it is necessary to classify the selections made within each programme in accordance with their origins and pedigrees. The genotypes within each set can then be compared and, subsequently, comparisons made between sets. Where several selections were made within the population cultivated on a farm (such as was done in Ecuador) these should be grouped together and compared with the sets of selections from other farms.

Although the analysis of the molecular markers has provided some information of value in determining the genetic constitution of the populations and the relationships among them there are several questions that remain to be answered. Perhaps the applications of the newer procedures in the field would supply answers to the areas where ambiguities have been found. Obviously, there are several populations from which living material exists, which have not been included in the analyses performed so far and the determinations of these would provide a means of completing the picture of the diversity. Among the populations that have not been included in the analysis of their genotypes are those from the R. Içá (lower R. Putumayo), the Parintins area (lower R. Madeira) and the cultivated population of the Óbidos–Alenquer district. Of particular importance are the varieties cultivated in the Amazon Extension Zone, where, apart from very few specimens, no genotypes have been analysed. The determination of the origins of these varieties, and those of the other types in areas where cultivation is practised in the Amazonian Region, is of special interest.

Of course, the populations that are not represented in living collections and, therefore, do not provide specimens for the determination of their DNA constitute an unknown element in the picture of the diversity.

Another phase in the study of the inter-population relationships would be the analysis of the genotypes from inbred and hybrid descendants. The analysis of these families would contribute to the identification of the loci involved in the expressions of the distinguishing characteristics of the populations and the action of their alleles.

9 The Utilization of the Genetic Resources

The knowledge of the diversity of the species and the understanding of the origins, evolution and characteristics of its component elements are by themselves matters of considerable interest and scientific value. The acquisition of this knowledge is complementary to the process involved in the conservation of this diversity since it is facilitated by the availability of specimens of the populations in a form that permits detailed observations to be carried out. However, the main justification for the efforts in the conservation of the genetic resources is the application of the variability of character expressions contained in the diversity for improvement of the varieties that are cultivated as a source of income for the growers.

In this chapter we shall examine the needs of the growers of the crop and the manner in which the diversity has been used for its genetic improvement as well as the prospects for the development of cultivars to be used in the future.

Principles and Procedures Applied to Cacao Improvement Related to the Specific Properties of the Species

In its natural state cacao is reproduced sexually by the production of fruits that contain a fairly high number of seeds. In some cases, especially under natural growing conditions, as described in Chapter 4, the trees are able to reproduce vegetatively by means of the production of roots from the adventitious shoots that arise in certain circumstances. In such a manner a genotype has the capacity to enlarge its presence in its environment. At the beginning of the 20th century efforts were made to develop artificial methods of vegetative reproduction following the system by which temperate fruit species were being reproduced. In cacao, artificial vegetative propagation has become an established process. The availability of two reasonably easy methods of reproduction has important advantages both from the point of view of

© B.G.D. Bartley 2005. *The Genetic Diversity of Cacao and its Utilization*
(Basil G.D. Bartley)

conservation of the individual genotypes that constitute the diversity as well as in the processes of genetic improvement.

In general, the great advances that have been made during the past 50 years in the accumulation of knowledge of the diversity and its conservation in genetic resources programmes have not been accompanied by the same rate of its utilization in improvement programmes. Some of the reasons for this disparity are related to the nature of the species. However, the main obstacles to progress are the inadequate approaches to the organization and execution of improvement programmes and their adaptation to the needs of the producers. These aspects result in the existence of constraints regarding the ability to obtain an adequate utilization of the resources that are available. The subject of the adequacy of the research on improvement is described in greater detail in the summary of the history of cacao genetic improvement in Bartley (1995).

At the outset it is appropriate to describe some of the conditions that affect the ability to utilize the diversity in improvement programmes. One basic fact is that the species is a perennial tree. This means that several years are required for an individual tree to reach the adult productive stage. Consequently, the generation cycle is prolonged, which implies that any activities related to improvement must be on a long-term basis. The nature of the tree requires that each individual occupies a relatively large area and, in order for meaningful results to be obtained, the effective conduct of an improvement programme demands that a sufficient area of land should be available. These factors have important effects on the continuity of research on variety development since they determine the attractiveness of the research to human resources regarding the conditions for achieving results. In order for a programme to function adequately the time and space factors must be taken into account in the provision of financial and other resources that will guarantee the effective operation that will lead to the attainment of a programme's objectives.

The above aspects have a considerable bearing on how the diversity can be used. In the preceding chapters the ample diversity that exists in the species has been described. During the last half of the 20th century a large proportion of this variability became available for use in improvement programmes. It is probable that the diversity is much larger than the existing facilities and resources allocated to research and development can provide for the utilization of more than a fraction of it.

The availability of both sexual and vegetative methods of reproduction provides the breeder with a choice of the type of cultivars that can be developed. The trees that develop from reproduction by seed form a natural architecture. Since the quantities of seed contained in the fruits can be relatively large and modest numbers of trees are required to establish a given area, the production of the planting material is relatively easy under adequate conditions of management.

The reproduction of clone cultivars by vegetative propagation is somewhat more costly, usually requiring the application of special procedures, especially when large quantities of planting material are needed. Vegetative reproduction provides the advantage of being able to fix individual genotypes regardless of their genetic composition. The important application of vegetative reproduction in the improvement process is in the selection and reproduction of the superior individuals in heterogeneous populations and their maintenance as clones. From the point of view of cultivar development the degree of heterozygosity of an individual genotype is unimportant and clones can be selected from individual trees whose pedigrees may involve several parents. The tendency, especially when few alternatives are available, to form monoclonal cultivations may be a disadvantage in terms of effects of diseases and pests; but clones are no different from seed-produced varieties consisting of single genotypes. Another disadvantage that applies to clones is the lack of the flexibility necessary to counteract unfavourable circumstances and react to changes in environmental and cultural conditions.

Although it is easy to establish improvement programmes, one aspect concerning the future use of the cultivars that may be developed has to be taken into consideration. This concerns the fact that, since cacao is a long-lived perennial species, the incentives to replace existing varieties with other cultivars, which may have superior qualities, are frequently lacking. From the conceptual standpoint clones are easier to develop than seed-reproduced cultivars since they do not necessarily require any predetermined processes of development and are immediately obtainable.

Every individual genotype that comprises the diversity has a potential use in an improvement programme and in the development of cultivars. In addition to its capability to be selected as a clone, in which form it may be a potential cultivar, it can be used as a parent in the development of seed-propagated cultivars or the formation of populations on which further breeding activities may be based.

In the previous chapters mention was made of homozygous varieties that, being sexually reproduced, formed the basis of the cultivation practised in several countries during many decades, and even centuries in some cases. In situations where two distinct varieties occurred, usually as a consequence of introduction, hybridization between them would have produced a generation of offspring that were heterozygous but forming a homogeneous population. When such homogeneous sources of planting material are available there is little distinction between seed and vegetatively reproduced genotypes, the advantages of either type depending on practical considerations regarding the production of the planting material and management of the resulting fields.

In practical terms all improvement programmes whose purpose is to develop cultivars designed to provide the growers with determined advantages, such as better returns, would be based on the formulation of desired objectives and criteria. These objectives would comprise those sought in the cultivars, such as adaptation to the specific environments in which they are grown, higher yields, efficiency in cultivation on the basis of reduced management costs and elimination of factors that affect performance, such as the proneness to diseases and pests.

Consequently, the use of any genotype in the improvement process would depend on its possession of the characteristics that would satisfy the intended purposes in terms of the criteria required in the cultivars to be created. In some instances the attributes that are sought may be discerned in a genotype by its phenotypic expression. However, the genotype may possess useful attributes that are controlled by recessive alleles and, therefore, are not immediately identifiable. In addition, the objectives or attributes that are sought in a breeding programme are subject to change in such situations as the appearance of new diseases and pests and alterations in cultivation practices and industrial requirements. For these reasons no genotype should be discarded merely on the basis of its phenotype.

The first step in determining the role that a genotype or population can play in an improvement programme is to identify the characteristics that it possesses in terms of its genetic composition. Once the potentially useful characters that a genotype possesses have been identified, their application in variety improvement would be enhanced by the knowledge of the mode of inheritance of the loci that determine the expressions of their characters. This knowledge can be acquired by several means but can be obtained most effectively during the process of breeding in programmes that have been designed so that the information can be obtained concomitantly with the actions taken in the improvement process.

Since the possibility of reproducing individual trees by vegetative propagation is at the disposal of the plant breeder this may be applied at any stage of the breeding programme. In this way the process of improvement is facilitated since it can be based on individual genotypes selected according to the attributes they express. These attributes should be related to the criteria desired in the intended programme, rather than to family or population. The use of single genotypes selected in families is important in reducing the quantity of individual genotypes that need to be maintained at each stage of the improvement process.

With the aid of vegetative reproduction several alternative procedures for the utilization of any given element of the diversity become feasible. In addition to the objectives of the improvement programmes it is necessary to decide which form of cultivar is desired for the particular circumstances that prevail. The production of cultivars as clones is relatively straightforward since all that is required is to produce genotypes that combine the desired attributes in any possible manner and to evaluate them in terms of their performance and other attributes. On the basis of the results of the evaluation, the genotypes that best meet the requirements will be selected for reproduction as cultivars. There is no limit to the sources of genetic materials that can be combined to produce clone cultivars, but the process of their formation would be facilitated by the availability of knowledge about the characteristics possessed by the elements of the diversity and their modes of inheritance.

On the other hand, production of cultivars to be reproduced by seed requires that greater attention be paid to details of the fundamental principles of genetics. In the survey of principles and methods that follows it is assumed that the source genotypes and parents are maintained and reproduced in quantity in the form of progenies of clones. The exception may be made in the case of homozygous self-compatible genotypes, which will be reproduced in their progenies from self-fertilized fruits.

In the utilization of any given component genotype of the diversity the breeder is faced with three choices, one of which is the evaluation of its potential as a clone cultivar. The choices concerned with the use of seed reproduction are the production of families obtained by inbreeding or its employment as a parent in the development of hybrid cultivars. In addition to their potential use as cultivars, seed-reproduced families also provide new combinations in which individuals may be selected on the basis of their own traits with the aim of reproducing them as clones.

The development of cultivars by inbreeding can be employed effectively if the parental genotypes possess the requisite characteristics and are relatively homozygous for the desired attributes. In some instances families produced by inbreeding could provide a greater degree of adaptability when they are heterozygous for certain traits.

Cacao improvement methods that have developed during the past few decades have been restricted to the production of cultivars formed by single-cross hybrids by mating selected parents. This process has become an accepted procedure in cultivar development. This stage has involved the use of cloned parents that were selected on the basis of their assumed performance without sufficient attention being paid to their genetic constitutions. The consequence of this procedure has been the production of hybrid 'cultivars' that are often highly heterozygous in nature with average performances below the level of the best individuals that segregate in the hybrid families.

Future development of hybrid cultivars will require the use of parents specifically selected for the purpose. It is for this reason that the matter of the determination of the genetic constitution of all genotypes that are available assumes importance. In addition to the determination of the genetic content of genotypes by the application of molecular biology techniques, two possible procedures can be used. One is the use of inbreeding by self-fertilization and observations of the segregation that occurs for the properties exhibited in the resulting families. The other is by mating the candidate genotypes with tester parents that have been selected on the basis of possessing characteristics that will enable the genetic constitutions of the genotypes to be determined.

The employment of inbreeding, either by self-fertilization or by sib-mating, allows for the expression of deleterious and undesirable traits that result from the action of recessive alleles. The expression of these traits in the progeny may place a heterozygous genotype at a disadvantage when it is used as a parent. As indicated in the previous chapters the presence of deleterious factors is common in most cacao populations. In the first generation of inbreeding the genes causing deleterious and undesirable characters will be identified and eliminated. The remaining members of the family will contain individuals that are homozygous for the normal type and others that are heterozygous for the recessive allele in the ratio of 1:2. A second generation of inbreeding of each individual in the family will be needed to identify the homozygous and heterozygous genotypes, respectively. This will permit the heterozygous segregants to be eliminated, leaving the homozygous individuals on which selection will be practised for the desired traits. The more promising genotypes could be maintained as clones for further use.

Hybridization with selected tester genotypes permits the breeder to detect in the F_1 generation the dominance status of the characteristics of the candidate genotype. The most useful tester parents are those that are homozygous, or relatively so, and those whose breeding behaviour with regard to the desired characters is known. On the basis of the results obtained from the test crosses, the candidate genotypes will be selected for use as parents in future cycles of improvement involving the development and use of hybrid cultivars. In addition, the most promising test crosses may themselves be suitable cultivars and reproduced accordingly, thus saving a step in the application of the superior genotype for the purpose. Of course, the potential parents could be combined with other genotypes in the formation of hybrid cultivars.

It is necessary to take into consideration the fact that, in addition to the physical and natural circumstances of the species, the existence of incompatibility would play an important role in the determination of the procedures to which a given genotype would be subjected. Since incompatibility is common in the species it would constitute an important obstacle to the ability to use many genotypes in the breeding process. For a start, self-incompatible genotypes would not, in theory, be inbred by self-fertilization. Their utilization in hybridization would also be prevented in cases where both parents have the same incompatibility alleles. Families of hybrids between self-incompatible genotypes would consist of various proportions of self- and cross-incompatible progenies according to the genotypes of the parents. The various mating systems that apply to genotypes carrying different incompatibility alleles and the consequences that occur in the subsequent generations were worked out and presented by Bartley and Cope (1973).

As explained in Chapter 7 the incompatibility system of cacao is not a complete barrier to obtaining families resulting from self-fertilization. Various techniques could be used to overcome the restrictions of incompatibility although the procedures may entail the use of greater resources to obtain families of sufficient size that would permit selection to be carried out. Bartley and Cope (1973) described the techniques that may be applied in the production of inbred progeny from self-incompatible genotypes. Presumably, the same techniques could be used to produce hybrids between genotypes that carry the same incompatibility alleles.

The most reliable method involves the application of pollen from *Herrania* sp. at the same time as pollen from the genotype is applied to the flowers of the tree which is a candidate for inbreeding. This method can be used for all genotypes. The egg nuclei fertilized by *Herrania* gametes result in seeds that do not contain cotyledons and have thick testas. Thus they can be readily identified and eliminated leaving the S_1 seeds to be used to form the inbred families. The number of normal seeds obtained would vary according to the genetic status of the incompatibility loci for each genotype. The proportion of these seeds could be very low which implies that large numbers of pollinations may be required to obtain families of sufficient size.

However, not all research stations and germplasm collections are likely to have plants of *Herrania* species available and alternative methods need to be investigated. In the case of genotypes with green coloured fruits one technique is to use the pollen from self-compatible genotypes which are homozygous for the red fruit colour. All seeds obtained will be normal consisting of a mixture of self-fertilized and hybrid seeds. The latter may be identified during germination by the presence of pigment in the hypocotyls or by the pigmentation of the petioles in the young plants. To employ this technique it is necessary to obtain the genotypes with the marker characters or produce them by breeding.

In the case of red-fruited genotypes the procedures may be more complicated. Here it will be necessary to use the genotypes that carry easily identifiable recessive genes in the homozygous condition as the marker parents. One character which is suitable for the purpose is the anthocyanin inhibitor gene and its use is described as an example. Fruits resulting from a mixture of the candidate genotype's pollen and that of the genotype with the marker gene will contain normal seeds which may have pigmented cotyledons of varying degrees. All seeds would have to be sown and when the progeny trees reach the adult stage they should be backcrossed to the genotype carrying the marker gene. The trees which produce white cotyledons, or progenies with unpigmented leaves, will be those resulting from hybridization and the progenies whose fruits contain seeds entirely with the characteristics of the candidate genotypes would belong to the S_1 generation.

Because of the influence of self- and cross-incompatibility on the actions that may be applied in the utilization of most elements of the diversity breeders, geneticists and curators of gene banks would be faced sooner or later with the task of determining the compatibility status of the available genotypes. In the populations in which self-incompatibility is found to occur the determination of the individual alleles and the frequency of their occurrence will constitute an aspect of the research on population structure. The knowledge acquired concerning the genetic constitution of the individual genotypes and their relationships would make it possible to plan the utilization of the diversity in a feasible manner by allowing the resources to be used more economically.

Up to the present time the utilization of the genetic resources has been undertaken, with few exceptions, on a casual basis and in a mostly unscientific manner. Selection programmes have been pursued without any definite objectives, with the result that 'gene banks' that are no more than clone collections are occupied with cloned genotypes that are duplicates or for which the curators have no idea about their potential value. Similarly, when the notion of hybrid cultivars was established, the attitude seems to have been that the production of these combinations was an end in itself. However, the state of understanding of the subject of improvement resulted in a lack of comprehension of the fact that the hybrids produced could be merely the first step in the process of variety improvement and the utilization of the genes carried by the genotypes comprising the basic diversity. Several procedures and tools can be applied to use the genetic factors carried by the progenies of the hybrid families that have been shown to have the capacity to contribute to the development of cultivars.

Practical Applications for the Exploitation of Specific Traits Exhibited by the Diversity

One of the best ways to illustrate how these procedures could be applied in specific circumstances is to describe the actions taken in some programmes or those that are possible from currently existing families.

One of the most active lines of research has concerned the genotypes belonging to the group named 'Scavina', which originated on the R. Ucayali in Peru. Some of these genotypes were observed to be free of infection by the pathogen *Crinipellis perniciosa*. However, the commercial utilization of these genotypes as clone cultivars was not feasible on account of their possessing other undesirable characteristics, which included the fact that they carry specific self-incompatibility alleles. The next phase was to overcome the undesirable qualities of the 'SCA' genotypes by mating them with others that possessed superior agronomic qualities. From these hybrid combinations it became apparent that variation occurred among the 'Scavina' genotypes with regard to their reactions to the pathogen although they appeared to be phenotypically similar. Also, although the hybrids possessed excellent productive potential they exhibited other characteristics that reduced their fitness as cultivars.

The principal factor that affects the productive capacity of the hybrids is the fact that all progenies are self-incompatible. A programme of inbreeding starting from the F_1 generation produced by mating a self-compatible genotype with the clone 'SCA 6', considered to possess the highest level of resistance to the disease, was undertaken. This programme had the objective of producing families of homozygous individuals that would combine the best attributes of both parents.

The first barrier to be overcome was the consequences of the self-incompatibility status of 'SCA 6', which carries two co-dominant incompatibility alleles. Although the F_1 progenies obtained by mating to self-compatible parents are self-incompatible, they segregate for two genotypes (each carrying one of the self-incompatibility alleles) where the individuals with the same genotype are inter-incompatible but compatible with individuals of the other genotype. This knowledge was applied to produce the F_2 generation by mating selected individuals belonging to the opposite genotypes in the F_1 families. The families of the F_2 generations segregate for self-compatible individuals in the proportion of 25%. These individuals would constitute the material on which further inbreeding would be conducted, a procedure that could be continued for as long as required, with selection being practised to isolate the genotypes that possess the desired characteristics.

An additional adverse consequence of the constitution of the 'SCA' parent was the production in the subsequent generations of fruits with small seeds. The dominance of the small seed size characteristic of genotypes of the 'SCA' group was demonstrated in Chapter 7. In addition, the progenies produced in advanced generations of crosses involving 'SCA' parents, segregate for several undesirable growth traits, such as the 'dwarfing' character. This character was also described in Chapter 7. This situation resulted in a large proportion of the progenies being discarded and required that large families be produced in order to obtain a sufficient number of individuals that could be selected as potential parents. The products of the lines with potential for cultivation would be utilized as inbred cultivars, clones based on selection or as parents in other lines of cultivar development.

In the specific programme described above the principal criterion sought was a superior reaction to the disease in question. The isolation of individuals with this characteristic would involve the evaluation of the reactions of the individuals in each generation of the programme and the use of those with the potential in this respect to form the succeeding generations.

Similar procedures could be used in other populations in which self-incompatibility is present but where individual genotypes with potentially useful characteristics, excluding the specific disease problem, may be encountered but which are not themselves suitable for use as cultivars.

An alternative procedure in such circumstances involves backcrossing F_1 individuals to the self-compatible, large seeded tester parent. This method has been used with satisfactory results as it permits the more rapid fixation of the tester parent's genes. The progenies of the first backcross generation selected for the desired 'SCA 6' characters could be employed in a further programme of backcrossing or by sib-crossing, followed by a sequence of inbreeding according to the potential exhibited by each individual in terms of achieving the objectives of the programme.

Among the more recent contributions to the programme of conservation of the diversity that offer potential for the development of cultivars is that of the system of the R. Purús. F_1 hybrids involving several genotypes from the populations in this system show that they possess an above average combining ability for fruit production that could be used to advantage. This ability is exhibited throughout the genotypes tested regardless of their individual performance as clones.

On account of the small seed size associated with these populations the most beneficial method of utilizing the genes present in the R. Purús genotypes would involve backcrossing F_1 individuals to a self-compatible tester parent that has the ability to transmit its large seed size to its progenies. Inter-crossing among the selected progenies of the backcross would produce a population in which selection for the desired traits would be practised. This selection would involve determining the combining ability of individual genotypes of potential value in order to select those to be used as parents. The compatibility situation in the R. Purús populations is virtually unknown. Once this is determined it may be possible to conduct a programme of reciprocal recurrent selection in which the same, or similar genotypes, would be employed in parallel mating systems using backcrossing.

The availability of an ample range of diversity provides the plant breeder with the opportunity to identify and utilize specific traits that are expressed by incorporating the genotypes that exhibit these traits in the improvement programmes. A trait that has a potential application in the development of new systems of cultivation is the narrow canopy architecture of genotypes from the R. Tarauacá (as shown in Plate 1a and b). If the expression of this trait is genetically determined it would be possible to use it to develop cultivars that could be planted at high densities, in which the apparent efficiency of the plants in terms of fruit production related to canopy size would be an additional advantage.

Hybrids between genotypes from this population and with others of 'normal' architecture and habit develop into plants with an intermediate canopy size with fruits that are larger than the R. Tarauacá genotypes. Figure 51 shows a plot of one of the F_1 families descended from 'CAB 182'. This is another specimen from the R. Tarauacá population similar to those depicted in Plate 1a and b. While the F_1 hybrids may serve adequately as cultivars, further improvement may be achieved by a programme of reciprocal recurrent selection to develop clones that would be used as cultivars and as parents of seed reproduced varieties.

This is one example of an F_1 family that would serve as the precursor to a process of improvement through which a new trait could be incorporated into the development of cultivars. The results from the crosses involving the elements from the R. Tarauacá population suggest that they provide conditions for this purpose. Various avenues are available to accomplish the objective. Inbreeding in the F_1 generation, either by self-fertilization or by sib-mating, whether or not self-incompatiblity is present, would create new combinations of genotypes with the desired architecture and improved agronomic qualities. An alternative would be backcrossing to the parent with normal architecture in order to concentrate the factors for its attributes.

Fig. 51. The effect of the tree habit of the variety from the R. Tarauacá (see Plate 1) as shown by the progenies of the cross 'CAB 25' × 'CAB 182'.

During the 1960s, to implement concepts that had been advanced, I began to investigate the possibility of creating varieties whose characteristics of growth and architecture could be adapted to management by the application of specific cultural practices. The basic objective of this line of research was the reduction of management costs by reducing the inputs needed.

Among the avenues of research that presented themselves was the exploitation of the wide variability shown for tree architecture and habit to develop varieties of plants with low height. Such trees would be more efficient to manage compared with the large vigorous trees that are outstanding but frequently of low productive ability. This concept does not involve the use of the genotypes with the lowest heights and growth rates but those that are adequate for sustained development and production by possessing architectural features that facilitate cultural practices. Trees of low height with compact canopies offer a special advantage in the control of disease and pests. Such varieties would, of course, incorporate the other characteristics that are desirable for economic cultivation.

The first attempt to investigate the potential for the development of lines of trees with low stature involved mating the clone 'M 252' with 'ICS 6'. 'M 252' was identified many years ago as being a plant that developed a limited height. In retrospect it seems that the height is limited because of the inhibition of the vegetative initials in the orthotropic stem which would produce vertical extension growth. The plants of 'M 252' are not weak as they develop vigorous basal suckers quite readilty. The fruit has a green colour and is of medium to small size.

Figure 52 depicts a progeny of the cross, the photograph being taken about 15 years after its establishment. The contrast between the height of this plant and the trees in the background of the same age is evident. The height of

Fig. 52. Demonstration of the dominance of the dwarfing habit of the clone 'M 252' as shown by a (approx.) 15-year-old tree of the cross 'M 252' × 'ICS 6'.

the plant illustrated is perhaps too low for it to be utilized as a cultivar but it is a step along the way to developing genotypes possessing similar characteristics. The encouraging feature of the plant was the large size of the fruits. The elongate shape probably was inherited from a *Criollo* ancestor of 'M 252'.

The two aspects described above introduce another choice with regard to the type of cultivars that should be produced, a choice that has not been considered until recently. Eventually, a decision will have to be taken, especially in the breeding programmes that have a wide range of the diversity at their disposal, concerning the systems of cultivation that are most appropriate to obtain the best results from certain genotypes. Simply speaking, this boils down to choosing whether cultivars should be developed for specific cultural systems or whether the cultural systems have to be developed in order to take the best advantage of the specific habits of the genotypes.

Cacao is a species that has the intrinsic capability of producing flowers and fruits throughout the year. However, in most environments various factors may operate to limit fruiting, either entirely or partially, to certain seasons. In tropical environments seasonal effects may not be exactly defined. In spite of the environmental control of flowering and fruiting patterns, a degree of variability in this respect has been expressed over the range of diversity. For various reasons the cultivation of genotypes with the capacity to fruit outside the normal seasonal patterns in given environments may have some advantages. These include the ability to escape infection from fungal diseases and insect infestations as well as economic benefits resulting from lengthening the harvesting periods. Although some genotypes with such tendencies have been identified the incorporation of these characteristics has not been undertaken on account of the absence of conditions to make observations over the time period necessary to establish the existence of the traits.

The differential periodicity in the availability of mature fruits may be related to the differences between genotypes and populations with regard to the period that elapses between fertilization and the onset of maturity. In the diversity that is known there is a range of about 40 days in the length of the maturity period. These differences may have important consequences in terms of the reactions of varieties to diseases affecting the fruits and the development of traits such as seed size but the effects of the variability of this character have not been determined.

The traits that are of fundamental importance in the selection of cultivars for commercial production are those related to their yielding capacity. Basically, the yield of an individual or variety is measured in terms of the weight of seeds produced at the marketable stage. The yield is a function of two main components, namely, the numbers of fruits harvested per tree and the average weight of the marketable seeds contained in the fruits. The latter, in turn, has two components: the numbers of seeds per fruit; and the mean dry weight of the seeds. These components have relatively high heritabilities under normal circumstances and it is possible to select successfully for each component individually. The numbers of seeds produced by given genotypes are limited by the numbers of ovules contained in the ovaries of the flowers. Since ovule numbers show a wide range of variability this character is important because the maximization of seed content per fruit depends on the combination of high seed numbers (from genotypes with high ovule numbers) and seeds with large dry weights. Hence, the development of high yielding cultivars would be based on the identification of the genotypes with the highest values for these components.

However, selection for these traits is not straightforward, as genetic and environmental associations between the components have to be considered. Within certain ranges the association of seed weight with seed number tends to be negative. Similarly, over the potential range of fruit production a negative correlation may be encountered between fruit numbers per tree and the weights of their seed contents.

Although the attainment of high production levels is one of the main objectives of improvement programmes, high yields are not necessarily advantageous by themselves if the achievement of this goal is obtainable only by the application of expensive inputs. The attainment of economically viable levels of production on a commercial scale requires that the cultivars also must possess attributes that will give the growers returns that will be lucrative under most conditions. While selecting the varieties that would have the highest fitness for cultivation it is necessary to consider those that exhibit the highest productive efficiency. Within this concept the traits that would be considered as related to efficiency are those concerned with the relationship between fruit production and the architecture and physiological characteristics of the trees. The higher the ratio of fruits to canopy size and trunk diameter the greater the physiological efficiency of the trees. As described above the architecture could have important effects on the costs of management of a variety in cultivation. Another index of efficiency relates to harvesting costs in terms of the relation between the weight of commercial product per fruit and the total weight of the fruits at harvest. Accordingly, the characteristics connected with the efficiency of production should be taken into consideration during the process of evaluation of the performance of the genotypes and varieties.

Besides the prospects of enriching the improvement programmes by seizing the opportunities the ample diversity presents in terms of introducing the specific characteristics that may be beneficial to cultivation, there are many ways in which the diversity may be used to produce cultivars. In general, improvement programmes conducted since breeding of cacao began have been limited in scope by the restricted range of germplasm that has been available. As described above and in Chapter 4 much of the germplasm available has consisted of the selections made with their use as clones in view. It was these clones that provided the parents for the development of hybrid cultivars when this concept was introduced.

Few breeding programmes have gone beyond the production of single cross hybrids using the clones from the first cycle of selection and those from a few of the genotypes from the collection activities carried out in Peru in 1937–1938. In rare cases selections have been made in these hybrids with the intention of using them to develop advanced lines and combinations. Various procedures are available in which the characteristics that individual genotypes of these progenies possess can be employed in the development of cultivars.

These include triple crosses and double crosses, as well as backcrosses, in addition to the possible use of inbred lines. As mentioned at the beginning of the chapter, the families developed at any stage can also be subjected to selection to use the most outstanding and suitable individuals as clones, either as cultivars or as parents in further cycles of breeding.

One additional method involves the improvement of populations. In some ways the rather nebulous concept of 'germplasm enhancement' is a corollary to population improvement. It is understood that the concept relates to the manipulation of the germplasm contained in a given population in order to concentrate the most advantageous genes in a smaller number of genotypes. However, no procedures for doing this have been defined.

Any procedures that will be undertaken with this objective in mind have to consider such factors as the size of the population, the presence and distribution of incompatibility and the homozygosity status of its component individuals. Obviously, as explained previously, no progress can be expected if the population is composed of individuals that are homozygous for the desirable traits, which is, for example, apparently the case of the Mexican *Criollo* population.

The populations that are extremely heterogeneous and large would require a great deal of effort to produce any significant gain through the random combination of genotypes compared with that involving selected individual genotypes.

One application of the concept of population improvement is in the raising of levels of performance and fitness of cultivated populations that are heterogeneous in genetic content. It is envisaged that a programme of this nature would be conducted on a long-term basis and would involve several generations of selection and hybridization. The procedure concerns the inter-crossing of the most desirable phenotypes in the base population to produce large populations. In the first generation cycle of the programme, selection of the most desirable phenotypes would be carried out and these hybridized among themselves. The process would be repeated in the subsequent cycles of the programme. Observations would be recorded on the traits exhibited by the individuals comprising the families produced from each combination during the successive cycles, regardless of the manner in which the composite population is established. The analysis of these observations would provide information about the inheritance of the characters contributed by each selected parent, which would be used in the process of improving the performance level of the base population. The application of the procedure envisages the use of the plants produced in each generation in the cultivation system.

With the ample diversity now known to be present in the species the breeding programmes would be supplied with a wider genetic base. The procedures for utilization of the individual genotypes would be determined by their possession of the characteristics that satisfy the criteria sought for commercial purposes. Fundamentally, it is the genotypes and their intrinsic attributes that will determine the suitable breeding methods to be applied in individual circumstances.

10 Epilogue

Closing Remarks

The review of the present knowledge about the diversity of cacao presented in this work leads us to certain clear conclusions.

The first is that the cultivated populations, excluding those in the Amazonian Region and the eastern coast of Brazil, have evolved from a narrow genetic base, although they exhibit considerable variability with regard to their morphological characteristics. In this case we are considering the populations that had been created prior to 1950, which were mainly constituted by the remnants of the *Criollo* populations and hybrids between them and certain elements of the Amazonian Region's germplasm. Depending on the localities in which the hybrid populations had been established and the histories of their evolution their genetic constitution would vary according to the proportions of genes contributed by each of the base varieties from which they had evolved.

Interpretation of the molecular analyses of the *Criollo* populations, presented in Motamayor *et al.* (2002), leads to the conclusion that the genetic difference between these populations is virtually non-existent. If we accept this interpretation (albeit with reservations) then it does not matter which source of germplasm from the Circum-Caribbean Region contributed to the formation of the cultivated populations. From this evidence it may be concluded that, in spite of the historical importance of the *Criollo* group varieties, they represent an insignificant proportion of the total diversity in the species.

Similarly, there would have been minor genetic differences between two genotypes from the Amazonian Region, which were introduced into the Circum-Caribbean Region, namely the Trinidad 'Calabacillo' and the Martinique 'Créole' or 'Matina'. Also belonging to the same genetic background is the 'West African Amelonado', which, although it was cultivated on a large scale, became a component of hybrids with *Criollo* genotypes at a later period and to a lesser degree, principally in São Tomé and Cameroon.

The position of the Ecuador 'Nacional' is rather different since the hybrid populations that involve the genes from this source have been restricted to Ecuador itself and certain areas of Mesoamerica. The exception to this situation would be the possible hybridization involving 'Nacional' genotypes that were introduced into São Tomé and Cameroon. The resulting hybrids would be difficult to distinguish from those that involved the 'West African Amelonado' and hybrid types from Trinidad. The results of the DNA analyses performed indicate that the 'Nacional' variety is genetically distant from the Amazonian Region genotypes that were parents of the hybrid populations produced in countries such as Trinidad. Accordingly, the hybrids between the 'Nacional' and members of the *Criollo* group would not be genetically equivalent to the hybrids between the *Criollo* and the specific Amazonian genotypes. Therefore, they cannot be classified under the same term that is applied to the hybrids of the type originally formed in Trinidad and that made a major contribution to the variability in other cultivated populations.

It is abundantly clear that nearly all of the naturally occurring diversity of the species is to be found in the Amazonian Region, as defined in this book. However, this finding has to take into consideration the fact that our knowledge of the diversity is still incomplete. Even in the Brazilian Amazon, where the greatest activity has taken place, it is estimated that about 20% of the potential diversity has been explored. In Peru and Colombia a great deal more needs to be done to conserve the diversity that exists in those countries, especially in the areas that will be mentioned in due course.

As far as Bolivia is concerned, information on the diversity occurring in that country is virtually non-existent. Even if samples of the diversity in the various populations are available they have not been studied or analysed to the stage where conclusions can be reached regarding their position in the spectrum of the diversity.

The second finding is one that has become apparent from the explorations realized during the past 30 years and is that the major populations of naturally occurring diversity are located in the southwestern segment of the Amazonian Region. All of these populations are distributed along drainage systems that have their sources in what we have termed the 'Province of Caravaya'. Not only do each of these populations represent an ample diversity but they differ markedly with regard to their genetic structure. We will not be able to determine how this individuality of the populations evolved until we are able to examine those that are distributed in the region of the headwaters of the rivers that lie in Peru.

The principle that the diversity in the Amazonian Region is distributed in discrete populations along the drainage systems was confirmed by a study undertaken by Dias *et al.* (2002). These authors analysed several morphological traits of the fruits and seeds of a sample of genotypes from the R. Purús, R. Ji-Paraná and R. Japurá drainage systems. They also sampled the cultivated population in the Parintins area, as representative of a part of the diversity along the R. Amazon. Although the authors found a large inter-genotype variation within each population, which agrees with the descriptions of these populations in Chapter 5, the samples may not have encompassed all of the diversity in each population. While the results showed that the differences between drainage systems were significant for all attributes measured, no indication is given as to the actual values obtained for each attribute or to which the variability is dispersed in each system. The greatest phenotypic distance was found in the comparison of the R. Ji-Paraná and R. Japurá systems, but this result does not agree with the genetic distance that may be deduced from the results of the analyses performed by Marita (1998).

The variability expressed in the large sample that was collected in the Parintins area was undoubtedly extensive, a result that confirms the visual observations made, but it should be appreciated that the plants of the specimens that were established were still in the developing stage when the data were collected. This would have affected the ability to obtain a sample that reflected the range of variability in the genotypes that represent the population. This is one of the areas in the Amazonian Region and Amazon Extension Zone where cultivation has been practised for centuries. The scale of variability in each zone appears to be large and, like the naturally distributed populations, large differences occur between zones.

Similar analyses of the variability that occurs within the populations that derive from Equatorial Oriental Piedmont need to be made on the morphological and molecular levels. In order to complete the picture of the diversity in this region the germplasm collected should be amplified through collections made in the upper R. Marañon valley and along the stretches of its tributaries that flow within Peru. This would provide a basis for determining the origins of the Ecuador 'Nacional' variety, and perhaps elucidate the true nature of the variety situation in the littoral zone.

The third finding was that, from the point of view of the diversity that offers the greatest potential for use in improvement of the species, the Solimões and lower R. Japurá population is perhaps the most outstanding in this respect. However, the Solimões has not been sampled for its diversity throughout its entire length. In addition the lower reaches of the right-bank tributaries of the Solimões remain to be explored. This especially includes the lower reaches of the R. Purús, where observations made by travellers in the past have indicated the existence of extensive occurrences of cacao.

In view of the fact that the diversity in the Amazonian Region is distributed in discrete, apparently unrelated, populations it becomes inappropriate to identify the diversity in the region by a general term, such as applying the name *Forastero* indiscriminately to all genotypes from the region. It will be necessary to adopt a system of nomenclature, which would be applied to identify the populations individually.

As a result of the lack of comprehension of the subject, several misconceptions exist regarding the genetic composition of the diversity of the Amazonian Region. For example, Lanaud *et al.* (1999) wrote: 'Les Forastero de basse Amazonie et de la vallée de l'Orenoque sont largement cultivés dans tout le bassin amazonien. La forme la plus typique en est l'Amelonado' ['The 'Forastero' of the lower Amazon and the Orinoco valley are widely cultivated throughout the Amazon Basin. The most typical form is the "Amelonado"'].

As could be seen from the illustrations of fruits from the cultivated populations in the region, elongate fruits occur frequently and are probably just as numerous as those within the concept of 'Amelonado' of Lanaud *et al*. Even among the types to which this name for the fruit shape could be applied there are considerable differences in the genetic constitutions of the genotypes among and within populations. Similarly, because the short broad fruit shape is not the most common within the diversity of the Amazonian Region and there is no proof that it is the basic shape found in the natural populations, the application of a subspecies concept called '*Sphaerocarpum*' is inadmissible and could be abolished.

On the basis of the few samples from the cacao population that occurs in the upper R. Tapajós system, especially that of the R. Teles Pires, this variety may well prove to be one of the most unique in the range of the diversity. Also, the few samples from the R. Branco system indicate that a significant range of variability is to be found in this valley and perhaps the adjacent tributaries of the R. Negro. The importance of this population lies in the fact that it is isolated from the main areas of dispersal of the species, except for a possible connection with the upper R. Orinoco population.

Among the areas that should be investigated is that of the island of Gurupá. This site should be a priority area for future collecting of germplasm since it may provide evidence of the origin of the cultivated population in the Amazon Extension Zone.

One of the most notable results from the collecting activities carried out in recent years has been the evidence of the existence of several varieties with fruits having red pigmentation on their surfaces. None of these varieties could have been contaminated with genes from the hybrids with *Criollo* ancestry. These findings satisfy one of the criteria suggested by Cheesman (1944), by which it can be shown that the *Criollo* varieties introduced into the Circum-Caribbean Region could have originated in the Amazonian Region. It remains for investigations to be conducted to determine the genetic constitution of the red-fruited Amazonian genotypes in order to establish the similarity between them and the *Criollo* varieties.

In connection with the appearance of red-fruited genotypes in the Mesoamerican populations an alternative scenario may be envisaged. This would allow for the original introduction from South America to have consisted of green-fruited varieties. During the course of time, a mutation would have occurred that conferred the ability to synthesize anthocyanin pigmentation in the fruits and other plant organs. Since the mutation was caused by a dominant allele it would have become easily established. Subsequently, selection for this trait, as described by Martínez (1912), resulted in red-fruited genotypes becoming a dominant element in the Mesoamerican populations. This scenario prompts various questions concerning the manner in which these varieties were distributed as cultivation spread in the Region.

The occurrence of red-fruited genotypes in the Sub-Andean and Catatumbo/Maracaibo areas of Venezuela would also be ascribable to chance mutations of the same kind. Alternatively, these variants may have been introduced from Mesoamerica. In either case, a similar process involving selection for red fruits would have resulted in the formation of groups of homozygous individuals in the areas to which cultivation was extended.

The diversity originating in the Equatorial Oriental Piedmont has a genetic structure that also contains elements with phenotypic characteristics that resemble those of the *Criollo* genotypes. However, the molecular analysis indicates that the genetic distance between the two populations may be very wide. In the due course of time this subject may be properly investigated with regard to the mode of inheritance of the characters concerned and the relationship between the two groups.

The germplasm collected in Peru in 1937 and 1938 has been given a certain amount of space in this book. However, this should not be construed as intentionally implying that this germplasm is more important than the other elements of the diversity in the Amazonian Region. This germplasm contains the oldest specimens from the Region known outside it and has gained a degree of importance owing to its use in breeding programmes, albeit only a small proportion of the collections have actually been used. The insufficient information regarding the collections and the manner in which the individuals have been identified prompt an interest in the determination of the parentage, family structure and inter- and intra-family relationships. For these reasons information about the collections has been interpreted in the context of the knowledge acquired about the individual clones resulting from the collections. An attempt has been made to explain how germplasm collections could be managed to generate the information needed to elucidate situations of this nature.

One of the messages that we have attempted to put forward in this survey of diversity is that the differences among the populations and varieties are based on the variability expressed in all organs of the plants and their habits. Because of this situation it is old-fashioned to use only the fruit characters to differentiate between varieties. All of the previous classifications of cacao populations made on this basis should be discarded, as well as the terminology related to fruit shapes and Cuatrecasas' classification. These should be replaced by a more logical system of classification.

Once we can clear our heads of the old-fashioned systems of classification and apply a realistic system, it will become easier to understand the range and nature of the species' diversity.

In this survey we have also set out to demonstrate the importance of mutations in creating the differences between populations and the variability that exists within them.

At this time it is not convenient to speculate on the location of a centre of origin of the diversity in the species. It is now obvious that the greatest diversity occurs in the southwest and west of the Amazon Basin. The area of the Solimões and R. Japurá has been highlighted as the region where the most significant genetic diversity may be found. These regions are outstanding, not only on the basis of the extensive variability contained in them, but also in terms of the sizes of their cacao populations. The details of the dispersal of variability in the Amazonian Region given in this book may be applied in the formulation of hypotheses about the creation of the diversity and its nature. In this respect an attempt has been made to suggest the possible areas in which the diversity may have been created.

It is not intended to discuss the anthropological aspects of the subject but it is evident that some, perhaps on a significant scale, human activity in the pre-Columbian era as well as in more recent times, has been responsible for much of the distribution of the diversity. Since none of the history of these developments has been recorded it is impossible to provide exact indications as to what activities occurred in specific locations and the periods in which certain populations were established. If it is accepted that the varieties in the *Criollo* group originated in the Amazonian Region the transfer of the material from South America to Mesoamerica must have been undertaken by peoples in South America who had a use for the plant and, therefore, may have domesticated it.

Some indications have been given in this book regarding the areas where exploration would be beneficial, or at least add to our knowledge of the diversity. It may already be too late to sample some of the other areas in which the species is distributed and some of the collections made have already been lost. The conservation of the ample diversity requires that more effort be made than has been the case in the past and investigations carried out on the inheritance of the characters of the individual genotypes and populations as well as the determination of their genetic constitutions.

References

Acuña, C. (1641) *Nuevo Descubrimiento del Gran Río de las Amazonas.* Imprenta del Reyno, Madrid, 48 pp.

Allen, J.B. (1987) London Cocoa Trade Amazon Project, Final Report Phase 2. *Cocoa Growers' Bulletin* 39, 1–94.

Allen, J.B. (1988) Geographical variation and population biology in wild *Theobroma cacao.* PhD thesis, University of Edinburgh.

Almeida, C.M.V.C., Barriga, J.P., Machado, P.F.R. and Bartley, B.G.D. (1987) *Evolução do Programa de Conservação de Recursos Genéticos de Cacau na Amazônia Brasileira, Boletim Técnico* no. 5. Comissão Executiva do Plano da Lavoura Cacaueira, DEPEA, Brazil, 108 pp.

Almeida, C.M.V.C., Machado, P.F.R., Barriga, J.P. and Silva, F.C.O. (1995) *Coleta de Cacau (*Theobroma cacao L.*) da Amazônia Brasileira: Uma Abordagem Histórica e Analítica.* Comissão Executiva do Plano da Lavoura Cacaueira, SUPOC, Brazil, 83 pp. + appendices.

Anon (1903) Cacao in Fiji. *The Agricultural News* 2 (44), 409.

Anon (1904) Criollo cocoa in Jamaica. *Journal of the Jamaica Agricultural Society* 8 (5), 190–193.

Anon (1966) Historiadores de Indias, Alguns capítulos relacionados con Guatemala I. Relación breve y verdadera de algunas cosas de las muchas que sucedieron al Padre Fray Alonso Ponce. *Anales de la Sociedad de Geografía y Historia de Guatemala* 39 (1–4), 123–233.

Bailey, L.H. (1947) *The Standard Cyclopedia of Horticulture*, vol. III, Part 2. The MacMillan Company, New York, pp. 2423–3639.

Baker, R.E., Cope, F.W., Holliday, P.C., Bartley, B.G.D. and Taylor, D.J. (1954) The Anglo-Colombian Cacao Collecting Expedition. In: *A Report on Cacao Research 1953*. Imperial College of Tropical Agriculture, Trinidad, pp. 8–29.

Barrett, O.W. (1925) The food plants of Porto Rico. *The Journal of Agriculture of Porto Rico* 9 (2), 61–208.

Barriga, J.P., Machado, P.F.R., Almeida, C.M.V.C. and Almeida, C.F.G. (undated) *Preservação e Utilização dos Recursos Genéticos de Cacau na Amazônia Brasileira, Comunicado Técnico Especial,* no 5. Comissão Executiva do Plano da Lavoura Cacaueira, DEPEA, Brazil, 23 pp. + appendices.

Bartelink, E.J. (1884) *The Cacao Planters' Manual*, translated by H.J. Vogin. Kirkland and Cope, London, 57 pp.

Bartley, B.G.D. (1964a) Notes of the cacao of Maracaibo and the relationships of pod and cotyledon colours. *Cacao (Costa Rica)* 9(1), 8–12.

Bartley, B.G.D. (1964b) Genetic studies. *Annual Report on Cacao Research, 1963*. Imperial College of Tropical Agriculture, University of the West Indies, Trinidad, pp.30–33.

Bartley, B.G.D. (1967) Genetic studies. *Annual Report on Cacao Research, 1966*. Imperial College of Tropical Agriculture, University of the West Indies, Trinidad, pp. 28–29.

Bartley, B.G.D. (1994) A review of cacao improvement: fundamentals, methods and results. In: End, M.J. (ed.) *Proceedings of the International Workshop on Cocoa Breeding Strategies*. INGENIC, Kuala Lumpur, Malaysia, pp. 3–17.

Bartley, B.G.D. and Chalmers, W.S. (1971) Cacao germplasm collection at the University of the West Indies. *Plant Genetic Resources Newsletter* 26, 2–4.

Bartley, B.G.D. and Cope, F.W. (1973) Practical aspects of self-incompatibility in *Theobroma cacao* L. In: Moav, R. (ed.) *Agricultural Genetics – Selected Topics*. John Wiley & Sons, New York, pp. 109–134.

Bartley, B.G.D., Yamada, M.M., Castro, C.T. and Melo, G.R.P. (1983) Genética de *Theobroma cacao*: ocorrência do fator letal Luteus-PA na família Parinari. *Revista Theobroma (Brazil)* 13(3), 275–278.

Bartley, B.G.D., Machado, P.F.R., Ahnert, D., Barriga, J.P. and Almeida, C.M.V.C. (1988) Descrição de populações de cacau da Amazônia brasileira. I. Observações preliminares sobre populações de Alenquer, Pará. In: *Proceedings of the 10th International Cocoa Research Conference, Santo Domingo, 1987*. Cocoa Producers' Alliance, Lagos, Nigeria, pp. 665–672.

Bernoulli, G. (1869) Uebersicht der bis jetzt bekannten Arten von Theobroma. *Neue Denkschriften der Allgemeinen Schweizerischen Gelleschaft für die Gesammten Naturwissenschaften* 23(3), 1–15.

Blanco, M. (1837) *Flora de Filipinas*. Imprenta de Santo Thome, Manila, 887 pp.

Bondar, G. (1924) *O Cacao, Parte I, A Cultura e Preparo do Cacao*. Imprensa Official do Estado, Bahia, Brazil, 158 pp.

Bondar, G. (1929) *O Cacao, Parte I, A Cultura e Preparo do Cacao*, 2nd edn. Secretário da Agricultura, Industria, Commercio, Viação e Obras Públicas da Bahia, Brazil, 131 pp.

Chalot, C. (1901) Rapport sur le jardin d'essai de Libreville. *L'Agriculture Pratique des Pays Chauds*, 1° Anneé, no. 2, 168–181.

Chalot, C. and Luc, M. (1906) Le cacaoyer au Congo Française: Introduction du cacaoyer au Gabon. *L'Agriculture Pratique des Pays Chauds*, 6° Année, no. 37, 283–294.

Chandless, W. (1968) Notes on the ascent of the River Purús. *Journal of the Royal Geographical Society* 36(1), 86–118.

Charters, Y.M. and Wilkinson, M.J. (2000) The use of self-pollinated progenies as 'in-groups' for the genetic characterization of cocoa germplasm. *Theoretical and Applied Genetics* 100, 160–166.

Cheesman, E.E. (1944) Notes on the nomenclature, classification and possible relationships of cacao populations. *Tropical Agriculture (Trinidad)* 21(8), 144–159.

Chevalier, A. (1908) *Le cacaoyer dans l'Ouest Africain*. In: *Les Vegétaux Utiles d'Afrique Tropicale Française*, vol. 4. Challamel, Paris, 245 pp.

Cieza de León, P. (1923) *The War of Las Salinas*. Translated by C. Markham. The Hakluyt Society, London, 304 pp.

Ciferri, R. (1929) *Informe General Sobre la Industria Cacaotera de Santo Domingo.* Serie B Botanica, no. 16. Estación Agronomica de Moca, Dominican Republic, 190 pp.

Ciferri, R. (1933) Monografia delle varietá forme e razze di cacao coltivate in Santo Domingo. *Italia, Reale Accademia, Memorie della Classe di Scienza, Fisiche, Matematica e Naturali* 4(18), 589–676.

Ciferri, R. and Ciferri, F. (1949) *Reconocimiento de la Explotación Cacaotera de los Valles de Riego del Sector Central (Estado Aragua).* Dirección de Agricultura, Sección de Cacao, Caracas, Venezuela, 153 pp.

Ciferri, R. and Ciferri, F. (1950) *Reconocimiento de la Explotación Cacaotera de los Andes Venezolanos,* Serie 5, vol. G. Instituto Botanico della Universitá, Paiva, Italy, 38 pp. (Mimeographed.)

Clark, P.D.G. (1891) *Report on the Central Territory of Peru.* The Peruvian Corporation, London, 26 pp.

Coe, M.D. and Coe, S.D. (1996) *The True History of Chocolate.* Thames and Hudson, New York, 280 pp.

Cope, F.W. (1962) The mechanism of pollen incompatibility in *Theobroma cacao* L. *Heredity* 17, 157–182.

Crouzillat, D., Bellanger, L., Rigoreau, M., Bucheli, P. and Pétiard, V. (2001) Genetic structure, characterisation and selection of Nacional cocoa compared to other genetic groups. In: Bekele, F., End, M. and Eskes, A. (eds) *Proceedings of the International Workshop on New Technologies and Cocoa Breeding, INGENIC.* Kota Kinabalu, Malaysia, 2000, pp. 47–64.

Cruz, de la, M., Whitkus, R., Gómez-Pompa, A. and Mota-Bravo, L. (1995) Origins of cacao cultivation. *Nature* 375, 542–543.

Cuatrecasas, J. (1964) Cacao and its allies: a taxonomic revision of the genus *Theobroma: Contributions from the U.S. National Herbarium* 35(6), 379–614.

Dash, J.S. (1929) Historical cacao in British Guiana. *Agricultural Journal of British Guiana* 2(2), 26–28.

Demandt, E. (1914) Samoanische Kakaocultur. *Beihefte zum 'Der Tropenpflanzer'* 15 (2/3), 135–307.

Dias, L.A.S., Barriga, J.P., Kageyama, P.Y. and Almeida, C.M.V.C. (2002) How genetic variation of cacao populations from Brazilian Amazon is distributed. In: *Proceedings, 12th International Cocoa Research Conference, Kota Kinabalu, Malaysia, 2000.* Cocoa Producers' Alliance, Lagos, Nigeria, pp. 183–188.

Díaz del Castillo, B. (1908) *The True History of the Conquest of New Spain,* translated by A.P. Maudslay, 2 vols. Hakluyt Society, London.

End, M.J., Wadsworth, M.J. and Hadley, P. (1994) The International Cocoa Germplasm Database. In: de Lafforest, J. (ed.) *Proceedings, 11th International Cocoa Research Conference, Cote d'Ivoire, 1993.* Cocoa Producers' Alliance, Lagos, pp. 537–543.

Engels, J.M.M., Bartley, B.G.D. and Enriquez, G.A. (1980) Cacao descriptors, their states and modus operandi. *Turrialba (Costa Rica)* 30(2), 209–218.

Fernández Duro, C. (1890) Ríos de Venezuela y de Colombia: Relaciones Ineditas. *Boletin de la Sociedad Geográfica de Madrid* 28(1), 76–174.

Froidevaux, H. (1894) Explorations Françaises à l'intérieur de la Guyane pendant le second quart du XVIIIe siécle (1720–1742). *Bulletin de Géographie, Historique et Descriptive,* Année 1894, no. 2, 218–301.

Garcia, B.C. (1954) Contribución al estudio de la variabilidad de la población cacaotera de Colombia. *Cacao en Colombia* 3 (supplement), 1–8.

Gillett, D. (1949) Report to Cadbury Bros. Ltd, Bournville, on a visit to the Caribbean Area, February 1949. Cadbury Brothers Ltd, Bournville, 24 pp.

Gosselin, M. (1895) *Report for the Year 1893–1894 on the German Colonies in Africa and the South Pacific. Miscellaneous Series*, no. 352, 1895. Foreign Office, London, 94 pp.

Guérin, P. (1896) *Culture du Cacaoyer – Étude fait a la Guadeloupe.* Bibliothèque d'Agriculture Colonial, Challamel, Paris, 64 pp.

Harris, C.A. and de Villiers, J.A.J. (1911) *The Rise of British Guiana.* Hakluyt Society, London, vol. I, pp. 1–372, vol. II, pp. 373–703.

Hart, J.H. (1893) Report on journey to Nicaragua in 1893. *Trinidad, Royal Botanic Gardens, Bulletin of Miscellaneous Information* 19, 1–11.

Hart, J.H. (1908) The characters of Criollo cacao. *West Indian Bulletin* 9(2), 161–162.

Hart, J.H. (1910) Cacao notes. *Proceedings of the Agricultural Society of Trinidad and Tobago* 10(8), 303–304.

Harvey, C. and Harwood, L.W. (1958) A progress report on cocoa development in Fiji. In: *Report of the Cocoa Conference, 1957.* The Cocoa, Chocolate and Confectionery Alliance, London, pp. 26–28.

Hernández, F. (1959) *Obras Completas, História Natural de Nueva España*, vol 1. Universidad Nacional Autonoma de México, México, 980 pp.

Hernández, S.A. (1968) El Cacao. *Agronomía (Venezuela)* 9, 8–23.

Herndon, W.M.L. and Gibbon, L. (1853) *Exploration of the Valley of the Amazon*, vol. 1. Robert Armstrong, Public Printer, Washington, DC, 414 pp.

Herndon, W.M.L. and Gibbon, L. (1854) *Exploration of the Valley of the Amazon*, vol. 2. A.O.P. Nicholson, Public Printer, Washington, DC, 339 pp.

Howes, F.N. (1946) The early introduction of cocoa to West Africa. *Tropical Agriculture (Trinidad)* 23(9), 172.

Huber, J. (1906a) Sur l'indigénat du *Theobroma cacao* dans les alluvions du Purús et sur quelques autres especes du genre *Theobroma. Bulletin Herbier Boisssiere* (2me série) 6(24), 272–274.

Huber, J. (1906b) Materiaes para a flora Amazônica, VI. Plantas vasculares coligidas e observadas no baixo Ucayali e no Pampa de Sacramento, nos meses de outubro e dezembro de 1898. *Boletim do Museu Goeldi (Museu Paraense)* 4, 510–619.

Humboldt, A. von (1821) *Personal Narrative of Travels to the Equinoctial Regions of the New Continent during the Years 1803 and 1804*, 5 vols. Longman, Hurst, Rees, Orme and Brown, London.

Isschot, E.-Ch. van (1900) Le cacao et le caoutchouc à l'Equateur. *Revue des Cultures Coloniales* 7(60), 520–522.

Jennings, V.L.K., Molesworth, D., Thorn, S. and Watt, N. (also Hanks, D., Johnston, M.A., Lay, E., Morris, R. and Spink, S.) (1999) *Guyana Cocoa Research Initiative, 2 July – 15 September 1998, Final Report.* University of the West of England, Bristol, UK, 102 pp.

Johnson, W.H. (1912) *Cocoa, Its Cultivation and Preparation.* John Murray, London, 186 pp.

Lachenaud, Ph. and Sallée, B. (1993) Les cacaoyers spontanés de Guyane. *Café Cacao Thé* 37(2), 101–114.

Lachenaud, P., Mooleedhar, V. and Couturier, C. (1997) Les cacaoyers spontanés de Guyane: nouvelles prospections. *Plantations, Recherche, Developpement* Janvier–Fevrier, 25–30.

Lachenaud, P., Bonnot, F. and Oliver, G. (1999) Use of floral descriptors to study variability in wild cocoa trees (*Theobroma cacao* L.) in French Guiana. *Genetic Resources and Crop Evolution* 46, 491–500.

La Condamine, C.-M. (1745) *Relation Abrégée d'un Voyage Fait dans l'Interieur de l'Amérique Méridionale.* Veuve Pissot, Paris, 216 pp.

Lanaud, C. (1986) *Rapport de prospection au Venezuela, Fevrier 1986*. Mimeographed, CIRAD, France, 8 pp.

Lanaud, C., Motamayor, J.-C. and Sounigo, O. (1999) Le cacaoyer. In: Hamon, P., Seguin, M., Perrier, X. and Glaszmann, J.C. (eds) *Diversité Génétique des Plantes Tropicales Cultivées*. CIRAD, France, pp. 141–166.

Leal, O. (1893) As regiões de terra e agoa. *Revista da Sociedade de Geographia do Rio de Janeiro*, 19(1 and 2), 5–31.

Le Cointe, P. (1934) *A Cultura do Cacau na Amazônia*. Ministerio de Agricultura, Rio de Janeiro, Brazil, 35 pp.

Lerceteau, E., Quiroz, J., Soria, J., Flipo, S., Pétiard, V. and Crouzillat, D. (1997) Genetic differentiation among Ecuadorian *Theobroma cacao* L. accessions using DNA and morphological analyses. *Euphytica* 95, 77–87.

Lisboa, M.M. (1866) *Relação de uma Viagem a Venezuela, Nova Granada e Equador*. Brussels, 393 pp.

López, E.A. and Cuatrecasas, J. (1953) El género *Theobroma* en la 'Flora Peruviana et Chilensis de Ruíz et Pavón'. *Ciencia (Mexico)* 22(4), 85–92.

López-Baez, O., Mulato, B.J. and López, J.I. (1989) Recolección de germoplasma de cacao (*Theobroma cacao* L.) en México. *Plant Genetic Resouces Newsletter* 80, 23–24.

Lunan, J. (1814) *Hortus Jamaicensis*, vol. I. St Jago de la Vega Press, Jamaica, 508 pp.

Marita, J.M. (1998) Characterization of *Theobroma cacao* using RAPD-marker based estimates of genetic distance and recommendations for a core collection to maximize genetic diversity. MSc thesis, The University of Wisconsin, Madison, Wisconsin.

Martínez, L. (1912) *Cultivo y Beneficio del Cacaotero*, 2nd edn. Imprenta y Fototipia de la Secretária de Fomento, Mexico, 72 pp.

Morris, D. (1887) *Cacao – How to Grow and How to Cure It*. Aston W. Gardner and Co., Jamaica, 42 pp.

Motamayor, J.C., Risterucci, A.M., López, P.A., Ortiz, C.F., Moreno, A. and Lanaud, C. (2002) Cacao domestication: 1, The origin of the cacao cultivated by the Mayas. *Heredity* 89(5), 380–386.

Myers, J.G. (1930) Notes on wild cacao in Surinam and in British Guiana. *Kew Bulletin* 1, 1–10.

Myers, J.G. (1934) Observations on wild cacao and wild bananas in British Guiana. *Tropical Agriculture (Trinidad)* 9(10), 263–267.

Newland, H.O. (1916) *Sierra Leone. Its People, Products and Secret Societies*. John Bale, Sons and Danielson Ltd, London, 251 pp.

Nooten, B.H. van (1863) *Fleurs, Fruits et Feuillages Choisis de L'ile de Java*. 3rd edn. Merzbach and Falk, Brussels, 32 pp.

Ocampo, R.F. (1985) Informe sobre recolección de germoplasma de *Theobroma cacao* L. *El Cacaotero Colombiano* 31, 24–29.

Oviedo y Valdéz, G. (1855) *História General y Natural de las Indias*, 4 vols. Real Academia de la História, Madrid.

Pastorelly, D.M., Vera, J.S., Enriquez, G.A. and Bartley, B.G. (1994) Estudio da la compatibilidad y algunas caracteristicas del cacao tipo 'Nacional' de Ecuador. In: *Proceedings of the 11th International Cocoa Conference, 1993*. Cocoa Producers' Alliance, Lagos, Nigeria, pp. 887–900.

Patiño, V.M. (1963) *Plantas Cultivadas y Animales Domésticas en América Meridional*, vol. 1, Esterculiaceas, 53. *T. cacao*. Cali, Imprenta Departamental, Colombia, pp 268–340.

Pires, J.L., Marita, J.M., Lopes, U.V., Yamada, M.M., Aitken, W.M., Melo, G.P., Monteiro, W.R. and Ahnert, D. (2001) Diversity for phenotypic traits and molecular markers in CEPEC's germplasm collection in Bahia, Brazil. In: Bekele, F., End, M. and Eskes, A. (eds) *Proceedings of the International Workshop on New Technologies and Cocoa Breeding, INGENIC.* Kota Kinabalu, Malaysia, 2000, pp. 72–88.

Pittier, H. (1902a) Las variedades del cacaotero cultivadas en la zona Atlantica de Costa Rica. *Boletin del Instituto Fisico-Geográfico de Costa Rica* 2(18), 121–124.

Pittier, H. (1902b) Es el cacaotero indigena en Costa Rica? *Boletin del Instituto Fisico-Geográfico de Costa Rica,* 2(20), 193–196.

Pittier, H. (1926) A propos des cacaoyers du Vénézuela. *Revue de Botanique Appliquée et d'Agriculture Coloniale* 6(58), 345–346.

Pittier, H. (1934) El problema del cacao en Venezuela. *Publicación de la Camara de Caracas, Boletin* no. 251, 15 pp.

Pound, F.J. (1938) *Cacao and Witchbroom Disease of South America (*Marasmius perniciosus*). Report on a Visit to Ecuador, the Amazon Valley and Colombia, April 1937 – April 1938).* Yuille's Printerie, Port-of-Spain, Trinidad and Tobago, 58 pp.

Pound, F.J. (1943) *Cacao and Witches' Broom Disease (*Marasmius perniciosus*). Report on a Recent Visit to the Amazon Territory of Peru, September 1942 – February 1943.* A.L. Rhodes, Port-of-Spain, Trinidad and Tobago, 14 pp.

Pound, F.J. (1945) A note on the cacao population of South America. In: *Report and Proceedings of the Cocoa Research Conference held at Colonial Office, May–June, 1945.* The Colonial Office, His Majesty's Stationery Office, London, pp. 131–133.

Preuss, P. (1901) *Expedition nach Central- und Sudamerika 1899/1900.* Kolonial-Wirtschaftlichen Komitees, Berlin, 374 pp.

Ratnam, R. (1961) Introduction of Criollo Cacao into Madras State. *South Indian Horticulture* 9(4), 24–29.

Reyes, H.E., Moreno, A., Morillo, V., Pagnini, T. and Aristiguieta, C. (1993) *Catálogo de Cultivares del Cacao Criollo Venezolano,* Serie Especial no. 12. Fondo Nacional de Investigaciones Agropecuarias/Corporación Andina de Fomento, Venezuela, 219 pp.

Rivera de León, S. (1986) *Informe General Sobre el Proyecto de Recolección de Cacao Criollo en Guatemala. Report ABPG/IBPGR/86/156.* International Board for Plant Genetic Resources, Rome, Italy, 9 pp.

Roepke, W. (1917) *Onze Koloniale Landbouw XI. Cacao.* H.D. Yjeenk Willinck en Zoon, Haarlem, Netherlands, 164 pp.

Ronning, C.M. and Schnell, R.J. (1994) Allozyme diversity in a germplasm collection of *Theobroma cacao* L. *Journal of Heredity* 85(4), 291–295.

Rosenquist, E.A. (1950) Cocoa selection and breeding in Malaya. *The Malayan Agricultural Journal* 53(4), 181–193.

Sanchez, P. and Jaffé, K. (1989) El genéro *Theobroma* en el Territorio Federal: Amazonas (Venezuela). II. Distribución Geográfica. *Turrialba (Costa Rica)* 29(4), 24–29.

Schnee, L. (ed.) (1960) *Plantas Comunes de Venezuela.* Faculdad de la Universidad Central de Venezuela, Maracay, Venezuela, 663 pp.

Schomburgk, R.H. (1836) Report of the expedition into the interior of British Guyana in 1835–1836. *Journal of the Royal Geographical Society* 6, 224–284.

Schomburgk, R.H. (1837) Diary of an ascent of the River Berbice in British Guyana in 1836–7. *Journal of the Royal Geographical Society* 7(2), 302–350.

Sloane, H. (1707) *A Voyage to the Islands Madera, Barbados, Nieves, S. Christophers and Jamaica, with the Natural History of the Herbs and Trees, Four-Footed Beasts, Fishes, Birds, Insects, Reptiles etc. of the Last of those Islands,* 2 vols. London 1707–1725.

Soria, V.J. (1961) *Anotaciones Sobre un Viaje a las Zonas Productoras de Cacao en México (Marzo 6–18, 1961)*. Turrialba, IICA, Informe 44-E, Costa Rica, 18 pp.

Soria, V.J. (1965) Notes on the native cacaos in the vicinity of Iquitos (Peru) and Alto Beni (Bolivia). *Cacao (Turrialba)* 10(4), 14–16.

Soria, V.J. (1970) The latest cocoa expeditions to the Amazon basin. *Cacao (Turrialba)* 15(1), 5–15.

Sounigo, O., Christopher, Y., Ramdahin, S., Umaharan, R. and Sankar, A. (2001) Evaluation and use of the genetic diversity present in the International Cocoa Genebank (ICG.T) in Trinidad. In: Bekele, F., End, M. and Eskes, A. (eds) *Proceedings of the International Workshop on New Technologies and Cocoa Breeding, INGENIC*. Kota Kinabalu, Malaysia 2000, pp. 65–71.

Spix, J.B. von and Martius, C.F.P. von (1828) *Reise en Brasilien in den Jahren 1817 bis 1820*, Martius C.F.P. von (ed.) Vol II. I.J. Lehntner, Munich, pp. 415–884.

Stahel, G. (1924) *Verslag van een Dienstreis naar Central Amerika, Bulletin* no. 65. Department van Landbouw, Nijverheid en Handel, Suriname, 73 pp.

Stell, F. (1933) *Report of Mr. F. Stell, Mycologist, Department of Agriculture, on his Visit to Ecuador, to Study Witchbroom Disease (*Marasmius perniciosus*) of Cocoa, Council Paper* no. 137 of 1933. Government Printer, Port-of-Spain, Trinidad and Tobago, 12 pp.

Stockdale, F.A. (1928) Cacao research. *Tropical Agriculturist* 71(6), 327–342.

Thomson, R. (1894) *Colombia. Report on the Cultivation of Cacao, Bananas and India-rubber in Districts Surrounding the Sierra Nevada de Santa Marta. Miscellaneous Series*, no. 322. Foreign Office, London, 17 pp.

Toxopeus, H. and Giesberger, G. (1983) History of cocoa and cocoa research in Indonesia. *Archives of Cocoa Research* 2, 7–34.

Tyler, C.D. (1894) The River Napo. *The Geographical Journal* 3(1), 476–484.

Vernon, A.J. (1971) *Phytophthora palmivora* disease of cocoa in Fiji. In: Tetteh, E.K. and Owusu, G.K. (eds) *Proceedings of the IIIrd International Cocoa Research Conference, Accra, Ghana, 1969*. Cocoa Research Institute, Ghana, pp. 375–386.

Watkins, C., Knowles, O. and Johnston, M. (undated) *Guyana Cocoa Diversity Project, Report*. University of the West of England, Bristol, UK, 38 pp.

Wellensiek, S.J. (1931) De erfelijkeid van zaadlobkleur bij cacao als basis voor qualiteits-selectie. *Archief voor de Koffiecultuur in Nederlandsch-Indie* 5, 217–232.

Whitkus, R., Cruz, M. de la, Mota-Bravo, L. and Gómez-Pompa, A. (1998) Genetic diversity and relationships of cacao (*Theobroma cacao* L.) in southern Mexico. *Theoretical and Applied Genetics* 96(5), 621–627.

Wildeman, É. de (1908) *Les Plantes Tropicales de Grande Culture*, vol. I. Alfred Castaigne, Brussels, 387 pp.

Wohltmann, F. (1904) Pflanzung und Siedlung auf Samoa. *Beihefte zum Tropenpflanzer* 5(1 and 2), 164 pp.

Wright, H. (1907) *Theobroma cacao – or Cocoa: Its Botany, Cultivation, Chemistry and Diseases*. Messrs A.M. and J. Ferguson, Colombo, 249 pp.

Yamada, M.M., Bartley, B.G.D., Castro, G.C.T. and Melo, G.R.P. (1982) Herança do fator compatibilidade em *Theobroma cacao* L. Relações fenotípicas na família PA (Parinari). *Revista Theobroma (Brazil)* 12(3), 163–167.

Yamada, M.M., Bartley, B.G.D., Lopes, U.V. and Pinto, L.R.M. (1996) Herança do fator compatibilidade em *Theobroma cacao* L. I. Relações fenotípicas em genótipos adicionais do grupo Parinari (PA). *Agrotrópica* 8(2), 51–52.

Zehntner, L. (1914) *Le Cacaoyer dans l'Etat de Bahia*. Friedlander und Sohn, Berlin, 156 pp.

Appendix

Variety Names

Abbreviations and acronyms used to identify the genotypes and families mentioned in the text and illustrations, with their equivalents and the origins of the varieties and/or names.

BE	Belém (Brazil)
CA	Careiro (Island) (Brazil)
CAB	Cacau da Amazônia Brasileira (Brazil)
CAM	Cacau da Amazônia (Brazil)
CJ	Cachoeira do Jarí (Brazil)
C. SUL	Cruzeiro do Sul (Brazil)
EBC	Expedición Botánica Caquetá
EEN	Estación Experimental de Napo (Ecuador)
EET	Estación Experimental Tropical (Ecuador)
ICS	Imperial College Selection (Trinidad)
IMC	Isla Morona Cocha (Peru)
M	Museum (Trinidad)
MA	Manaus (Brazil)
NA	Nanay (Peru)
P	Peru (Peru) (Pound, Trinidad)
PA	Parinari (Peru)
RB	Rio Branco (Brazil)
SCA	Scavina (Peru)
SIAL	Seleção do Instituto Agronómico do Leste (Brazil)
SIC	Seleção do Instituto de Cacau (Brazil)
SPEC	Specimen – Colombia 1952–1953 (Trinidad)
TAP	Tapiche (Peru)
U	Ucayali (Peru)
UCA	Ucayali (Peru)
UF	United Fruit (Company) (Costa Rica)

Index

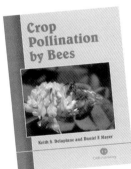

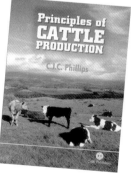